An Introduction to **Modern Algebra**

An Introduction to

BURTON W. JONES *The University of Colorado*

Modern Algebra

MACMILLAN PUBLISHING CO., INC.

New York

COLLIER MACMILLAN PUBLISHERS

London

Printed in the United States of America

Macmillan Publishing Co., Inc.
866 Third Avenue, New York, New York 10022

Collier-Macmillan Canada, Ltd.

Library of Congress Cataloging in Publication Data

Jones, Burton Wadsworth, (   )
  An introduction to modern algebra.

  Bibliography: p.
  1. Algebra, Abstract. I. Title.
QA162.J66   512'.02   74–3795
ISBN 0–02–361250–9

Printing: 1 2 3 4 5 6 7 8   Year: 5 6 7 8 9 0

To Marian

Preface

This text is intended for junior and senior students in colleges or universities who have the mathematical maturity that a beginning course or two in calculus should impart. It is written at a somewhat lower level than *A Survey of Modern Algebra* by Garrett Birkhoff and Saunders Mac Lane and still lower than *Topics in Algebra* by Israel Herstein and *Modern Algebra* by Bartel van der Waerden. The influence of these three great books on this text is evident.

In writing the book, I have been especially mindful that the student should have a sense of involvement in the development of the subject. To this end, in order to deal with a fundamental concept like a group, for instance, I do not begin by listing a set of postulates but rather show how the idea comes up as an abstraction of a set of properties common to two number systems. In general, if a term is to be defined, I want to make sure that the student has some prior acquaintance with ideas that make the concept meaningful and useful.

I do not usually introduce a new term until I am ready to use it. For instance, the idea of a normal subgroup is not defined when cosets are first introduced nor even when the kernel of a homomorphism is first dealt with. But as soon as it appears that we would be in trouble unless the left and right cosets of the kernel are the same, then the concept of a normal subgroup is forced upon us.

vii

From an algebraist's point of view, the role of ideals in ring theory is probably much more impressive than for unique decomposition into products of prime ideals. But from a less sophisticated point of view, it is when we are faced with the problem of unique decomposition that Dedekind's idea shines in all its brilliance. Similarly, Galois theory, although of fundamental importance in algebra, can be shown to be important to a beginner by its immediate usefulness in the solution of the classical problem of constructions and solutions by radicals.

Special effort is made to keep the student informed about where we are going and why. Tentative exploration often seems useful. Where a somewhat special result is derived, the student is told that it *is* special and where we shall use it later. This information will also guide those teachers who may want to omit certain topics and need to know the price to be paid for such omissions.

Much use is made of numerical examples to lead to ideas and to illustrate concepts that have already been formulated. There are also a number of places where a difficult proof is first shown in a numerical setting, then carried out for a special case in the pattern of the general proof, before finally proving the theorem completely. I feel no compulsion to state a theorem in its most general form.

The exercises are a very important part of the text. Some are routine, to help fix the abstract ideas in concrete situations. A number of proofs of theorems are left to the student so that he can test his knowledge and feel involved in the development of the results. Answers and partial answers are given to selected exercises. Sometimes the answer is merely "yes" or "no," to reassure the student that he is on the right track while leaving the reasons to him. At other times an answer to part of an exercise is given in some detail.

The first three chapters give the basic ideas of groups and fields. An average class should be able to acquire most of this material in a one-semester course that meets three times a week. I would, however, be disappointed not to take some of the topics in the last two chapters, because it is there that the results on groups and fields are really put to work. Furthermore, the last chapter, especially, shows the intimate connection between groups and fields. Students who have some prior knowledge of linear algebra should, in a semester's course, be able to get at least a taste of one of the last two chapters.

In Chapter I we lead up to the abstract idea of a group, then define it, and devote the rest of the chapter to important examples of groups from various parts of mathematics. The purpose is to show the student how ubiquitous the properties of a group are, to help him to become familiar with a group in many of its forms, and to acquire a backlog of experience and examples on which the concepts of the next two chapters can be built. The first four sections are closely interrelated. The fifth and sixth deal with functions and transformations. They all lead up to the idea of a subgroup and what it takes to prove that a subset of a group is a group. Sections 8 and 9 are intimately related. In Section 10 we not only have a rudimentary example of an ideal but derive a result that will be immediately useful. Even the little group from logic is not just a curiosity but concerns itself with the structure of proof. The idea of an abstract group is a simple one (I have taught it to freshmen) but to really appreciate it requires experience; providing this experience is an important aim of this chapter.

Chapter II begins with the idea of an isomorphism. Logically, it might seem more efficient to introduce first the concept of a homomorphism and then specialize it to an isomorphism. But the latter is a simpler concept and much more readily applicable to examples at hand. Cyclic groups are dealt with in detail but only after the student has had experience with more general groups. Since normal subgroups and homomorphisms go together, they are introduced in juxtaposition.

Chapter III begins with the idea of a field because, in my opinion, this is a simpler concept for the beginning student than that of an integral domain or a ring. Another reason for postponing the idea of a ring is that the most accessible example of a ring that is not an integral domain is the set of matrices of some order; and a student without linear algebra in his background might not feel at home with such an illustration. The section on derivatives and separability is introduced, first, because there may be some interest in seeing how a derivative can be defined for polynomials without the idea of a limit, and, second, this section is needed in parts of Chapter VI. It should be pointed out that we treat algebraic extensions only partially in Section 14 because we lack some ideas of a vector space. When a class is familiar with vector spaces, one could give extensions more adequate treatment by introducing at this stage part of Section 2 of Chapter VI.

Chapter IV is a service chapter for those students without any background in linear algebra. The point is that certain topics in linear algebra are needed in the last two chapters, and it would seem awkward to have to refer students to other texts for these results. The chapter begins with the basic idea of a vector space and linear dependence and independence. These are the only parts of Chapter IV that are needed in Chapter VI. Matrices and determinants are used crucially in Section 6 of Chapter V. Otherwise, except for the use of matrices as examples of rings, we do not need them in the rest of the book. Thus the choice of material in this chapter depends on the training of the class and what additional topics in the book they will study.

Chapter V justifies the concept of an ideal by posing the problem of unique decomposition and then solving it. The latter involves, as is there noted, two rather difficult theorems about algebraic fields, which neither the student nor the teacher may want to tackle. In such a case I would prefer, myself, to assume these two theorems and then proceed to solve the problem at hand, since, as I noted earlier, I think the student would be much more impressed with the use of ideals in this connection than for homomorphisms of rings.

But it is true that Section 8 through Theorem 8.7 and Section 9 could be taken up without much of the rest of the chapter.

Chapter VI uses the first three sections of Chapter V but is otherwise independent of its predecessor. Hence a class could omit all but those three sections before proceeding to Chapter VI. The latter is concerned with the related problems of constructions and solutions by radicals. It develops as needed the tool of Galois theory to solve these problems. This chapter ties together most of what has previously been covered in the book: groups, fields, and algebraic extensions of fields.

My indebtedness to the four authors mentioned in the first paragraph of this preface is very great. I especially admire the exploratory point of view from which Herstein wrote his book, and this has influenced my approach to some topics. I should also acknowledge that I have adopted with some enthusiasm the practice of George Simmons in his book on differential equations and have given brief biographical sketches of those mathematicians whose names are linked with the subject and whose ideas underlie the basic theory.

I should like to record my special appreciation of the late Carl B. Allendoerfer, who in many ways encouraged the writing of this book and whose careful and perceptive comments were a crucial influence in its development. I am also grateful to Charles Brase, who, in reading the semifinal version of the manuscript, gave many helpful suggestions. Furthermore, I want to acknowledge the contributions of Miss Kanda Kunze of the University of Arizona and Mrs. Mae Jean Ruehlman of the University of Colorado, who typed the manuscript at various stages. Thanks could not be complete without including the Macmillan staff in their meticulous attention to many details of production.

Boulder, Colorado   BURTON W. JONES

Contents

III. Fields, Integral Domains, and Rings   *111*

IV. Vector Spaces and Matrices   *185*

V. Ideals   *217*

VI.   Constructions and Galois Theory   *272*

1

Definition and Examples
of Groups

1. Introduction

In the process of studying any subject, one should stop from time to time to correlate what he has learned. Such coordination is especially useful in mathematics, where the body of knowledge increases rapidly and emphasis shifts. Without such periodic reassessment, what needs to be learned can quickly become unmanageable by sheer volume alone. To achieve such correlation, we shall in this book deal with some fundamental mathematical structures—sets of objects from various parts of the subject that have certain properties in common. When we look at these common properties, we may see relationships not previously perceived. Such a look will increase our insight into known mathematics and will lead into realms that are new to us, although they have important properties in common with the old.

The subject matter of this book is thought of as algebra, but we shall see that it has much in common with parts of analysis and geometry as well. In fact, it is the interplay of various parts of mathematics that lends importance to much of the material presented.

We start with the technical idea of a group because it can be described briefly and has a wealth of application. To work toward this fundamental concept, we first point out a list of properties common to two sets of numbers, and from this abstract the definition of "group." In the rest of the chapter we explore various examples from different parts of mathematics not only to show how being a group can serve as a unifying concept, but also to provide a source for the development of properties of groups that are discussed in later chapters. We postpone until Chapter II consideration of most of the general properties of groups.

2.   Definition of a Group

First, let us consider certain properties of two sets of numbers: **Z**, the set of all integers, and **R***, the set of all nonzero real numbers (**R** denotes the real numbers, including zero). To emphasize the relationships, we list the properties in parallel columns.

Addition of integers, **Z**	Multiplication of nonzero real numbers **R***
1. Closure	
If b and c are in **Z**, then $b + c$ and $c + b$ are in **Z**.	If b and c are in **R***, then $b \cdot c$ and $c \cdot b$ are in **R***.
2. Associativity	
For all a, b, and c in **Z**, we have $(a + b) + c = a + (b + c)$.	For all a, b, and c in **R***, we have $(a \cdot b) \cdot c = a \cdot (b \cdot c)$.
3. Existence of an identity	
There is a number 0 in **Z** such that $0 + a = a + 0 = a$ for all a in **Z**.	There is a number 1 in **R*** such that $1 \cdot a = a \cdot 1 = a$ for all a in **R***.
4. Existence of an inverse	
For each a in **Z**, there is a number $(-a)$ in **Z** such that $a + (-a) = (-a) + a = 0$.	For each a in **R***, there is a number a^{-1} such that $a^{-1} \cdot a = a \cdot a^{-1} = 1$. (We also write $1/a$ for a^{-1}.)

To be sure, these are not the only properties common to addition of integers and multiplication of nonzero real numbers. But we select these because there is a striking parallelism between these two sets of four properties and because we shall see later in this chapter that many other examples of sets also have these properties.

Notice that for multiplication of integers, property 4 fails to hold, since, for example, there is no integer x such that $3x = 1$. On the other hand, the four properties of addition hold if **Z** is replaced by **Q**, the set of rational numbers, **R**, the set of real numbers, or **C**, the set of complex numbers.

Furthermore, the properties of the right-hand column hold if **R*** is replaced by **Q*** or **C***, the respective sets of **Q** and **C** with zero excluded.

It is also true that the four properties of addition hold if **Z** is replaced by **Z**[x], the set of polynomials $f(x)$ in x with coefficients in **Z**. In fact, the coefficients of the polynomials in x could be rational, real, or complex and the properties listed for addition would still hold.

In order to gradually develop a concept that includes both sets with addition and multiplication, respectively, let us look more carefully at the two sets above. Each integer is an element of the set **Z** and each nonzero real number an element of **R***. So we start with a set. Then there is a means of combining two elements of the set to get a third, a process we call an *operation*. For the first set above, the operation is addition and, for the second, multiplication. If we denote the operation by ∘, we can write

$$s \circ s' = s'',$$

where s, s', and s'' are in the set S. For addition of integers, ∘ is $+$, and for multiplication of real numbers it is a raised dot ·, the symbol for multiplication. Property 2 affirms that the operation is associative.

You may well wonder why we did not include in our list the commutative properties: $a + b = b + a$ and $a \cdot b = b \cdot a$. We have, in fact, purposely avoided this requirement in order to enlarge the scope of the concept of a group. So if the result of the operation is to be allowed to depend on the order, we must consider not just the pair of elements but an *ordered pair*, that is, a pair in which the order makes a difference. Thus we have led up to the following formal definition of a binary operation ("binary" because it combines two elements).

Definition.   *Given a set S, we call ∘ a binary operation on S if it assigns to each ordered pair of elements s and s' of S a unique element s'' of S.*

This operation can be written in at least two ways:

$$s \circ s' = s'' \qquad \text{or} \qquad (s, s') \xrightarrow{\circ} s'',$$

where s, s', and s'' are elements of S.

Since the idea of a binary operation includes closure, we need not mention this property in the formal definition of a group, which follows.

Definition.   *Given a nonempty set of elements S and a binary operation ∘ on S, then we call S a group "under" the operation ∘ if the following properties hold:*

1. *The binary operation is associative; that is, for every a, b, and c in S,*

$$a \circ (b \circ c) = (a \circ b) \circ c.$$

2. *There is an element e of S such that e ∘ a = a ∘ e = a for every element a of S. Such an element e is called an* identity *element.* (We shall prove that it is unique.)

3. *To each element b of S there corresponds an element $\bar{b}$ of S such that $b \circ \bar{b} = \bar{b} \circ b = e$, where e is the identity element. Such an element is called an* inverse *of b.* (It can be proved that $\bar{b}$ is uniquely determined by b.)

If you let S be **Z** and ∘ be $+$, you will see that the integers form a group under addition, whereas if S is **R*** and the operation is multiplication, it follows that the nonzero real numbers form a group under multiplication. Similarly, the sets **Q**, **R**, and **C** (rational, real, and complex numbers), as well as **Z**[x], form a group under addition; **Q*** and **C*** form a group under multiplication.

There are five other properties common to **Z** under addition and **R*** under multiplication, which we list as follows, using the terminology that we have used for the definition of a group.

4. The identity of a group is unique; that is, there is only one identity element in a group.

5. The inverse of any element of a group is unique; that is, each element b has exactly one inverse.

6. If a and b are any elements of a group, then there are unique elements x and y of the group for which $a \circ x = b$ and $y \circ a = b$.

7. The cancellation properties hold: (a) if $a \circ b = a \circ c$, then $b = c$; (b) if $b \circ a = c \circ a$, then $b = c$.

8. The inverse of $a \circ b$ is $\bar{b} \circ \bar{a}$.

First, let us see what these mean for addition of integers. Property 4 affirms that 0 is the only integer z for which $z + a = a + z = a$ for all integers a, and property 5 affirms that every integer has only one negative. Both of these are very obvious. Slightly less obvious is property 6, which maintains that $a + x = b$ and $y + a = b$ are solvable for x and y, no matter what integers a and b are. Property 7 states that $a + b = a + c$ or $b + a = c + a$ implies that $b = c$. Property 8 affirms that the inverse of $a + b$ is $(-b) + (-a)$. You should follow through these same properties for **R*** under multiplication.

Now, these five properties are different from the first three, in that they can be deduced from the first three. This means that whenever we have verified that a system is a group, these properties also hold as extra dividends. So now we proceed to show how property 4 and half of property 6 follow from the group properties, leaving the rest of the proofs as exercises.

To prove property 4, suppose that there are two identity elements e and e' of a group. Then $e \circ e' = e'$ since e is an identity, and $e \circ e' = e$ since e' is an identity. Hence $e = e'$.

To show property 6, first suppose that there is an element x of S such that $a \circ x = b$. Then $\bar{a} \circ b = \bar{a} \circ (a \circ x)$. Using the associative property and $\bar{a} \circ a = e$, we have

$$\bar{a} \circ b = \bar{a} \circ (a \circ x) = e \circ x = x.$$

So if there is an element x such that $a \circ x = b$, then x must be $\bar{a} \circ b$. Thus we have proved that there is no more than one element x such that $a \circ x = b$. To show that $x = \bar{a} \circ b$ *is* a solution, perform the computation:

$$a \circ (\bar{a} \circ b) = (a \circ \bar{a}) \circ b = e \circ b = b,$$

using in turn group properties 1, 3, and 2. Thus there is one and only one element of S, namely, $\bar{a} \circ b$, such that $a \circ x = b$. Note that we have assumed a property of equality and the operation: that if $x = y$, then $a \circ x = a \circ y$. This could be thought of as the converse of the cancellation property. Similarly, one can show that the unique solution of $y \circ a = b$ is $b \circ \bar{a}$. Note that $b \circ \bar{a}$ is not necessarily equal to $\bar{a} \circ b$, the solution of $a \circ x = b$.

When the operation is addition, we customarily write the identity as 0 and the inverse of b as $-b$. When the operation is multiplication, the identity is often written as 1 and the inverse of a as a^{-1} or $1/a$. When the operation is neither addition nor multiplication, we usually write $a \circ b$ as ab and the inverse of a as a^{-1} in place of $\bar{a}$, just as if it were multiplication; we usually denote the identity by e.

Note, however, that there is one notation, b/a, that cannot well be carried over to groups. If the set is the set of nonzero real numbers, b/a is used to denote the solution of $ax = b$, that is, $b(1/a)$. But if multiplication is not commutative, $b(1/a)$ and $(1/a)b$ might be different, that is, $a(1/a)b = b$ by the associative property, but $(1/a)ba$ might not be b. So in place of dividing we multiply by the inverse, being careful on which side we multiply.

Note that for multiplication of nonzero real numbers, the fact of $\mathbf{R}^*$ being a group implies that if a and b are in $\mathbf{R}^*$, then ab is in $\mathbf{R}^*$, and hence the product of any two nonzero real numbers is not zero. In other words, the group property for $\mathbf{R}^*$ implies that if $ab = 0$ for a and b real numbers, then one of a and b must be zero. This, as we know, is an important property.

If it happens that a group also has the commutative property, $a \circ b = b \circ a$, for all a and b in S, we say that it is *commutative* or *Abelian*. (The name Abelian is in honor of the Norwegian mathematician Niels Henrik Abel. See Section 13.) Sometimes when a group is Abelian, we use the symbol $+$ instead of $\circ$, even if ordinary addition is not involved. The commutative property cannot be deduced from the first three properties of a group, since, as we shall see shortly, there are examples of groups that are not commutative.

Referring back to the examples at the beginning of this section, we can see that the three sets of numbers $\mathbf{Q}$, $\mathbf{R}$, and $\mathbf{C}$ have the following properties:

1. $\mathbf{Q}$, $\mathbf{R}$, and $\mathbf{C}$ form Abelian groups under addition, with identity 0.
2. $\mathbf{Q}^*$, $\mathbf{R}^*$, and $\mathbf{C}^*$ are Abelian groups under multiplication, with identity 1.
3. The distributive property holds: $a(b + c) = ab + ac$.

Any set that has two or more elements with two operations that satisfy these three conditions is called a *field*. We shall deal with this notion later in the book and give other examples of fields. Until that time there will be no harm in thinking of the term "field" as denoting one of the sets **Q**, **R**, and **C**, if you prefer. But whenever a result is true for a field *F*, we shall state it in this more general form.

If a group has *n* elements, we say that its *order* is *n* and call it a *finite group*. If it has infinitely many elements, we call it an *infinite group* and say that its order is infinite. A convenient notation is $o(G)$ for the order of a finite group *G*. All the groups considered in this section are infinite. We shall have a number of examples of finite groups later in this chapter.

The properties of a group can be thought of as a codification of ideas. Instead of saying that a set and operation have such and such properties, we can be more brief about it by merely characterizing them as a group. Furthermore, as we saw above, once we have a group, other properties hold as a consequence.

The few examples we have given so far are not sufficient to give one an adequate feeling for what a group is in the abstract. For this reason the following set of exercises is especially important. The object of most of them is to help in acquiring an understanding of what a group is. Further examples are the chief content of the rest of the chapter.

You are asked to show in Exercise 22 that an equivalent definition of a group can be given by replacing properties 2 and 3 by seemingly less restrictive ones. But usually there is no great advantage in doing so.

Exercises

1. In which of the following is ∘ a binary operation on *S*? Which are commutative and which are not?

    (a) $S = \mathbf{Z}$ and $a \circ b = a + b$.　　(b) $S = \mathbf{Q}^*$ and $a \circ b = a/b$.

    (c) $S = \mathbf{R}$ and $a \circ b = a$.　　(d) $S = \mathbf{Q}^*$ and $a \circ b = \sqrt[a]{b}$.

    (e) $S = \mathbf{Q}$ and $a \circ b = a/b$.　　(f) $S = \mathbf{Z}$ and $a \circ b = a^b,\, a \neq 0$.

    (g) $S = \mathbf{Q}^*$ and $a \circ b = a^b$.　　(h) $S = \mathbf{Q}$ and $a \circ b = a + b - ab$.

2. In those cases of Exercise 1 where ∘ is a binary operation, which are groups? When they are groups, point out which is the identity element and which is the inverse.

3. If $S = \mathbf{Z}$, does $a \circ b = a - b$ define a group?

4. Let $S = \{1, i, -1, -i\}$, where $i^2 = -1$, and define $a \circ b$ to be the product ab for each pair of elements of *S*. Is *S* a group? Is it Abelian?

5. Suppose that a group has exactly two elements, *e* and *a*, where *e* is the identity. What is the inverse of *a*?

6. Let C mean "move to the other side of the river." How can we define $C \circ C$ so that C will be an element of a group having two elements?

7. Let c be a root of $x^2 + x + 1 = 0$ and define the binary operation to be multiplication. Show that the set $S = \{1, c, c^2\}$ is a group.

8. Let R denote a rotation of $120°$ about the origin in the coordinate plane. Let $R \circ R$ denote "rotate through $120°$ and then rotate again through the same angle in the same direction." Do the elements R and $R \circ R$ form a group? If not, what is the smallest group containing both R and $R \circ R$?

9. Suppose that R denotes a rotation of 1 radian about the origin in the coordinate plane and let $R \circ R$ denote "rotation twice in succession through an angle of 1 radian." What is the smallest group that contains R?

10. What are the additive and multiplicative identities for $\mathbf{Z}[x]$? What polynomials in $\mathbf{Z}[x]$ have the property that their reciprocal is in $\mathbf{Z}[x]$?

11. Answer Exercise 10 for $\mathbf{Q}[x]$, the set of polynomials with rational coefficients.

12. Prove properties 5, 7, and 8 of a group.

13. Does the set of polynomials in x with real coefficients form a group under multiplication? Explain your answer.

14. Suppose that a group G contains all the nonzero polynomials with rational coefficients and that the operation is multiplication. What other elements must G contain?

15. Let $S = \{0, 1\}$. Is S a group under one or both of addition and multiplication? Explain.

16. For a group, show that $(\overline{\overline{a}}) = a$.

17. Let $G = \{x^i\}$, where i is in $\mathbf{Z}$. Define $x^i \circ x^j = x^{i+j}$ for all integers i and j and $x^0 = 1$. Prove that G is an Abelian group.

18. Let G be a group and H the subset of all elements of G with the property that $h \circ g = g \circ h$ for every h in H and every g in G. Prove that H is a group.

19. Show that if there is a binary operation $\circ$ on a set S that is associative and satisfies conditions 6, then S is a group.

$\star$**20.** Suppose that every element of a group has the property that its square is the identity. Prove that the group is Abelian.†

† Here and elsewhere in this book, starred exercises are those that seem to the author to be more difficult.

*21. Prove that if $(ab)^2 = a^2b^2$ for all elements of a group, then it is Abelian.

*22. Suppose that in the definition of a group we replace properties 2 and 3 by

2′. There exists an element of S such that $e \circ a = a$ for all a in S.
3′. For every element a in S there is an element $\bar{a}$ in S such that $\bar{a} \circ a = e$.

That is, we confine ourselves to the identity and inverse on the left. Show that 2′ and 3′ imply properties 2 and 3.

3.   *Vectors*

We know that in analytic geometry, every point in the plane may be located with reference to two perpendicular axes by an *ordered* pair of real numbers (x, y) called the *coordinates* of the point. Every ordered pair determines a point, and every point determines an ordered pair. Notice that (x, y) is an *ordered* pair, since (x, y) and (y, x) represent different points unless x and y are equal. In fact, the two points (a, b) and (a', b') are the same if and only if $a = a'$ and $b = b'$.

Each ordered pair of numbers also defines a vector from the origin to the point in question. From this point of view we call the ordered pair a *vector*. Just as for points, the two vectors (a, b) and (a', b') are said to be equal if and only if $a = a'$ and $b = b'$. An advantage of thinking of ordered pairs as vectors is that we can then add vectors by "completing the parallelogram," as is shown in Figure I.1. That is, if $A = (a_1, a_2)$ and $B = (b_1, b_2)$, we get the vector that is the sum of these two by forming the parallelogram, two of whose sides are OA and OB. Then if C is the fourth vertex, the sum of the two vectors is OC. Algebraically, this amounts to

(3.1) $$(a_1, a_2) + (b_1, b_2) = (a_1 + b_1, a_2 + b_2).$$

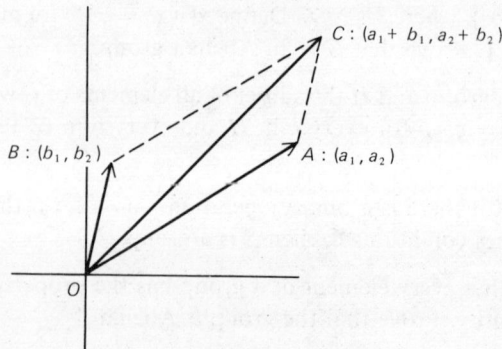

Figure I.1

Notice that the plus sign on the left refers to the addition of two vectors and those on the right to the addition of numbers. Thus we are using the same symbol for two different operations.

Now let S be the set of all ordered pairs of real numbers (vectors if you will) (x_1, x_2) and let the operation be vector addition as in (3.1). We show part of the following theorem.

Theorem 3.1. The set of all vectors (x_1, x_2), where x_1 and x_2 are real numbers, under (or with respect to) the operation of vector addition, forms a group.

Partial proof: Equation (3.1) shows that vector addition is a binary operation. To show that vector addition is associative, we perform the following calculations:

$$[(a_1, a_2) + (b_1, b_2)] + (c_1, c_2) = (a_1 + b_1, a_2 + b_2) + (c_1, c_2)$$
$$= [(a_1 + b_1) + c_1, (a_2 + b_2) + c_2],$$
$$(a_1, a_2) + [(b_1, b_2) + (c_1, c_2)] = (a_1, a_2) + (b_1 + c_1, b_2 + c_2)$$
$$= [a_1 + (b_1 + c_1), a_2 + (b_2 + c_2)].$$

Since addition of real numbers is associative, $(a_i + b_i) + c_i = a_i + (b_i + c_i)$ for $i = 1$ and 2 and we see that

$$[(a_1, a_2) + (b_1, b_2)] + (c_1, c_2) = (a_1, a_2) + [(b_1, b_2) + (c_1, c_2)].$$

That is, addition of vectors is associative.

We leave as Exercise 15 after Section 4 the proof that there is an identity and inverse for vector addition. This will complete the proof.

We should take the opportunity to recall a method of representing graphically complex numbers that you have probably already seen. If x and y are real numbers, the point (x, y) can be considered to be the point associated with the single complex number $x + iy$, where $i^2 = -1$. This representation is often called the *Argand diagram*. Instead of adding vectors, we can add complex numbers; that is, corresponding to (3.1) we have

$$(a_1 + ia_2) + (b_1 + ib_2) = (a_1 + b_1) + i(a_2 + b_2).$$

We shall exploit this idea further in Section 8 and also in Chapter II.

Theorem 3.1 is also true for ordered triples instead of ordered pairs of numbers. Indeed, it holds for n-tuples. [$(a_1, a_2, \ldots, a_n)$ is an n-tuple.] Also, the numbers in the pairs, triples, or n-tuples need not be real; they could be rational, complex, or indeed any elements of a field.

Can one form a group by considering in some sense a product of vectors? One could look at the *inner product* or *dot product*:

$$(x_1, x_2) \cdot (y_1, y_2) = x_1 y_1 + x_2 y_2.$$

But this does not define a binary operation, since combining two vectors in this way does not yield a vector but a number. You may know that for three dimensions there is a *vector product* or *cross product*. Although such an operation is a binary operation, vectors do not form a group for this operation; for one thing, the associative property does not hold. A third possibility might be multiplication by a scalar defined by

$$c(x_1, x_2) = (cx_1, cx_2).$$

Although this is a useful concept, it does not yield a group, since here a number and a vector are being combined instead of two vectors.

We shall, however, find it helpful to use scalar multiplication, partly because it implies that

$$x_1(1, 0) + x_2(0, 1) = (x_1, x_2).$$

We say that every vector in two dimensions is a *linear combination* of the vectors $(1, 0)$ and $(0, 1)$. We shall see later that this is true for any pair of vectors in place of $(1, 0)$ and $(0, 1)$, provided that neither is a scalar multiple of the other.

4. Matrices of Order 2

A rectangular array of numbers is called a *matrix*. For instance, a vector is a matrix that has one row. In this section we shall deal with groups of matrices of *order* 2, that is, matrices with two rows and two columns. (In Chapter IV we consider matrices in general.) Two matrices are called *equal* if corresponding elements are the same. For instance,

$$\begin{bmatrix} 1 & 2 \\ 3 & 4 \end{bmatrix} = \begin{bmatrix} \frac{3}{3} & \frac{8}{4} \\ 3 & 4 \end{bmatrix} \quad \text{but} \quad \begin{bmatrix} 1 & 2 \\ 3 & 4 \end{bmatrix} \neq \begin{bmatrix} 1 & 3 \\ 2 & 4 \end{bmatrix}.$$

The first group of this section is obtained by letting S be the set of matrices of order 2 and defining addition of matrices just as we defined it for vectors, by adding corresponding elements:

$$(4.1) \qquad \begin{bmatrix} a_1 & a_2 \\ b_1 & b_2 \end{bmatrix} + \begin{bmatrix} a_1' & a_2' \\ b_1' & b_2' \end{bmatrix} = \begin{bmatrix} a_1 + a_1' & a_2 + a_2' \\ b_1 + b_1' & b_2 + b_2' \end{bmatrix}.$$

Again, the plus sign between the two matrices indicates a different kind of addition from that of numbers. The additive identity is the *zero matrix*, that is, the matrix of order 2 all of whose elements are zero. We leave as Exercise 5 the proof that the set of matrices of order 2 with elements in a field forms a group under addition as defined in (4.1). We could carry through the same process for square matrices of any size except that all matrices of any such set S must be of the same order.

Vector addition extends to matrix addition with little alteration. But here, contrary to the situation with vectors, we shall see that a matrix product can be defined so that some subset S^* of matrices of order 2 is a multiplicative group. The most useful definition of the product of two matrices is the following:

$$(4.2) \qquad \begin{bmatrix} a_1 & a_2 \\ b_1 & b_2 \end{bmatrix} \cdot \begin{bmatrix} a_1' & a_2' \\ b_1' & b_2' \end{bmatrix} = \begin{bmatrix} a_1 a_1' + a_2 b_1' & a_1 a_2' + a_2 b_2' \\ b_1 a_1' + b_2 b_1' & b_1 a_2' + b_2 b_2' \end{bmatrix}.$$

(Motivation for this definition is given in Section 6.) To get the number in the first row and first column of the product, find the sum of products of corresponding elements in the first row of the left-hand matrix and elements in the first column of the right-hand matrix. To get the element in the first row and second column of the product, use in a similar manner the first row of the left matrix with the second column of the right matrix. The other two elements of the product can be calculated similarly. It is apparent from (4.2) that multiplication of such matrices is a binary operation.

Does the set of matrices of order 2 with coefficients in a field form a group under multiplication? Let us check the properties. The associative property could be verified by direct calculation, but the algebra involved is somewhat tedious. Since we shall have in Section 6 a better method of showing associativity, we assume it for the present.

The identity matrix for multiplication turns out to be

$$I = \begin{bmatrix} 1 & 0 \\ 0 & 1 \end{bmatrix},$$

since multiplication shows that $A \cdot I = I \cdot A = A$ for every matrix A of order 2. It will follow from property 4 of a group that this is *the* identity, once we have proved that we indeed have a group.

The third property, that of having a multiplicative inverse, presents some problems. We notice, for instance, that

$$\begin{bmatrix} x_1 & x_2 \\ y_1 & y_2 \end{bmatrix} \begin{bmatrix} 1 & 0 \\ 5 & 0 \end{bmatrix} = \begin{bmatrix} 1 & 0 \\ 0 & 1 \end{bmatrix}$$

is impossible for any values of x_1, x_2, y_1, and y_2, since the product on the left must have all zeros in the second column. Even though the matrix is different from the zero matrix, it has no multiplicative inverse. This means that if we are to have a group under multiplication, we must be content with a subset of the matrices of order 2. To determine this subset, we set out to solve the following matrix equation:

$$(4.3) \qquad \begin{bmatrix} x_1 & x_2 \\ y_1 & y_2 \end{bmatrix} \begin{bmatrix} a_1 & a_2 \\ b_1 & b_2 \end{bmatrix} = \begin{bmatrix} 1 & 0 \\ 0 & 1 \end{bmatrix}.$$

If we carry out the multiplication in accordance with (4.2), we get the following four equations, which must be satisfied if there is to be a solution:

(4.4)
$$x_1a_1 + x_2b_1 = 1, \qquad x_1a_2 + x_2b_2 = 0,$$
$$y_1a_1 + y_2b_1 = 0, \qquad y_1a_2 + y_2b_2 = 1.$$

To solve the pair of equations in x_1 and x_2 for x_1 by the usual process, we multiply the first by b_2, the second by b_1, and subtract. Similarly, we can solve for x_2, y_1, and y_2. This yields

(4.5)
$$x_1(a_1b_2 - b_1a_2) = b_2, \qquad x_2(a_1b_2 - b_1a_2) = -a_2,$$
$$y_1(a_1b_2 - b_1a_2) = -b_1, \qquad y_2(a_1b_2 - b_1a_2) = a_1.$$

We see that if there is a solution, $d = a_1b_2 - b_1a_2 = 0$ would imply that a_1, a_2, b_1, and b_2 are all zero, which contradicts (4.4). So if there is a solution, $d \neq 0$ and x_1, x_2, y_1, and y_2 are determined by (4.5).

On the other hand, if $d \neq 0$, substitution shows that the following is a solution of (4.4):

(4.6)
$$x_1 = \frac{b_2}{d}, \qquad x_2 = -\frac{a_2}{d}, \qquad y_1 = -\frac{b_1}{d}, \qquad y_2 = \frac{a_1}{d}.$$

Thus for the matrix

(4.7)
$$A = \begin{bmatrix} a_1 & a_2 \\ b_1 & b_2 \end{bmatrix},$$

we have almost proved the following theorem.

Theorem 4.1. The matrix A in (4.7) has a multiplicative inverse if and only if $d \neq 0$, where $d = a_1b_2 - b_1a_2$. In that case we denote the inverse of A by A^{-1} and have

$$A^{-1} = \begin{bmatrix} \dfrac{b_2}{d} & -\dfrac{a_2}{d} \\ -\dfrac{b_1}{d} & \dfrac{a_1}{d} \end{bmatrix}.$$

The proof is not quite complete, since we must show that $AA^{-1} = I$ also. This can be done by direct multiplication. Henceforth in referring to the multiplicative inverse of a matrix, we omit the adjective "multiplicative," since we call the additive inverse of A the negative of A, or $-A$.

Notice that for matrices of order 2, the inverse of A can be obtained by interchanging a_1 and b_2, changing the signs of the other two elements, and dividing each element by d. The number $d = a_1b_2 - b_1a_2$ is called the *determinant* of A in (4.7) and is written det(A). Any matrix whose determinant is not zero, that is, which has a multiplicative inverse, is called *nonsingular*. Otherwise it is called *singular*.

Since the above result is an important one, it is useful to look at it also from a vector point of view. So we prove, using vectors, another theorem, which turns out to be equivalent to Theorem 4.1.

Theorem 4.2.   Let $A = (a_1, a_2)$ and $B = (b_1, b_2)$ be two vectors whose corresponding points are not collinear with the origin. Then for every vector (c_1, c_2), the following equation is solvable for x_1 and x_2:

$$(4.8) \qquad\qquad x_1(a_1, a_2) + x_2(b_1, b_2) = (c_1, c_2).$$

Proof: Designate the points A, B, and C as in Figure I.2. We want to express the vector OC as the resultant (that is, the vector sum) of some multiple of OA and some multiple of OB. To this end, let the line through C, parallel to OB, intersect OA in the point A' and the line through C, parallel to OA, intersect OB in the point B'. Then OC is the vector sum of the vectors OA' and OB'. Moreover, $A' = x_1(a_1, a_2)$ and $B' = x_2(b_1, b_2)$ determine a solution x_1, x_2 of (4.8). This construction can always be carried through, provided that the points O, A, and B are not collinear. This completes the proof.

In Theorem 4.1 we showed that equations (4.4) are solvable if and only if $d = a_1b_2 - b_1a_2$ is different from zero. In Theorem 4.2 we showed that if the points $(0, 0)$, (a_1, a_2), and (b_1, b_2) are not collinear, then (4.8) is solvable for every pair (c_1, c_2), in particular for the two pairs $(1, 0)$ and $(0, 1)$. Thus Theorem 4.2 affirms that if the points are not collinear, equations (4.4) are solvable. This suggests the following result, which correlates the two theorems.

Theorem 4.3.   The points $(0, 0)$, (a_1, a_2), and (b_1, b_2) are collinear if and only if $d = a_1b_2 - a_2b_1 = 0$.

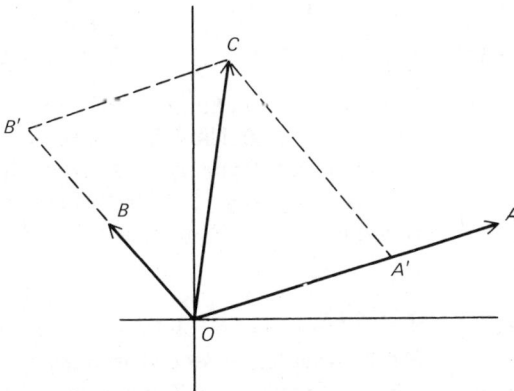

Figure I.2

Proof: Notice first that the points O, A, and B are collinear if and only if the vector (b_1, b_2) is a scalar multiple of (a_1, a_2). Now if the points are collinear, then $(b_1, b_2) = m(a_1, a_2)$ for some number m. Thus $b_1 = ma_1$ and $b_2 = ma_2$, which imply that

$$d = a_1 b_2 - b_1 a_2 = a_1 m a_2 - m a_1 a_2 = 0.$$

If $d = 0$, then

$$b_1(a_1, a_2) - a_1(b_1, b_2) = (0, b_1 a_2 - a_1 b_2) = (0, 0).$$

If a_1 and b_1 are both zero, the points A and B are collinear with O on the y-axis. Otherwise one of (a_1, a_2) and (b_1, b_2) is a scalar multiple of the other. In both cases the points are collinear and the proof is complete.

In this connection it is interesting to note that the absolute value of $d = a_1 b_2 - a_2 b_1$ is twice the area of the triangle whose vertices are the three points O, A, and B. You are asked to prove this in Exercise 24 after Section 8. Theorem 4.3 follows almost immediately from this result, since if three points determine a triangle, the area cannot be zero.

Now, with all this information, we can prove the following theorem.

Theorem 4.4. The set S^* of nonsingular matrices of order 2 is a multiplicative group.

Proof: First, we must show that the operation of multiplication is a binary operation for the subset S^* of S, the set of matrices of order 2. Now, multiplication *is* such an operation for the set S; that is, if A and B are in S so is their product. To show that it is a binary operation for S^*, we need to prove that if A and B are in S^*, their product is also. So, suppose that matrices X and Y both have inverses. We want to show that XY has an inverse. Taking our cue from property 8 of a group, we can try $Y^{-1}X^{-1}$ to see if it is an inverse of XY. We have

$$(XY)(Y^{-1}X^{-1}) = X(YY^{-1})X^{-1} = X \cdot I \cdot X^{-1} = X \cdot X^{-1} = I,$$

using the associative property and the definition of the inverse. Similarly, it can be shown that $(Y^{-1}X^{-1})(XY) = I$. Thus the binary operation is closed over the set of nonsingular matrices. Property 2 of groups holds, since I is nonsingular (its determinant is 1). Finally, X^{-1} is nonsingular, since it has X as its inverse by the definition of inverse. [Note that $(X^{-1})^{-1} = X$.] Thus the proof is complete.

Now, if A is nonsingular, $AX = B$ has a solution $X = A^{-1}B$ and $YA = B$ has the solution $Y = BA^{-1}$, whether or not B is nonsingular. (Compare property 6 of a group.) Also, as for property 7 of a group, $AB = AC$ implies that $B = C$ if A is nonsingular, since $A^{-1}AB = A^{-1}AC$.

Notice that the multiplicative group of nonsingular matrices of order 2 is not Abelian, as is shown by the following example:

$$A = \begin{bmatrix} 1 & 1 \\ 0 & 1 \end{bmatrix}, \quad B = \begin{bmatrix} 1 & 0 \\ 0 & 2 \end{bmatrix}, \quad AB = \begin{bmatrix} 1 & 2 \\ 0 & 2 \end{bmatrix}, \quad BA = \begin{bmatrix} 1 & 1 \\ 0 & 2 \end{bmatrix}.$$

Exercises

1. For matrices A and B, find A^{-1}, B^{-1}, $(AB)^{-1}$, and $(BA)^{-1}$ and solve the equations $AX = B$ and $YA = B$.

$$A = \begin{bmatrix} 1 & 2 \\ 3 & 4 \end{bmatrix}, \quad B = \begin{bmatrix} 5 & 3 \\ 3 & 2 \end{bmatrix}.$$

2. Find the matrices asked for in Exercise 1 when

$$A = \begin{bmatrix} 3 & -5 \\ 5 & -8 \end{bmatrix}, \quad B = \begin{bmatrix} 4 & -7 \\ 7 & -12 \end{bmatrix}.$$

3. If in Exercise 2, B is replaced by the singular matrix

$$C = \begin{bmatrix} 1 & 2 \\ 2 & 4 \end{bmatrix},$$

does the inverse of CA exist? If the inverse exists, find it. Are the equations $AX = C$ and $YA = C$ solvable? If the equations are solvable, find the solutions.

4. In each case there is a given set S of matrices with elements in the number system noted. Which form additive groups, which multiplicative groups, which both, and which neither?

(a) $S = \begin{bmatrix} 1 & a \\ 0 & 1 \end{bmatrix}$, a in **R**.   (b) $S = \begin{bmatrix} a & b \\ 0 & c \end{bmatrix}$, $ac \neq 0$, a, b, c in **Q**.

(c) $S = \{I, A, A^2, A^3, \ldots\}$, where $A = \begin{bmatrix} 0 & -1 \\ 1 & 0 \end{bmatrix}$.

(d) $S = \{I, A^i\}$, where i is in **Z** and A is as in (c). Note that A^{-c} for c a positive integer is defined to be $(A^{-1})^c$.

★(e) $S = a_1 I + a_2 J + a_3 K + a_4 L$, where the a_i are real and

$$J = \begin{bmatrix} i & 0 \\ 0 & -i \end{bmatrix}, \quad i^2 = -1, \quad K = \begin{bmatrix} 0 & 1 \\ -1 & 0 \end{bmatrix}, \quad L = JK.$$

Note that $(a_1 I + a_2 J + a_3 K + a_4 L)(a_1 I - a_2 J - a_3 K - a_4 L) = (a_1^2 + a_2^2 + a_3^2 + a_4^2)I$. The numbers of S are called *quaternions*.

5. Prove that the set of all matrices of order 2 forms an additive group.

6. Let Z be the zero matrix of order 2 (its elements are all zero). Show that if A is a nonsingular matrix, then $AX = Z$ implies that $X = Z$.

7. Find all matrices X such that

$$\begin{bmatrix} 1 & 2 \\ 2 & 4 \end{bmatrix} X = Z,$$

where Z is the zero matrix.

8. Let A be a singular matrix. Show that $AX = Z$, for Z as defined in Exercise 6 has more than one solution.

9. Prove that if AB is nonsingular for two matrices of order 2, then both A and B are nonsingular.

10. If A, B, and C are nonsingular matrices of order 2, express the inverse of ABC as a product of the inverses of A, B, and C.

11. If A is a nonsingular matrix of order 2, show that $\det(A) \cdot \det(A^{-1}) = 1$.

12. Find a necessary and sufficient condition that the matrix A in (4.7) have the property that $A^2 = I$, the identity matrix.

13. Does the set of nonsingular matrices of order 2 form a group under addition? Give reasons for your answer.

14. Show that if A, A', and A'' are three matrices of order 2, then $A(A' + A'') = AA' + AA''$. This is the distributive property. Also show that $(A' + A'')A = A'A + A''A$.

15. Complete the proof of Theorem 3.1.

16. Does Theorem 3.1 hold if the numbers x_1 and x_2 are complex numbers?

17. Does Theorem 3.1 hold if the numbers x_1 and x_2 are restricted to be integers?

18. Does Theorem 3.1 hold if x_1 and x_2 are polynomials in z with real coefficients?

19. Define the product $(x_1, x_2)B$, where B is a matrix of order 2, in accordance with (4.2). Let C be a matrix of order 2. For this definition show that the distributive property holds; that is,

$$(x_1, x_2)(B + C) = (x_1, x_2)B + (x_1, x_2)C.$$

20. If the letters have the same meaning as in Exercise 19, which of the following hold?

(a) $[(x_1, x_2) + (y_1, y_2)]B = (x_1, x_2)B + (y_1, y_2)B$.
(b) $(B + C)(x_1, x_2) = B(x_1, x_2) + C(x_1, x_2)$.

21. Suppose that we define $(x_1, x_2) \cong (y_1, y_2)$ to mean $x_1 y_2 = x_2 y_1$ just as for fractions x_1/x_2 and y_1/y_2. Then give an example to show that $(x_1, x_2) \cong (y_1, y_2)$ does not imply that

$$(a_1, a_2) + (x_1, x_2) \cong (a_1, a_2) + (y_1, y_2).$$

Is the operation of addition with reference to the relationship $\cong$ a binary operation? Why or why not?

22. If the matrix A is nonsingular and B is singular, what can be said about the existence of solutions of $AX = B$ and $YA = B$? That is, do such solutions always exist? If not, under what conditions will they exist?

23. Show that the product of the determinants of two matrices of order 2 is equal to the determinant of their product. Show how this can be used to prove that the product of two nonsingular matrices of order 2 is non-singular.

24. Prove that if A and B are two matrices of order 2 such that

$$A^2 = I = B^2 = (AB)^2,$$

then $AB = BA$. (Compare Exercise 21 of Section 2.)

25. Use the definition of Exercise 19 to show that the matrix B is nonsingular if and only if $(x_1, x_2)B = (0, 0)$ implies that $(x_1, x_2) = (0, 0)$.

5. Groups of Functions Under Composition

Here we introduce a certain group of functions that is of fundamental importance in projective geometry. Initially, we identify a function with an algebraic expression. Then at the close of the section, we shall point out how a more careful treatment could be made on the basis of a more precise definition of a function.

Consider the following two functions or algebraic expressions:

(5.1) $$f(x) = 1 - x \quad \text{and} \quad g(x) = \frac{1}{x},$$

where x is a real number. If $g(x)$ is to be a function of x, we must exclude $x = 0$. We could combine these two functions by addition or multiplication, but this would not be very different from what we have considered previously. Another fruitful way of combination is by composition. That is, $f[g(x)] = f[1/x] = 1 - 1/x$ and $g[f(x)] = g[1 - x] = 1/(1 - x)$. For the latter to be a function of x, we must exclude $x = 1$. These can be written more briefly as

(5.2) $$fg(x) = 1 - \frac{1}{x} \quad \text{and} \quad gf(x) = \frac{1}{1 - x},$$

where by fg we mean the function obtained by applying g first and f second; for gf we apply f first. It is understood that x can be any real number except 0 and 1. Notice that fg and gf are not the same function. It should be emphasized that fg is "f of g"—not the product of $f(x)$ and $g(x)$. Also, the four functions exhibited above exist for all real numbers x except 0 and 1.

Now if we let $e(x) = x$ denote the identity function, we already have two groups of order 2:

$$(5.3) \qquad G_1 = \{e(x), f(x)\}, \qquad G_2 = \{e(x), g(x)\},$$

since $ff(x) = e(x)$ and $gg(x) = e(x)$. We customarily write ff as f^2, but we must keep in mind that it is "f of f" and not f squared. Since each of f and g is its own inverse, there is no problem with associativity.

However, if composition is to be a binary operation on the elements of a group G that contains both g and f, we must have closure. That is, the group G must contain not only fg and gf but all other functions that can be written as a sequence of compositions of f's and g's. Computation shows that

$$(5.4) \qquad gfg(x) = \frac{x}{x - 1} = fgf(x),$$

which is a sixth function in G. Let us list them:

$$(5.5)$$

$e(x)$	$f(x)$	$g(x)$	$fg(x)$	$gf(x)$	$gfg(x) = fgf(x)$
x	$1 - x$	$\dfrac{1}{x}$	$1 - \dfrac{1}{x}$	$\dfrac{1}{1 - x}$	$\dfrac{x}{x - 1}$

Are there more? In view of (5.4) and $f^2 = g^2 = e$, it follows that every "product" of four or more f's and g's is equal to a "product" of three or fewer, assuming that composition is associative [see (5.6)]. For instance, $gfgf(x) = g(gfg)(x) = fg(x)$. Hence every "product" of a sequence of f's and g's is equal to one of the functions in (5.5). Thus the operation of composition is closed; that is, it is a binary operation over the six functions. Now we can write a "multiplication table," that is, a table for the operation of composition (Table I.1). The operation is a binary operation, since only the six functions appear in the body of the table, that is, since the set is closed under the operation of composition. To verify the associative property, write the following sequence of equations:

$$(5.6) \qquad fg[h(x)] = f\{g[h(x)]\} = f[gh(x)] = \{f[gh]\}(x).$$

So the expression $fgh(x)$ is not ambiguous. It means, apply first h, then g, and then f. The table shows that every function has an inverse, since e occurs in each row and column. Hence these six functions form a group under composition, provided that $x = 0$ and $x = 1$ are excluded.

This is an important group in projective geometry. In fact, let a, b, c, and d be four distinct real numbers and

$$(5.7) \qquad \frac{(a - c)(b - d)}{(b - c)(a - d)} = x.$$

Table I.1

∘	e	f	g	fg	gf	$fgf = gfg$
e	e	f	g	fg	gf	fgf
f	f	e	fg	g	fgf	gf
g	g	gf	e	fgf	f	fg
fg	fg	fgf	f	gf	e	g
gf	gf	g	fgf	e	fg	f
fgf	fgf	fg	gf	f	g	e

For instance, if a and c as well as b and d are interchanged, the ratio x is not changed, whereas if a and b are interchanged, x is replaced by $g(x) = 1/x$. It can be shown that if one considers all the 24 permutations of a, b, c, and d one gets exactly the six functions dealt with above, each one occurring four times. The expression (5.7) is called the *cross ratio*.

Now let us look more carefully at the idea of a function. Recall that a function f is a correspondence from a set A to a set B with the property that every element of A corresponds under f to a unique element of B, called its *image* under f. The set A is called the *domain* of f, and the set of images is called the *range* of f. We say that f is a function or transformation of A *onto* B or *into* any set that contains B. If the composition fg of two functions is to have meaning, in the sense that g is first and f second, then the range of g must be in the domain of f. If both fg and gf are to have meaning, the range of each must be in the domain of the other.

For instance, suppose that the function $k(x) = x^2$ is over all rational numbers. Then $\mathbf{Q}$, the set of rational numbers, is its domain, and its range is the set of (nonnegative) squares of rational numbers. This function is a transformation or mapping of $\mathbf{Q}$ onto the set of squares of rational numbers and, for instance, into the set of rational numbers. For this function the image of a rational number is its square. Suppose that $p(x) = \sqrt{x}$ for the set of nonnegative real numbers x. Then the range of $p(x)$ is the same as its domain. Now the function $kp(x)$ does not exist, since the range of p contains irrational numbers, and such numbers are not in the domain of k. But $pk(x)$ does exist, since every number in the range of k, being a nonnegative rational number, is in the domain of p. In fact, $pk(x) = p(x^2) = \sqrt{x^2} = |x|$ for all rational numbers x, and its range is the set of nonnegative rational numbers.

Consider the functions of (5.1) in the light of the two previous paragraphs. The domain of $g(x)$ is $\mathbf{R}^*$, the set of nonzero real numbers. For gf to have meaning, the range of f must be in the domain of g. Since 0 is not in the domain of g, it cannot be in the range of f. Thus we must exclude $x = 1$ from the domain of f. Furthermore, if fg is to have meaning, we must exclude $x = 1$ from the range of g; that is, $x = 1$ cannot be in the domain of g. Thus if fg and gf are both to have meaning, we must exclude both $x = 0$ and $x = 1$. On the other hand, if we let S be the set of real numbers with 0 and 1 excluded, S will be the domain and range of f and g. In fact, for all six functions the domain and range is S.

There is another example of a group of functions in the next section, and others in the exercises that follow.

6. Transformations on Vectors

There are three terms in mathematics that have much the same fundamental meaning but arise from different contexts. These are the terms "function," "transformation," and "mapping." The first of these, which was described in Section 5, is most popular in analysis and we think of it as a correspondence. For instance, there is a *correspondence* between the set of numbers x and the set x^2. This same *function* can also be thought of as a *transformation* that "takes x into x^2" or transforms a number into its square. We could also express this relationship by calling it a *mapping* of x onto x^2, just as, in a map, points correspond to towns. So a binary operation can be thought of as a *transformation*, or a *function*, or a *mapping* of pairs of elements of a set S into S itself.

In this section we call a function a transformation because the context is algebraic. The term "mapping" is equally relevant. We emphasize the change in point of view from function to transformation or mapping by our notation. Let (x, y) be a pair of coordinates of a point in a plane. Then the equations

$$(6.1) \qquad\qquad x' = x + h, \qquad y' = y + k$$

define another point (x', y') obtained from (x, y) by adding h to the first coordinate and k to the second. We say that $(x + h, y + k)$ is obtained from (x, y) by a horizontal *translation* of h units and a vertical one of k units, where h and k may be any real numbers. This can be written

$$(6.2) \quad (x, y) \overset{\tau}{\to} (x + h, y + k) \qquad \text{or} \qquad (x, y)\tau = (x + h, y + k),$$

where τ denotes the translation that "takes the vector on the left into the vector on the right." A translation is a special kind of transformation. In general we shall use Greek letters for transformations.

Let τ' be another translation defined by

$$(6.3) \qquad\qquad (x, y)\tau' = (x + h', y + k').$$

Then we can combine the two translations by composition as follows:

$$(6.4) \quad [(x, y)\tau]\tau' = (x + h, y + k)\tau' = (x + h + h', y + k + k'),$$

and we can write the combination of the two translations of (6.4) as $\tau\tau'$. This means "do τ first and τ' second"; so the order of operation is from left to right.

It is a problem to choose which order to use in combining transformations. Here, where $(x, y)\tau$ denotes the image of (x, y), $\tau\tau'$ means "do τ first and then τ'." If we had used the functional notation as in the previous section, $fh(z)$, then h is applied first and f second. There are advantages and disadvantages in either choice. In this book, except where functional notation is used, we write the symbol for the transformation or mapping on the right. When this is done, whenever two transformations are combined, it is the one on the left that is applied first. This seems to be in accord with standard practice in most algebra books, but in reading any book one should bear in mind that mathematical usage is not uniform in this respect.

In this particular case, $\tau\tau' = \tau'\tau$ and the order is immaterial, but in most cases it does make a difference. As you would suspect by this time, our next question is: Does the set of all translations form a group? The answer is in the affirmative, and we leave the proof as an exercise.

Your first question at this point might well be: Why this obsession with showing that something is a group? We reply that we have already shown that it is a fruitful idea. Also, it points up relationships among quite dissimilar parts of mathematics. Furthermore, if you want to deal with a set of transformations, you need to be able to combine them in a meaningful manner—that means a binary operation. It is hard to get very far in combining unless you have associativity. It is convenient to be able to "undo" what you have done. And by that time you practically have a group.

A transformation of vectors of special importance is the *linear transformation* λ, defined as follows:

(6.5) $$(x_1, x_2)\lambda = (x_1a_1 + x_2b_1, x_1a_2 + x_2b_2).$$

The vector on the right is the image of the vector on the left under the transformation λ. Let L be the matrix

(6.6) $$L = \begin{bmatrix} a_1 & a_2 \\ b_1 & b_2 \end{bmatrix}.$$

If we define multiplication of a vector by a matrix on the right as we did for a product of matrices in (4.2), we can write (6.5) in matrix form as follows:

(6.7) $$(x_1, x_2)L = (x_1a_1 + x_2b_1, x_1a_2 + x_2b_2).$$

This can be written more compactly as $XL = X'$ for a vector X and its image X'. Notice that every linear transformation takes $(0, 0)$ into itself.

Each linear transformation has its matrix, and each matrix determines a linear transformation. That is, we have a one-to-one correspondence between the sets of linear transformations λ and the set of matrices L of order 2. This correspondence has at least two advantages. One is that we can use our knowledge of matrices to get information about linear transformations. Another is that we can prove associativity of multiplication of matrices from our knowledge that transformations are associative under composition.

But before doing this, we must show that this correspondence is such that if λ corresponds to L and λ' to L', then the "product" $\lambda\lambda'$ (that is, λ followed by λ') corresponds to LL'. This we do by working each out separately as follows. Let λ' be the transformation obtained from λ by replacing a_i and b_i by a_i' and b_i', respectively:

$$
\begin{aligned}
(x_1, x_2)&\lambda\lambda' \\
&= (x_1a_1 + x_2b_1, x_1a_2 + x_2b_2)\lambda' \\
&= ([x_1a_1 + x_2b_1]a_1' + [x_1a_2 + x_2b_2]b_1', [x_1a_1 + x_2b_1]a_2' + [x_1a_2 + x_2b_2]b_2') \\
&= (x_1[a_1a_1' + a_2b_1'] + x_2[b_1a_1' + b_2b_1'], x_1[a_1a_2' + a_2b_2'] + x_2[b_1a_2' + b_2b_2']) \\
&= (x_1a_1'' + x_2b_1'', x_1a_2'' + x_2b_2''),
\end{aligned}
$$

where

$$
\text{(6.8)} \qquad
\begin{aligned}
a_1'' &= a_1a_1' + a_2b_1', & a_2'' &= a_1a_2' + a_2b_2', \\
b_1'' &= b_1a_1' + b_2b_1', & b_2'' &= b_1a_2' + b_2b_2',
\end{aligned}
$$

since the components of the vectors are real (or, indeed, in any field). So if

$$
\text{(6.9)} \qquad (x_1, x_2) = X, \qquad (x_1, x_2)\lambda = X', \qquad X'\lambda = X'',
$$

we have the following corresponding matrix products:

$$
XL = X', \qquad X'L' = X''.
$$

Now if we define L'' by

$$
L'' = \begin{bmatrix} a_1'' & a_2'' \\ b_1'' & b_2'' \end{bmatrix},
$$

whose elements are defined in (6.8), we see, by comparison with (4.2), that $L'' = LL'$. (In fact, this outcome is perhaps the principal reason for defining the product of matrices as we did in the first place.) Thus

$$
\text{(6.10)} \qquad X(LL') = X''.
$$

Thus we have shown that

$$
\text{(6.11)} \qquad L \to \lambda, \quad L' \to \lambda' \qquad \text{implies that } LL' \to \lambda\lambda'.
$$

We can use this correspondence in two ways to prove that multiplication of matrices of order 2 is associative.

Theorem 6.1. Multiplication of matrices of order 2, whose elements are in a field, is associative.

Proof 1: From (6.11) we see that if λ, λ', and λ'' are three linear transformations and L, L', and L'' are three corresponding matrices, then

$$
(\lambda\lambda')\lambda'' \to (LL')L'' \qquad \text{and} \qquad \lambda(\lambda'\lambda'') \to L(L'L'').
$$

Since composition of linear transformations is associative, $(\lambda\lambda')\lambda'' = \lambda(\lambda'\lambda'')$. But there is a one-to-one correspondence between such transformations and matrices of order 2; hence $(LL')L'' = L(L'L'')$.

Proof 2:   Consider the product

$$\left\{\begin{bmatrix} x_1 & x_2 \\ y_1 & y_2 \end{bmatrix} L \right\} L' = \begin{bmatrix} X' \\ Y' \end{bmatrix} L' = \begin{bmatrix} X'' \\ Y'' \end{bmatrix},$$

where X' and X'' are defined as in (6.9) and Y' and Y'' are defined analogously. Also,

$$\begin{bmatrix} X \\ Y \end{bmatrix}(LL') = \begin{bmatrix} X'' \\ Y'' \end{bmatrix}.$$

If we let M be the matrix whose first row is X and second row $Y = (y_1, y_2)$, the above becomes $(ML)L' = M(LL')$, which is the associative property for multiplication of matrices of order 2. So we have two proofs of Theorem 6.1.

The correspondence between matrices and transformations also carries over for inverses, for suppose that the transformation λ has an inverse λ^{-1}. Then if L^{-1} is the matrix corresponding to λ^{-1}, we have that $X\lambda = XL = X'$ is equivalent to $X = X'\lambda^{-1} = X'L^{-1}$. That is, if L corresponds to λ and if λ has an inverse, then there is a matrix corresponding to λ^{-1} and this matrix is the inverse of L. So if we call a linear transformation *nonsingular* if it has an inverse, we see that a nonsingular transformation corresponds to a nonsingular matrix. Using Theorem 4.4 for matrices, we have immediately the corresponding theorem for transformations.

Theorem 6.2.   The set of nonsingular linear transformations of two-dimensional vectors is a group.

Using the result of Exercise 25 after Section 4, we deduce the following theorem.

Theorem 6.3.   A linear transformation λ of two-dimensional vectors is nonsingular if and only if $(0, 0)$ is the only vector whose image under λ is the zero vector.

Without much labor, what we have discussed in this section would carry over for transformations of vectors with n components, but we postpone until Chapter IV the discussion of such transformations.

7.   Subgroups

Before giving additional examples of groups, we should pay a little attention to the idea of a subgroup. There is nothing very complicated about this concept. A subset H of the elements of a group G, together with the operation of the group, is called a *subgroup* of G if it is itself a group. Note that under this definition, a group G is a subgroup of itself. When we wish to exclude *this* possibility, we use the term *proper subgroup*; that is, H is called a proper subgroup of G if H is a subgroup but $H \neq G$. Using the notation of sets, write $H \subseteq G$ or $G \supseteq H$; or if H is a proper subgroup, write $H \subset G$ or $G \supset H$.

We already have some examples of subgroups. The set of nonzero rational numbers under multiplication is a subgroup of the multiplicative group of nonzero real numbers. In Exercise 18 after Section 2 you were asked to show that if H is the subset of all elements of a group G such that $g \circ h = h \circ g$ for all g in G, then H is a subgroup of G. This important subgroup is called the *center* or centrum of G. If G is Abelian, it is its own center; if not, its center is a proper subgroup.

How does one determine whether or not a subset of a group is a group? The obvious answer is: See if the properties of a group hold for the subset. But we do not have to check *all* the properties. In particular, if H is a subset of a group G and if the binary operation for G is also a binary operation for H, then it must be associative in H, since all "products" in H are products in G. To be definite about this, we state and prove the following theorem.

Theorem 7.1. If H is a nonempty subset of elements of a group G, then H is a subgroup of G if

1. The binary operation of G takes each pair of elements of H into an element of H; that is, the operation is closed with respect to H.
2. When h is an element of H, then h^{-1} is also in H.

Proof: We use the multiplicative notation for the operation; that is, instead of $a \circ b$ we write ab. Since the operation is a binary operation for G, then ab is in G for all a and b in G. If this is to be a binary operation for H, then ab must be in H for all a and b in H. So requirement 1 affirms that the binary operation for G is also one for H. As we have noted, the binary operation is associative over H, since it is over G.

Requirement 2 is that the inverse of every element h of H is to be in H. Then requirement 1 implies that $hh^{-1} = e$ is in H. This completes the proof.

NOTE: If H is a finite group, requirement 2 can be deduced from requirement 1 as follows: Let h be some element. If h is the identity element of H, it is its own inverse. Let h be some element not the identity and $S = \{h^n\}$ over all positive integers n. Since H is finite, two powers of h must be equal, say $h^r = h^s$, for $r > s$. Hence, by the cancellation property, $h^{r-s} = e$, the identity element, and h^{r-s-1} is the inverse of h.

8.   The Orthogonal Group

An important subgroup of the set of nonsingular linear transformations depends on the idea of distance between two points. For this purpose we restrict the field to be that of real numbers and define the distance between two points (x_1, y_1) and (x_2, y_2) in the usual way as

$$\sqrt{(x_1 - x_2)^2 + (y_1 - y_2)^2}.$$

Euclidean geometry is concerned with properties of figures (set of points) that are unchanged under "rigid motions"—that is, transformations that preserve distance. In other words, we call a transformation a *rigid motion* or *Euclidean transformation* if the distance between any two points is equal to the distance between their images. Not all Euclidean transformations are linear; for instance, a translation is Euclidean, but it does not leave the origin $(0, 0)$ unchanged. We here deal with a subset of the Euclidean transformations defined as follows.

Definition.   *An* orthogonal transformation *is a Euclidean transformation that is linear.*

Theorem 8.1.   Let S be the set of vectors (x, y), where x and y are real numbers. The set of orthogonal transformations of S onto S is a subgroup of the multiplicative group of nonsingular linear transformations of S.

Proof: First, we must show that the set of orthogonal transformations is a subset of the nonsingular transformations. Now since an orthogonal transformation ρ preserves the distance of each point from the origin, and since the origin is the only point at distance 0 from the origin, it follows that $(x_1, x_2)\rho = (0, 0)$ implies that $(x_1, x_2) = (0, 0)$. Use of Theorem 6.3 then affirms that ρ must be nonsingular.

Since the set of nonsingular transformations on two-dimensional vectors is a multiplicative group, we need by Theorem 7.1 only show closure and the existence of an inverse. Let ρ and ρ' be two orthogonal transformations. Then the distance between points P_1 and P_2 is equal to that between $P_1\rho$ and $P_2\rho$; the latter is equal to the distance between $P_1\rho\rho'$ and $P_2\rho\rho'$. Hence the distance between P_1 and P_2 is equal to that between $P_1\rho\rho'$ and $P_2\rho\rho'$, which shows that $\rho\rho'$ is an orthogonal transformation, since the product of two linear transformations is linear.

To show that the inverse of an orthogonal transformation is orthogonal, note that $P_1\rho = Q_1$ and $P_2\rho = Q_2$ implies that $P_1 = Q_1\rho^{-1}$ and $P_2 = Q_2\rho^{-1}$; in both cases the distance between P_1 and P_2 is equal to that between Q_1 and Q_2. So we have proved the theorem.

Definition. *The group of all (nonsingular) linear transformations that preserve distance is called the* orthogonal group *and denoted for two dimensions by* O_2.

We now find explicitly the matrices of orthogonal transformations in two dimensions. Looking back at (6.5), we see that if the distance of (x_1, x_2) from the origin is to be preserved, then

(8.1) $$x_1^2 + x_2^2 = (a_1x_1 + b_1x_2)^2 + (a_2x_1 + b_2x_2)^2$$

for all pairs of real numbers (x_1, x_2). If this condition holds, then the distance between any two points and their images is the same, since the distance between the points (x_1, x_2) and (y_1, y_2) is the same as the distance between the origin and $(x_1 - y_1, x_2 - y_2)$.

If in (8.1) we take (x_1, x_2) to be in turn $(1, 0)$, $(0, 1)$, and $(1, 1)$, we get the following set of three equations:

(8.2) $$a_1^2 + a_2^2 = 1, \qquad b_1^2 + b_2^2 = 1, \qquad a_1b_1 + a_2b_2 = 0.$$

Geometrically, this means that the points (a_1, a_2) and (b_1, b_2) are on the circle with radius 1 and center at the origin—that is, the *unit circle*. Now let φ be the angle from the x-axis to the point (a_1, a_2) and see that $a_1 = \cos \varphi$ and $a_2 = \sin \varphi$. In general the solutions of the equation $a_1x_1 + a_2x_2 = 0$ are given by $x_1 = -ka_2$ and $x_2 = ka_1$, for some k. Thus $0 = a_1b_1 + a_2b_2$ implies that $b_1 = -ka_2$ and $b_2 = ka_1$. But

$$1 = b_1^2 + b_2^2 = k^2(a_1^2 + a_2^2) = k^2$$

implies that $k = 1$ or -1. Hence

(8.3) $(b_1, b_2) = (-\sin \varphi, \cos \varphi)$ or $(b_1, b_2) = (\sin \varphi, -\cos \varphi)$.

[Note in Figure I.3 that the vector (b_1, b_2) is in both cases perpendicular to the vector (a_1, a_2).] Thus, in the respective cases, the following are the matrices of the orthogonal transformations:

(8.4) $$L_1 = \begin{bmatrix} \cos \varphi & \sin \varphi \\ -\sin \varphi & \cos \varphi \end{bmatrix}, \qquad L_2 = \begin{bmatrix} \cos \varphi & \sin \varphi \\ \sin \varphi & -\cos \varphi \end{bmatrix}.$$

We leave as an exercise showing that, for $a_1 = \cos \varphi$, $a_2 = \sin \varphi$, and b_i as in (8.3), Equation (8.1) holds.

You will probably recognize L_1 as the matrix of the transformation for the rotation of axes. The second is probably less familiar. To determine the geometrical meaning of these two transformations, write (x_1, x_2) in polar form: $(r \cos \theta, r \sin \theta) = r(\cos \theta, \sin \theta)$. Then computation shows that

(8.5) $$r(\cos \theta, \sin \theta)L_1 = r[\cos(\varphi + \theta), \sin(\varphi + \theta)].$$

(8.6) $$r(\cos \theta, \sin \theta)L_2 = r[\cos(\varphi - \theta), \sin(\varphi - \theta)].$$

Equation (8.5) shows that L_1 is a rotation about the origin through an angle of φ in a counterclockwise direction if φ is positive. To interpret (8.6),

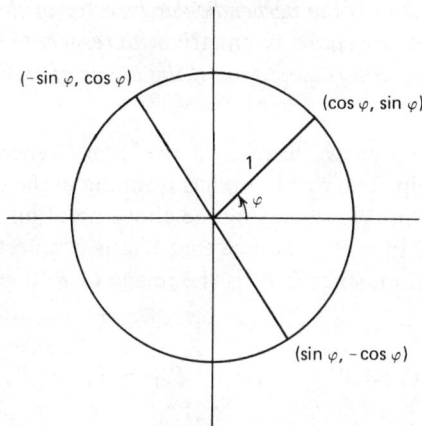

Figure I.3

see that the image of (x_1, x_2) under L_2 is on the line through the origin whose angle of inclination is $(\varphi - \theta)$ and at a distance of r from the origin. But $(\theta + \varphi - \theta)/2 = \varphi/2$ is independent of θ; that is, the bisector of the angle from the point to its image is $\varphi/2$. Thus the image of (x_1, x_2) (see Figure I.4) is a reflection in the line through the origin whose angle of inclination is $\frac{1}{2}\varphi$. An important property of L_2 is that its square is the identity. This can be seen geometrically or by squaring the matrix of L_2.

What really is a reflection? We think of it, quite properly, as an image in a mirror. When we write of a "reflection in a line," we consider the line as a mirror, which is in accord with the following definition.

Definition.   *Given a line m and a point P, if P is on m, it is its own* reflection *in m. If P is not on m, the reflection of P in m is the point P', called the* image

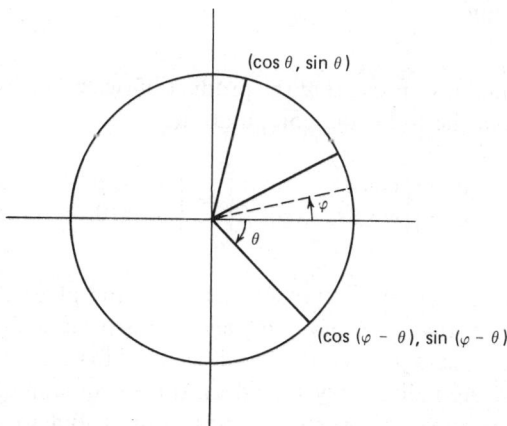

Figure I.4

of P in m such that m is the perpendicular bisector of the line segment PP'.
The points P and P' are called symmetric *with respect to m. A set S of points*
is called symmetric *with respect to m if the image of every point of S is in S.*

In the above definition we have used the term "reflection" in the static
sense of a relationship. The word also has meaning in the dynamic sense, that
is, as a transformation. We say, using the above notation, that a reflection in
the line m "takes P into P'." Notice that if ρ is a reflection, then ρ^2 is the
identity transformation, since if P' is the image of P, then P is the image of
P'. In notation,

$$P\rho = P' \quad \text{and} \quad P'\rho = P\rho^2 = P.$$

There are also reflections in points and planes, but here we confine ourselves
to reflections in lines of plane figures.

One can show geometrically that L_1 is the product of two reflections. But
it is simpler to show it algebraically as follows:

$$L_1 = \begin{bmatrix} 1 & 0 \\ 0 & -1 \end{bmatrix} \begin{bmatrix} \cos\varphi & \sin\varphi \\ \sin\varphi & -\cos\varphi \end{bmatrix} = RL_2,$$

where R is a reflection in the x-axis. We have proved part of the following
theorem.

Theorem 8.2. Every transformation of the orthogonal group in two dimensions is either a rotation about the origin or a reflection in a line through the
origin as shown in (8.4). A rotation is a product of two reflections in lines
through the origin, and every product of two reflections in lines through the
origin is a rotation.

Proof: It remains to prove that the product of two reflections is a rotation.
This follows from the following computation:

$$(8.7) \quad \begin{bmatrix} \cos\alpha & \sin\alpha \\ \sin\alpha & -\cos\alpha \end{bmatrix} \begin{bmatrix} \cos\beta & \sin\beta \\ \sin\beta & -\cos\beta \end{bmatrix} = \begin{bmatrix} \cos(\beta-\alpha) & \sin(\beta-\alpha) \\ -\sin(\beta-\alpha) & \cos(\beta-\alpha) \end{bmatrix}.$$

It can be shown that every transformation of the plane into itself which
preserves distance, that is, every Euclidean transformation, is either in the
orthogonal group, is a translation, or is a combination of these types of
transformations. Actually, every Euclidean transformation can be expressed
as a sequence of three or fewer reflections in lines. But we shall not prove
these here.

Exercises

1. Let

$$f(x) = \frac{3x + 4}{2x + 3} \quad \text{and} \quad g(x) = \frac{2x + 1}{3x + 2}.$$

Find $fg(x)$ first by substitution and second by multiplication of matrices.

2. For $f(x)$ as in Exercise 1, find $f^{-1}(x)$ and $gf(x)$.

3. Do Exercise 1 for the two functions

$$f(x) = \frac{2x + 1}{-3x + 2}, \quad g(x) = \frac{4x + 3}{3x + 2}.$$

4. Find $g^{-1}(x)$ and $gf(x)$ in Exercise 3.

5. Let the linear transformation τ be defined by

$$(x_1, x_2)\tau = (x_2, -x_1 - x_2).$$

Show that e, τ, τ^2 is a multiplicative group.

6. Find the inverse of L_1 in (8.4).

7. For what values of φ is L_1^2 in (8.4) the identity matrix?

8. Find the smallest group that contains the two functions $f(x) = -1 - x$ and $g(x) = 1/x$, where the binary operation is composition. Note what values of x are excluded. Write the "multiplication table" for the group.

9. Show that if a and d are interchanged in (5.6), then x is replaced by $1 - x$.

10. If L and L' are rotations with angles φ and φ', show that LL' is a rotation with angle $\varphi + \varphi'$.

11. Show by multiplication that the square of a reflection is the identity.

12. Let $A = L_1$, where A is defined in (4.7) and L_1 in (8.4). Prove that (8.1) holds.

13. Do Exercise 12 with L_1 replaced by L_2.

14. Prove that the set of translations (6.2) forms an Abelian group.

15. Show that if S and T are subgroups of a group G, then their inter-section,† $S \cap T$, is a subgroup of both S and T. Is the same conclusion justified when S and T are not subgroups of another group?

16. Show by an example that if S and T are subgroups of a group G, their union,† $S \cup T$, is not necessarily a group.

† $S \cap T$ is the set of elements common to S and T, and $S \cup T$ is the set of elements in S or T or both.

17. Does the set of rotations in O_2, that is, matrices L_1 in (8.4), form a multiplicative group? If it is a group, is it Abelian?

18. Does the set of reflections in O_2, that is, matrices L_2 in (8.4), form a multiplicative group? If it is a group, is it Abelian?

19. Let L be a matrix of order 2 whose square is the identity and whose determinant is -1. Must L be the matrix of a reflection?

20. Let ρ_1 and ρ_2 be two reflections and $\rho_1\rho_2 = \sigma$, a rotation. Show that $\rho_1\rho_2 = \rho_2\rho_1$ if and only if σ^2 is the identity.

21. Prove geometrically that if ρ and ρ' are two reflections in lines m and m' that intersect in a point P, then $\rho\rho'$ (ρ first and ρ' second) is a rotation about P whose angle is twice the angle from m to m'.

22. Prove algebraically the result of Exercise 21.

23. Verify Equation (8.7).

24. Prove that the area of the triangle $(0, 0)$, (a_1, a_2), (b_1, b_2) is

$$\tfrac{1}{2}|a_1b_2 - b_1a_2|.$$

   (*Hint:* Write the coordinates of the points in polar coordinates.)

25. Prove, by use of equations (8.4) or otherwise, that a product of a rotation and a reflection about a line through the origin in the plane is a reflection about a line through the origin.

*26. Use Exercise 25 or by other means prove that in the plane a product of an odd number of reflections is a reflection and a product of an even number of reflections is a rotation.

*27. Show that the orthogonal group O_2 is generated by the reflections in lines through the origin.

*28. Show geometrically or algebraically that a reflection in a line through the origin is determined by the image of one point, not the origin. That is, given points A and A', which may be the same or different, show that there is exactly one reflection in a line through the origin that takes A into A'.

9.   Symmetries and Permutations

Two important kinds of groups have to do with geometrical figures on the one hand and permutations on the other. Their interrelation is so intimate that it is best to discuss them together. Consider a triangle ABC. If sides AB and AC are of equal length, the triangle is symmetric with respect to the altitude through A. This static property is equivalent to the dynamical one that a reflection in this altitude (see Section 8) takes the triangle into itself. This reflection is not only a Euclidean transformation, since it preserves distances, but it also preserves the position of the triangle. However, it does change something: the labels of the vertices. That is, vertex B moves into the place where vertex C was before. Intuitively, we can think of the triangle as being made of cardboard and the reflection as rotating it out of the plane about the altitude and back onto where it was before. Thus, by taking a triangle or any other figure into itself, we mean leaving it unchanged in size and position without regard to labels.

If a triangle is equilateral, any one of the three altitudes can serve as an axis of symmetry. Using Figure I.5, we designate three reflections:

(9.1)
    α_1: reflection in the altitude through the lower right vertex.
    α_2: reflection in the vertical altitude.
    α_3: reflection in the altitude through the lower left vertex.

Each of these is a transformation that takes the triangle into itself. If we perform any one of these twice, we not only take the triangle into itself but preserve the labeling as well. So we think of α_i^2 as the identity transformation and call it ϵ. Thus we have so far three groups of two elements each:

$$(9.2) \qquad \{\epsilon, \alpha_1\}, \{\epsilon, \alpha_2\}, \{\epsilon, \alpha_3\}.$$

We can also describe each of these reflections in terms of what it does to a set of labels. Let us label the vertices 1, 2, and 3 in a counterclockwise direction, starting at the lower right corner. Then we can describe the reflections by the following symbols:

$$(9.3) \qquad \alpha_1 = \begin{pmatrix} 1 & 2 & 3 \\ 1 & 3 & 2 \end{pmatrix}, \qquad \alpha_2 = \begin{pmatrix} 1 & 2 & 3 \\ 3 & 2 & 1 \end{pmatrix}, \qquad \alpha_3 = \begin{pmatrix} 1 & 2 & 3 \\ 2 & 1 & 3 \end{pmatrix}.$$

The first symbol means: Leave vertex 1 unchanged and interchange 2 and 3, and the others have corresponding meanings. In the first row of these symbols

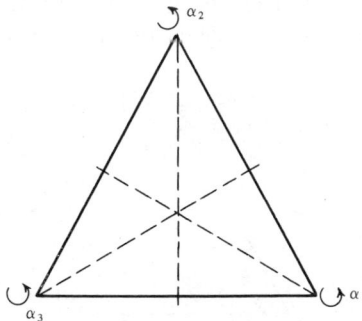

Figure I-5

we usually write the numbers in natural order, but that is not necessary. For instance, two other ways of representing α_1 are

$$\begin{pmatrix} 3 & 2 & 1 \\ 2 & 3 & 1 \end{pmatrix}, \qquad \begin{pmatrix} 2 & 3 & 1 \\ 3 & 2 & 1 \end{pmatrix}.$$

If each of the two transformations takes a triangle into itself, then certainly their "product," that is, first one and then the other, will have the same property. What is α_1 followed by α_2? One can find this geometrically, as in Figure I.6. This means that 1 moves into the position formerly occupied by 3, 2 into position 1, and 3 into position 2.

It is probably a little easier to find the product by means of the symbols in (9.3). Note that there α_1 takes 1 into 1 and α_2 takes 1 into 3; hence the sequence of two reflections has the net effect of taking 1 into 3. Computing in this way with the other two symbols, we have

$$(9.4) \qquad \rho_1 = \alpha_1\alpha_2 = \begin{pmatrix} 1 & 2 & 3 \\ 3 & 1 & 2 \end{pmatrix}, \qquad \rho_2 = \alpha_2\alpha_1 = \begin{pmatrix} 1 & 2 & 3 \\ 2 & 3 & 1 \end{pmatrix} = \rho_1^2.$$

Note that $\alpha_1\alpha_2$ means α_1 first and α_2 second. As we would expect from Section 8, each of these, being a product of two reflections, is a rotation. In fact, ρ_1 can be described as a rotation in a clockwise direction about the center of the triangle through an angle of 120° and ρ_2 is a rotation through 240° in the same direction. Also, ρ_2 can be thought of as a rotation through 120° in a counterclockwise direction. Note that the product $\alpha_1\alpha_2$ is not commutative.

If we compare the set (9.2) with (9.4), we note that the square of each α_i is the identity and the cube of each ρ_i is the identity. Furthermore, α_i changes the orientation of the triangle; that is, if initially the labels 1, 2, and 3 appear in counterclockwise order, after the reflection they appear in clockwise order. But the rotations preserve the orientation. This is only natural, since to get a rotation one changes the orientation twice. This idea of orientation is difficult to pin down rigorously, and hence it is often described in terms of the evenness or oddness of the number of reflections. Connected with this is the other intuitive idea that, for operations (9.1), one must move the triangle out of the plane, whereas for transformations (9.4), one can keep the triangle in the plane.

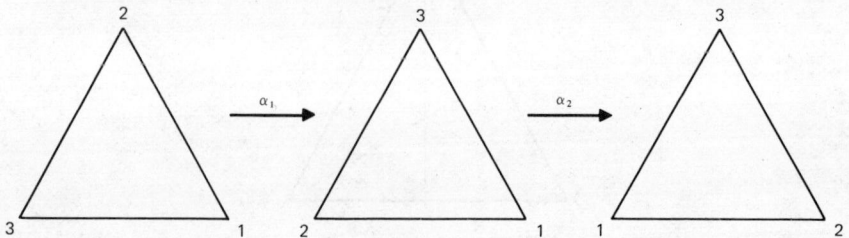

Figure I-6

Each of these six transformations, being either a reflection or a rotation, is, by Section 8, an orthogonal transformation. These six are called *symmetries* because they are not only orthogonal transformations but they take the triangle into itself in the sense described above. Of course, not all orthogonal transformations are symmetries; for instance, a rotation through an angle of 60° about the center of the equilateral triangle is an orthogonal transformation, but it does not take the triangle into itself. We shall see that the set of symmetries of a triangle is a subgroup of the orthogonal group O_2.

We have so far six symmetries of an equilateral triangle, including the identity transformation. Can there be any more? In this case it is easy to see that there are no more, since every symmetry of a triangle permutes the labels of the three vertices and there are exactly six permutations of three numbers. So every product of two of the six symmetries must be one of them again; that is, the operation of product is closed and thus is a binary operation. But just what these products are can best be exhibited by a multiplication table.

To compute the multiplication table, it helps to exploit as much as possible relationships among the elements. This can be done in various ways. Partly to pave the way for showing a relationship with another group, we start with α_1 and α_2. (In Exercise 1 you will be asked to form the multiplication table, starting with another pair of elements.) We have expressed in terms of α_1 and α_2 all six elements, except α_3. Since α_3 changes orientation, it cannot be a product of two reflections. So we try a product of three: $\alpha_1\alpha_2\alpha_1$. Geometrically or algebraically, we see that

$$\alpha_3 = \alpha_1\alpha_2\alpha_1.$$

It is also true that $\alpha_3 = \alpha_2\alpha_1\alpha_2$. Thus we have expressed every element of the set as a "product" involving only α_1 and α_2. We show below that these six elements form a group under "multiplication." We say that α_1 and α_2 *generate* the group. (We give a formal definition of "generator" below.) Table I.2 is the "multiplication table" expressed in terms of these generators. With this table at hand, we can establish the following theorem.

Table I.2

	ϵ	α_1	α_2	$\alpha_1\alpha_2$	$\alpha_2\alpha_1$	$\alpha_1\alpha_2\alpha_1 = \alpha_2\alpha_1\alpha_2 = \alpha_3$
ϵ	ϵ	α_1	α_2	$\alpha_1\alpha_2$	$\alpha_2\alpha_1$	$\alpha_1\alpha_2\alpha_1$
α_1	α_1	ϵ	$\alpha_1\alpha_2$	α_2	$\alpha_1\alpha_2\alpha_1$	$\alpha_2\alpha_1$
α_2	α_2	$\alpha_2\alpha_1$	ϵ	$\alpha_1\alpha_2\alpha_1$	α_1	$\alpha_1\alpha_2$
$\alpha_1\alpha_2$	$\alpha_1\alpha_2$	$\alpha_1\alpha_2\alpha_1$	α_1	$\alpha_2\alpha_1$	ϵ	α_2
$\alpha_2\alpha_1$	$\alpha_2\alpha_1$	α_2	$\alpha_1\alpha_2\alpha_1$	ϵ	$\alpha_1\alpha_2$	α_1
$\alpha_1\alpha_2\alpha_1$	$\alpha_1\alpha_2\alpha_1$	$\alpha_1\alpha_2$	$\alpha_2\alpha_1$	α_1	α_2	ϵ

Theorem 9.1. The six symmetries of an equilateral triangle form a group.

Proof: By Theorem 7.1 we need merely show closure and that each element has an inverse. Since every entry is one of the six, the table shows closure. Since ϵ occurs in each row and column of the table, each symmetry has an inverse. (Actually, from the note after Theorem 7.1, it is only necessary to show closure.)

Notice that there is one subgroup of order 3: $\{\epsilon, \rho_1, \rho_1^2\}$. This is generated by a single element, since the group consists of the powers of ρ_1. Whenever a group is generated by a single element, we call it a *cyclic group*. Groups (9.2) are also cyclic but of order 2. Correspondingly, we say that the *order* of the element ρ_1 is 3, and the order of α_1, as well as α_2, is 2. In general we call the *order of an element* the order of the cyclic group that it generates. That is, if σ is an element of a group, and if $\sigma^u = \epsilon$ holds for a positive integer u and for no smaller positive integer, we call u the order of σ. If there is no such u for σ, we say that σ is of *infinite order*. For instance, the group consisting of all integral powers of x is a cyclic group G. The order of x, as well as the order of G, is infinite.

Returning to the idea of generators, we give a formal definition and make a comment before going on to the symmetries of another geometrical figure.

Definition. *A set S of elements of a group is said to* generate *the group if every element of the group can be expressed as a "product" that contains only elements of S. In this case S is called a* set of generators *of the group.*

REMARK: There may be many sets of generators of a group. Also, a set of generators does not *have* to be minimal. In fact, the set of all the elements of a group is a set of generators. However, in practice we usually try to make a set of generators as small as possible.

Now consider the symmetries of a square. Here there are four reflections, as indicated in Figure I.7. If we number the vertices of the square 1, 2, 3, and 4 in a counterclockwise order starting with 1 in the lower right corner, we have

$$\alpha_1 = \alpha = \begin{pmatrix} 1 & 2 & 3 & 4 \\ 2 & 1 & 4 & 3 \end{pmatrix}, \qquad \alpha_2 = \begin{pmatrix} 1 & 2 & 3 & 4 \\ 3 & 2 & 1 & 4 \end{pmatrix},$$

$$\alpha_3 = \begin{pmatrix} 1 & 2 & 3 & 4 \\ 4 & 3 & 2 & 1 \end{pmatrix}, \qquad \alpha_4 = \begin{pmatrix} 1 & 2 & 3 & 4 \\ 1 & 4 & 3 & 2 \end{pmatrix}.$$

Let $\rho = \alpha_1 \alpha_2$ and compute

$$\rho = \begin{pmatrix} 1 & 2 & 3 & 4 \\ 2 & 3 & 4 & 1 \end{pmatrix}.$$

Thus ρ is a rotation through a right angle in a counterclockwise direction about the center of the square. Under this transformation, 1 moves into the

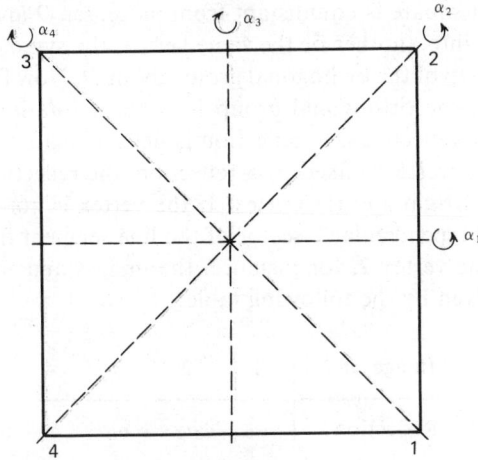

Figure I-7

place formerly occupied by 2, and so forth. We now show that α and ρ together generate the group of symmetries of a square.

First, calculation shows that $\alpha_2 = \alpha\rho$, $\alpha_3 = \alpha\rho^2$, and $\alpha_4 = \alpha\rho^3$. Also from the definitions of α and ρ it follows that $\alpha^2 = \epsilon = \rho^4$. In calculating the multiplication table for these generators, it is convenient to use the following, which can be obtained by direct calculation:

$$\rho\alpha\rho = \alpha, \qquad \text{that is,} \quad \rho\alpha = \alpha\rho^3.$$

The second can be obtained from the first by multiplying on the right by ρ^3. By this means we can calculate the multiplication table for the *octic group*, Table I.3. Are the table entries *all* the symmetries of a square? It certainly appears that there are no more, but there is room for a lingering doubt, since so far we have only 8 of a possible 24 permutations of four numbers. Here we can make some use of the orthogonal group discussed in Section 8. Since

Table I.3

	ϵ	ρ	ρ^2	ρ^3	α	$\alpha\rho = \alpha_2$	$\alpha\rho^2 = \alpha_3$	$\alpha\rho^3 = \alpha_4$
ϵ	ϵ	ρ	ρ^2	ρ^3	α	$\alpha\rho$	$\alpha\rho^2$	$\alpha\rho^3$
ρ	ρ	ρ^2	ρ^3	ϵ	$\alpha\rho^3$	α	$\alpha\rho$	$\alpha\rho^2$
ρ^2	ρ^2	ρ^3	ϵ	ρ	$\alpha\rho^2$	$\alpha\rho^3$	α	$\alpha\rho$
ρ^3	ρ^3	ϵ	ρ	ρ^2	$\alpha\rho$	$\alpha\rho^2$	$\alpha\rho^3$	α
α	α	$\alpha\rho$	$\alpha\rho^2$	$\alpha\rho^3$	ϵ	ρ	ρ^2	ρ^3
$\alpha_2 = \alpha\rho$	$\alpha\rho$	$\alpha\rho^2$	$\alpha\rho^3$	α	ρ^3	ϵ	ρ	ρ^2
$\alpha_3 = \alpha\rho^2$	$\alpha\rho^2$	$\alpha\rho^3$	α	$\alpha\rho$	ρ^2	ρ^3	ϵ	ρ
$\alpha_4 = \alpha\rho^3$	$\alpha\rho^3$	α	$\alpha\rho$	$\alpha\rho^2$	ρ	ρ^2	ρ^3	ϵ

every vertex of a square is equidistant from its center O and each symmetry takes one vertex into another or the same vertex, the symmetries of a square must be a subgroup of the orthogonal group about O. Now from Theorem 8.2 every element of the orthogonal group is either a rotation or a reflection. First, look at the reflections. A reflection is determined by the image of any vertex, since if the vertex is fixed by a reflection, the reflection must be in the line through the origin and the vertex. If the vertex is not fixed, the line of reflection is the perpendicular bisector of the line segment from the vertex to its image. For the vertex 2, for instance, the images and the accompanying reflections are given by the following table:

Image of 2	1	2	3	4
Reflection	α_1	α_2	α_3	α_4

This accounts for all the reflections.

Second, the only rotations that are symmetries are those which take the vertex 1, for instance, into 1, 2, 3, and 4, respectively. This accounts for all the rotations. We have proved the following theorem.

Theorem 9.2.   The symmetries of a square form a group of order 8. (It is called the *octic group*.)

It is instructive to consider the symmetries of other quadrilaterals, the rhombus, for instance. It is quite obvious that the group of symmetries of a rhombus must be a subgroup of the symmetries of a square; but partly to illustrate another method of attack, let us not assume this. We can start by noting that in any symmetry of a rhombus (not a square) the longer diagonal, say that connecting vertices 2 and 4, must be fixed. This means that either vertices 2 and 4 are both fixed or they are interchanged; for each of these, vertices 1 and 3 are both fixed or interchanged. Hence there are exactly four symmetries, as follows:

$$\begin{pmatrix} 1 & 2 & 3 & 4 \\ 1 & 2 & 3 & 4 \end{pmatrix} = \epsilon, \qquad \begin{pmatrix} 1 & 2 & 3 & 4 \\ 3 & 2 & 1 & 4 \end{pmatrix} = \alpha_2,$$

$$\begin{pmatrix} 1 & 2 & 3 & 4 \\ 1 & 4 & 3 & 2 \end{pmatrix} = \alpha_4, \qquad \begin{pmatrix} 1 & 2 & 3 & 4 \\ 3 & 4 & 1 & 2 \end{pmatrix} = \alpha_2\alpha_4 = \rho^2.$$

In the exercises you will be asked to discuss symmetries of other figures.

The groups of symmetries of a figure can be used to describe it at least partially. For instance, the only quadrilateral that has the octic group as its set of symmetries is a square. Crystals are in the form of solids that are "regular" in a certain sense, and the symmetries are helpful in crystallographic classification.

We have shown that each symmetry of an equilateral triangle or a square can be represented by a permutation (that is, a reordering) of three or four symbols. (We chose 1, 2, 3 or 1, 2, 3, 4, respectively, but any other symbols would do equally well.) We shall leave as an exercise the proof of the following theorem.

Theorem 9.3.   The set of all permutations of n symbols forms a group containing $n!$ elements.

Definition.   *The group of all permutations of n symbols is called the* symmetric group *and denoted by* S_n.

Not only are the symmetries expressible as permutation groups, but as we shall see in Chapter II, there is a theorem of Cayley which affirms that every finite group is, except for the notation, representable as a group of permutations. Even in case the group is infinite, there is a certain sense in which this is true. By exploiting this connection, one can get useful information about groups. So we shall be discussing permutation groups in more detail in Chapter II.

Exercises

1. Show that α_1 and ρ_1 form a pair of generators of the group of symmetries of the equilateral triangle and form the multiplication table in terms of these two transformations as generators.

2. Find all the subgroups of S_3.

3. Show that α_2 and α_4 (symmetries of the square) do not generate the octic group.

4. Do α_2 and ρ (symmetries of the square) generate the octic group?

5. Which of the transformations of the octic group leave the orientation of the square unaltered? Show that these form a subgroup of order 4.

6. Find the group of symmetries of a rectangle that is not a square. Is it the same group as that described in Exercise 5?

7. Find the group of symmetries of a parallelogram.

8. What elements of the octic group are commutative with all elements of the group?

9. Find all the subgroups of orders 2 and 4 of the octic group, and by use of a diagram show which are subgroups of which.

10. Suppose that we number the vertices of a square in the order 1, 3, 2, 4 in

a counterclockwise direction. What effect, if any, would this have on the symmetries of the figure?

11. Suppose, having numbered the vertices of a square as in Exercise 10, we connect them in the order 1, 2, 3, 4 to give a figure in the form X. How would this affect the group of symmetries?

12. Show that the set of integers is a cyclic additive group.

13. Is the set of even integers a cyclic additive group? Will the same be true if "even" is replaced by "odd"?

14. Do the integers form a cyclic multiplicative group? Why or why not?

15. Describe the group of symmetries of a circle.

16. Find the group of symmetries of the regular pentagon.

17. Find the group of symmetries of the regular hexagon.

18. Prove Theorem 9.3.

*19. Find the group of symmetries of the regular polygon of n sides.

*20. Find the group of symmetries of the regular octahedron.

*21. Find the symmetries of a cube that leave one vertex fixed.

*22. Find a set of four matrices that generate the additive group of matrices of order 2 with integral elements.

*23. Let T be a regular tetrahedron. Show that its symmetries fall into five categories:

1. The identity.

2. Those in which just one symbol is fixed, e.g. $\begin{pmatrix} 1 & 2 & 3 & 4 \\ 1 & 4 & 2 & 3 \end{pmatrix}$.

3. Those in which just two symbols are fixed, e.g. $\begin{pmatrix} 1 & 2 & 3 & 4 \\ 1 & 2 & 4 & 3 \end{pmatrix}$.

4. Those in which two pairs are interchanged, e.g. $\begin{pmatrix} 1 & 2 & 3 & 4 \\ 2 & 1 & 4 & 3 \end{pmatrix}$.

5. None of the above, e.g. $\begin{pmatrix} 1 & 2 & 3 & 4 \\ 2 & 4 & 1 & 3 \end{pmatrix}$.

Find how many there are in each category and how they can be described geometrically. Which, intuitively, can be accomplished in three dimensions with a tetrahedron made, say, of wood?

10.   *An Additive Group*

The example of this section is introduced for three reasons. Most of our applications so far have been multiplicative, or at least phrased in this terminology; this group is definitely additive. Also, this leads into the notion of an "ideal," which we shall see is of great importance in algebra. Furthermore, in the process we shall derive a result about integers which we will need at the beginning of Chapter II.

Let a and b be any two integers and consider the set S of all numbers of the form $ax + by$, where x and y are integers. That is, S is the set of all linear combinations of a and b with integer coefficients (compare Section 3). Since the numbers of S are all integers, S is a subset of the additive group of integers **Z**. So, from Theorem 7.1, to prove that S is a group we need to show the following:

1. If c and d are in S, then so is $c + d$.
2. If c is in S, then so is $(-c)$.

These two properties follow immediately, since

$$ax + by + ax' + by' = a(x + x') + b(y + y'),$$
$$-(ax + by) = a(-x) + b(-y).$$

So we have proved the following theorem.

Theorem 10.1.   If a and b are two integers, then the set of all numbers of the form $ax + by$, where x and y are integers, is an additive group.

We now proceed to show that the group described in Theorem 10.1 is generated by one integer; that is, S consists of the set of all integral multiples of a properly chosen integer c.

Theorem 10.2.   Given two integers a and b, let $S = \{ax + by\}$ for all integers x and y, and let g be the least positive integer in S. Then every element of S is an integral multiple of g.

Proof: We need to show that a and b are multiples of g, for if they are, certainly every element of S will be a multiple of g, which is itself an element of S. So divide a by g and get a quotient q and a remainder r; that is,

(10.1) $a = gq + r, \qquad 0 \le r < g.$

Since g is in S, we know that for some integers x_0 and y_0, $g = ax_0 + by_0$. Then

$$r = a - gq = a - (ax_0 + by_0)q = a(1 - x_0 q) + b(-y_0 q).$$

This shows that r is in S. But we assumed that g was the least positive integer in S and r is less than g. Hence the only possibility is that $r = 0$, which implies that g is a factor of a. Similarly, it can be shown that g is a factor of b. We have shown that every element of S is a multiple of g. Since also every integral multiple of g is in S, the proof is complete.

Notice that the number g determined in the theorem is a common factor of a and b. Also, since $ax_0 + by_0 = g$, any common factor of a and b is a factor of g. Hence we have the following corollaries.

Corollary 1.   The number g of Theorem 10.2 is the greatest common divisor (g.c.d.) of a and b. It can be expressed in the form $g = ax + by$, where x and y are integers.

Corollary 2.   The g.c.d. of a and b is 1 if and only if $ax + by = 1$ for some integers x and y. Here a and b are said to be *relatively prime*.

Corollary 3.   Consider the line whose equation is $ax + by = c$, with a, b, and c integers. The least positive value of c for which the line has points with integer coordinates is the greatest common divisor of a and b.

Another way to state the theorem is: The set of all linear combinations of a and b with integral coefficients is a cyclic additive group whose generator is the greatest common divisor of a and b.

Since the following theorem can be proved as was Theorem 10.2, we leave the proof as an exercise.

Theorem 10.3.   Given n integers $a_1, a_2, \ldots, a_n$, let

$$S = \{a_1x_1 + a_2x_2 + \cdots + a_nx_n\}$$

for integer values of x_i, and let g be the least positive integer in S. Then every element of S is an integral multiple of g.

In the proof of Theorem 10.2 we took as obvious two results that we shall deal with more carefully in Chapter III. The first was the existence of q and r satisfying (10.1). The second was that since S consists of integers some of which are positive, it contains a least positive integer. These results hold for integers and with some modification for polynomials but not for all real numbers. However, the set S, which we call an *ideal*, is useful when a and b are irrational numbers. We shall deal extensively with ideals in Chapter V.

11.   A Little Group from Logic

Here is an example of a little group in logic that we introduce for curiosity's sake and to show the ubiquity of the idea of a group. Let "*A* implies *B*" be a certain implication. For example, in the implication:

If he is a friend, then I shall smile at him,

"he is a friend," denoted by *A*, is the hypothesis and "I shall smile at him" is *B*, the conclusion. Three other statements can be obtained by modifying the implication as follows:

The implication, *I*:   If *A*, then *B*.
The contrapositive, *C*: If not *B*, then not *A*.
The converse, *V*:   If *B*, then *A*.
The obverse, *O*:   If not *A*, then not *B*.

(The obverse is sometimes called the *negative* or *inverse*.) The contrapositive for instance, is obtained by interchanging the conclusion and the hypothesis of the implication and negating both. The contrapositive of the implication above is: "If I shall not smile at him, then he is not a friend."

With a little hedging any pair of the four implications can be combined by the preposition "of." For instance, the contrapositive of the converse is the obverse, since, starting with "If *B*, then *A*," interchanging *A* and *B* and negating both gives "If not *A*, then not *B*," which is the obverse. Similarly, we can combine most of the others. Hedging is necessary when "the implication" comes before "of." For instance, "the implication of the converse" has no meaning. So for our purposes we have the understanding that "the implication of *I*, *C*, *V*, or *O*" shall mean *I*, *C*, *V*, or *O*, respectively. With this understanding we have a table of combination, Table I.4. Thus the four statements form a group under the operation implied by the preposition "of."

Table I.4

	I	*C*	*V*	*O*
I	*I*	*C*	*V*	*O*
C	*C*	*I*	*O*	*V*
V	*V*	*O*	*I*	*C*
O	*O*	*V*	*C*	*I*

Exercises

1. Prove Theorem 10.1 for three integers, *a*, *b*, and *c*, rather than for only two.

2. Suppose in Theorem 10.1 that the numbers *a* and *b* were allowed to be rational, but that *x* and *y* are to be integers; would the theorem still hold?

3. Does the group of Section 11 have a single generator? What are its subgroups?

4. Does Theorem 10.1 hold if x and y are restricted to be nonnegative integers?

5. Prove Theorem 10.3.

6. State Theorem 10.3 in terms of linear combinations and a cyclic additive group.

7. Let ρ_1 and ρ_2 be two rotations in the plane in a clockwise direction about the origin through angles of $a_1\pi$ and $a_2\pi$, respectively, where a_1 and a_2 are integers. Prove that the group generated by ρ_1 and ρ_2 is cyclic.

8. Prove Corollary 2 of Theorem 10.2.

*9. Given a set U. Let S be all the subsets of U. Define $R \circ T$, for R and T in S, to be the set of all elements of U that are in R or T but not both. Show that S is a group under the operation $\circ$.

*10. Let $g, g^2, \ldots, g^b = e$ be a multiplicative group G of order b. Let a be any positive integer less than b such that 1 is the greatest common divisor of a and b. Prove that g^a generates G. (*Hint:* Use Theorem 10.2 or Corollary 2.)

*11. Assume that π is irrational. Does the conclusion of Exercise 7 hold if a_1 and a_2 are rational numbers? Does it hold if they are irrational numbers? Explain.

12. Summary

In this chapter we defined the abstract concept of a group. To do this, we started with a set S of numbers, polynomials, figures, or any one of a wide variety of other entities. Then there was a way of combining two elements of S to get a unique element of S. We called this way of combining a *binary operation*. [It could also be thought of as a function or mapping of $S \times S$ (pairs of elements of S) into S.] Implicit in the definition of a binary operation was the property that the resulting element was independent of the symbols used to represent the elements of S. Then if the operation with respect to S had the three properties specified, we called S a *group* under the operation.

In the course of dealing with specific groups, we found two especially useful concepts: those of *subgroup* and *generator*. It should be pointed out that not all groups have a finite number of generators; two related examples are the multiplicative group of real numbers excluding zero and the group of all rotations in the plane about the origin.

In Chapter II we shall study properties of groups in general, using the examples of this chapter as a base.

13.   *Abel*

In the park surrounding the King's Palace in Oslo there is, on a modest pedestal, a bust of a young man bearing the name Niels Henrik Abel and the dates 1802–1829. Nothing more is needed, for the Norwegians know their greatest mathematician. Abel's father, a pastor in Findö, died in 1820, by which time Abel had already read Gauss's famous *Disquisitiones Arithmeticae*. The mathematical world of his day was slow to believe that he had proved that not all equations of the fifth degree are solvable by radicals (see Chapter VI). However, he did have influential friends who helped him in 1825 to procure a grant to study in Germany and France. His most notable work was a memoir on elliptic integrals and the proof of a theorem that bears his name. For this, one year after his death from tuberculosis, the French Academy jointly awarded its Grand Prize in Mathematics to him and Jacobi. (See Oystein Ore's *Niels Henrik Abel, Mathematician Extraordinary*, Minneapolis, University of Minnesota Press, 1957.)

II

Groups: Fundamental Concepts and Properties

1. Isomorphism

In this chapter, with a background of examples from the previous one, we shall look at groups as abstract entities, find some of their fundamental properties, and discuss what these properties mean in specific cases.

First, you have probably noticed certain similarities of structure among the various groups considered in Chapter I. Let us start with three of these: G_1, the multiplicative group $\{1, c, c^2\}$, where $c = (-1 + \sqrt{-3})/2$; G_2, the group of rotations through $0°$, $120°$, $240°$, a subgroup of the symmetries of an equilateral triangle; and the group G_3 of functions e, fg, and gf in Section 5. For G_1 the operation is multiplication; for G_2, addition of angles; and for G_3, composition of functions. To compare them, we write their "multiplication tables" (Table II.1).

Table II.1

G_1				G_2				G_3		
1	c	c^2		$0°$	$120°$	$240°$		e	fg	$(fg)^2 = gf$
1	1	c	c^2	$0°$	$0°$	$120°$	$240°$	e	e	fg	$(fg)^2$
c	c	c^2	1	$120°$	$120°$	$240°$	$0°$	fg	fg	$(fg)^2$	e
c^2	c^2	1	c	$240°$	$240°$	$0°$	$120°$	$(fg)^2$	$(fg)^2$	e	fg

The comparison between G_1 and G_3 is most marked. If in the multiplication table for G_1 we replace 1 by e and c by fg, we get table G_3. But notice that though the operation for G_3 is written as if it were multiplication, it is really composition of functions.

Comparing G_1 and G_2, we see that multiplication corresponds to addition: 1 to 0°, c to 120°, and c^2 to 240°. Not only do we have such a correspondence, but it carries over to the table, that is,

$$c \leftrightarrow 120°, \qquad c^2 \leftrightarrow 240°,$$

and

$$c(c^2) = 1 \text{ corresponds to } 120° + 240° = 0°.$$

Thus all three groups are the same except for the notation used.

In each case we have two properties of the comparison to check. First, there is a one-to-one correspondence between the elements of the two groups; that is, each element of one group corresponds to a unique element of the other group, in both directions. Second, the correspondence carries over to the multiplication tables. Now let us state this relationship formally and give it a name.

Definition.   *Let G and G' be two groups with operations* $\circ$ *and* $\circ'$. *We call them* isomorphic *if there is a one-to-one correspondence between the elements of the two groups such that*

(1.1)   *If* $g_1 \leftrightarrow g_1'$ *and* $g_2 \leftrightarrow g_2'$,   *then* $(g_1 \circ g_2) \leftrightarrow (g_1' \circ' g_2')$.

A correspondence with these properties is called an isomorphism.

If σ denotes the isomorphism, we write the correspondence (1.1) as

(1.2) $g_1\sigma = g_1'$ and $g_2\sigma = g_2'$

implies that $g_1 g_2 \sigma = (g_1\sigma)(g_2\sigma)$.

There is one respect in which the notation (1.2) slurs over a distinction that (1.1) emphasizes. In (1.2), $g_1 g_2$ is an abbreviation for $g_1 \circ g_2$ and $(g_1\sigma)(g_2\sigma)$ is an abbreviation for $(g_1\sigma) \circ' (g_2\sigma)$. In (1.2) we wrote both operations as if they were products; that is, we used the same notation for two operations that might be different. In spite of this we shall generally use the product notation, since it is more compact and often the two operations are the same in any case. In terms of (1.2) we could say that an isomorphism is a one-to-one correspondence that "preserves products."

One can think of (1.1) in terms of two calculations. First, one may use the operation $\circ$ to combine g_1 and g_2 and then find the image of the combination; second, one can combine the images of g_1 and g_2 using the operation $\circ'$. Condition (1.1) affirms that the two calculations give the same element of G'.

Precisely, "one-to-one correspondence" means that to each element g of G

corresponds a unique element g' of G' and that for each element g' of G' there is a unique element g of G which corresponds to it. Thus the uniqueness of the correspondence extends in both directions—hence the notation ↔ with two arrows. So an isomorphism σ between G and G' can be thought of as a transformation of G *onto* G', which has an inverse and for which (1.1) holds. The word "onto" signifies that every element of G' is the image of an element of G. (See Section 5 of Chapter I.)

There are three properties of an isomorphism that should be spelled out even though they may seem quite obvious:

1. A group G is isomorphic to itself.
2. If G is isomorphic to G', then G' is isomorphic to G.
3. If G is isomorphic to G' and G' to G'', then G is isomorphic to G''.

Since isomorphism has these three properties, it is called an *equivalence relation*. We shall discuss this concept in more detail in Section 2 because it occurs many times inside and outside mathematics. But at this point we merely notice that since an isomorphism has these three properties, it behaves much like equality. That is, we need not speak of G being isomorphic *to* G' but say that G and G' *are* isomorphic. The isomorphism is *between*, rather than from one to the other, since it acts in both directions.

We have talked around the subject enough. Let us now put the concept of isomorphism to some use. First, certain correspondences follow from an isomorphism. Let e be the identity element of a group G and σ an isomorphism between G and G'. Then $(e\sigma)(x\sigma) = (ex)\sigma = x\sigma$, which shows that $e\sigma$ is the identity e' for G'. Let x be an element of G; then since its inverse x^{-1} has the property that $xx^{-1} = e$, it follows that

$$(x\sigma)(x^{-1}\sigma) = (xx^{-1})\sigma = e\sigma = e'.$$

Thus $(x^{-1}\sigma)$ in G' is the inverse of $x\sigma$; that is, $x^{-1}\sigma = (x\sigma)^{-1}$. Furthermore, writing $xxx \cdots x(k$ times$)$ as x^k, we see that $(x\sigma)^n = x^n\sigma$ for all positive integers n. In fact, $(x\sigma)^n = x^n\sigma$ for $n = 0$ if we interpret x^0 to be e, and for $n = -c$, where c is a positive integer, if we interpret x^{-c} to be $(x^{-1})^c$. We collect these results in the following theorem.

Theorem 1.1.   If σ is an isomorphism between groups G and G', then the image of the identity of G is the identity of G' and, for the interpretation immediately above, $(x\sigma)^n = x^n\sigma$ for all integers (including zero, and negative and positive integers).

Corollary.   The order (see Section 9 of Chapter I) of any element of G is equal to the order of its image. (See Exercise 25 after Section 3.)

Now we use what we have learned to deal with two abstract groups. First, we leave as Exercise 24 after Section 3 proof of the following theorem.

Theorem 1.2.   Every group of order 3 is isomorphic to every other group of order 3. That is, there is only one group of order 3 as far as the structure is concerned, apart from notation.

What about groups of order 4? In Section 9 of Chapter I we found two subgroups of order 4 of the octic group. One was the group $H_1 = \{\epsilon, \rho, \rho^2, \rho^3\}$, rotations through multiples of 90°. Another was the group $H_2 = \{\epsilon, \alpha_2, \rho^2, \alpha_2\rho^2\}$ of the symmetries of a rhombus not a square. These two groups cannot be isomorphic, since the square of every element of H_2 is the identity, and this is not so for H_1. You will be asked to show in Exercise 13 after Section 3 that every group of order 4 is either a cyclic group (see Section 9 of Chapter I) isomorphic to H_1 or is isomorphic to the Abelian group $\{e, a, b, ab\}$, where the square of every element is the identity. The latter group is called the *Klein four-group*.

In Chapter I we had examples of two non-Abelian groups with six elements: the group of symmetries of the equilateral triangle in Section 9 and the six functions associated with the cross ratio in Section 5. Now if you replace f by α_1 and g by α_2 in the multiplication table for the six functions, you get the multiplication table of the symmetries of a triangle. Thus σ is an isomorphism if it is defined by $f\sigma = \alpha_1$ and $g\sigma = \alpha_2$. It may at first seem strange that two groups with such different backgrounds should be isomorphic. But once we show that there is, up to an isomorphism, only one non-Abelian group of order 6, it will not seem as strange. For this reason and since it will illustrate some ideas that we shall find useful later, we prove the following theorem.

Theorem 1.3.   Any two non-Abelian groups of order 6 are isomorphic.

Proof: Let G be a non-Abelian group of order 6. If the square of every element were the identity, Exercise 20 of Section 2 in Chapter I would imply that the group is Abelian. This follows since if $(ab)(ab) = e$ with $a^2 = b^2 = e$, we have $abba = e = abab$, which, by the cancellation property, implies that $ab = ba$.

So G must have an element whose square is not e. Let g be such an element and write its powers: $g, g^2, g^3, \ldots, g^r = e$, where $r > 2$ and r is the least positive power of g which is e; that is, r is the order of g. If $r = 6$, then G consists of the powers of g and is cyclic—therefore Abelian. If $r = 4$, there would be an element b in G not a power of g and hence G would have at least the following elements: $e, g, g^2, g^3, b, bg, bg^2$, and bg^3. No two of these eight are equal, for suppose that

$$b^i g^k = b^t g^u \qquad \text{for } 1 \geq i \geq t \geq 0$$

and k and u nonnegative integers less than 4. Then it follows that $b^{i-t} = g^{u-k}$. Since b is not a power of g, it follows that $i = t$ and $u = k$. So G would have at least eight elements, which is not true. One can deal similarly with $r = 5$. So we are left with $r = 3$. Then G has the following six elements:

$$(1.3) \qquad\qquad e, \quad g, \quad g^2, \quad b, \quad bg, \quad bg^2.$$

These are distinct, as above. There are no more, and since gb is also in G, it must be one of (1.3). Now $gb = g^i$ ($i = 0, 1, 2$) would imply that $b = g^{i-1}$ by the cancellation property, and this is false. Furthermore, $gb = b$ is impossible, since $g \neq e$; and $gb = bg$ would imply that G is Abelian. Hence the only possibility for gb is bg^2. By similar reasoning it can be shown that the only one of (1.3) that b^2 can be is e. Thus we had no choice at any point in the development, and G must have elements (1.3) with

$$g^3 = e = b^2, \qquad gb = bg^2, \qquad g^2 b = bg.$$

This determines the group and proves the theorem. (There are more efficient ways of proving this theorem in the light of later results.)

An immediate consequence of this theorem is the following result, which we have already proved by comparison of multiplication tables.

Corollary.   The group of symmetries of an equilateral triangle is isomorphic to the cross ratio group of functions of Section 5 in Chapter I.

It should be emphasized that the groups referred to in the corollary are not really the same group—they just have the same structure. By analogy, suppose that one person is interested only in the *price* of a car; then two cars with the same price would be isomorphic for him, although they might be very different in other aspects. Since for groups we are interested in structure, we often call two isomorphic groups the "same," but we must remind ourselves from time to time that structure is not *all important*. If two groups are isomorphic, we sometimes say that they are the same "up to an isomorphism."

Finally, we state and give an indication of the proof of a modification of the definition of an isomorphism, which is sometimes useful.

Theorem 1.4.   If in the definition of an isomorphism G' is not assumed to be a group, then the isomorphism implies that G' is a group. That is, let G be a group with a binary operation $\circ$, and G' a set with a binary operation $\circ'$; if there is a one-to-one correspondence between G and G' satisfying condition (1.1), then G' is a group and G and G' are isomorphic.

Indication of the proof: Essentially, the theorem holds since if G is a group and (1.1) holds, then $\circ'$ in a sense "behaves as $\circ$ does." We show this for the associativity property. From this it should be clear how the other properties of a group hold equally well for G'. Now

$$g_1 \circ (g_2 \circ g_3) \leftrightarrow g_1' \circ' (g_2' \circ' g_3'),$$
$$(g_1 \circ g_2) \circ g_3 \leftrightarrow (g_1' \circ' g_2') \circ' g_3'.$$

Since G is a group $g_1 \circ (g_2 \circ g_3) = (g_1 \circ g_2) \circ g_3$. Thus $g_1' \circ' (g_2' \circ' g_3')$ and $(g_1' \circ' g_2') \circ' g_3'$ correspond to the same element of G and hence are equal.

2.   *Equivalence Relations and Classifications*

The concept of equivalence relation that occurred in Section 1 is such a fundamental idea in mathematics (and elsewhere) that we should take time to consider it in some depth. First, what do we mean by a relation? (Outside mathematics we often use the word "relationship.") For instance, in answer to the question, "What is a relationship between 2 and 3?" the most obvious answer might be "2 is less than 3." For the question "What is a relationship between John and Bill?" we might answer that they are brothers, or "John is older than Bill," or taller, or more pleasant. (The question is so vague that many answers are possible.) Two things are "related" by some attribute in which they are alike or different. A relation is a kind of correspondence or comparison. But the more one tries to pin down what a relation is in the abstract, the more it slips out from under. In any context it is easy to say what a relation is. What chiefly matters is that if ρ is a relation, A is in the relation ρ to B, or A is not in this relation to B. So, with this circumlocution, we leave the concept undefined and use the notation $x \rho y$ to mean that, whatever relation ρ is, x is in the relation of ρ to y.

From this point on we can be more precise and give the following definition.

Definition.   *Given two elements a and b of some set S and a relation ρ, we use the notation $a \rho b$ to mean that a is in the relation ρ to b. Then ρ is called an* equivalence relation *if it has the following three properties, for all a, b, c, in S:*

1. *$a \rho a$ (the reflexive property).*
2. *If $a \rho b$, then $b \rho a$ (the symmetric property).*
3. *If $a \rho b$ and $b \rho c$, then $a \rho c$ (the transitive property).*

We noted in Section 1 that isomorphism is an equivalence relation between two groups. It is the symmetric property that makes it possible to use the preposition "between" since, so to speak, the action is in both directions. For instance, equality, congruence, "living in the same house as," "having the same father as" are all equivalence relations. But being the brother of, inequality, and inclusion are not.

Whenever one has an equivalence relation, one also has a classification, that is, a partition into classes. For instance, since "living in the same house as" is an equivalence relation, we can classify persons according to the house they live in. We can classify triangles according to congruence. We can classify groups according to isomorphism. But we cannot classify persons according to the relation "is the brother of." In general, we have the following definition.

Definition.   *We say that a set S can be* partitioned *or separated into classes if* $S = S_1 \cup S_2 \cup \cdots \cup S_k \cdots$, *where no pair* S_i, S_j *has an element in common, that is, if S is the union of disjoint subsets. Under these circumstances, each* S_i *is called a* class.

This means that if S can be partitioned into classes, the classification has three properties:

1c. Each element of S is in its own class.
2c. If a is in the same class as b, then b is in the same class as a.
3c. If a is in the same class as b, and b in the same class as r, then a is in the same class as r.

These three properties look surprisingly like the three properties of an equivalence relation. In fact, the two concepts are related by the following theorem.

Theorem 2.1.   Given a set S and an equivalence relation ρ among the elements of S, if we call two elements a and b in the same class whenever $a \rho b$, then the set S is partitioned with respect to ρ into classes. Conversely, if S can be partitioned into classes, let ρ be the relation "being in the same class as"; then ρ is an equivalence relation.

Partial proof: Suppose that an equivalence relation ρ is given. Then choose one element a of S and define as S_1 the set of all x in S such that $a \rho x$. If $S_1 \neq S$, let a_2 be an element of S not in S_1. Then define S_2 to be the set of all elements x of S such that $a_2 \rho x$. Continue in this fashion through all the elements of S. Then the properties which ρ has in virtue of being an equivalence relation imply that S is partitioned into classes. (Why?) (See the note after this theorem.)

Suppose that S is partitioned into classes. Let ρ be the relation "is in the same class as." Then the three properties 1c, 2c, and 3c translate directly into the three properties of an equivalence relation (Why?) Answering the question will complete the proof.

NOTE: There is some logical difficulty in the above argument if the partition of S yields an infinite number of classes, since in this case one would never finish the process of classifying. One may meet this difficulty in two ways. First, one could assume that the number of classes is finite—most of the examples in this book have this property. Or one can avoid the constructive approach and merely note that for each element a, the set of all elements x of

S such that $a \, \rho \, x$ forms a class; and that given a partition into classes we can say that two elements are in the relation ρ if they are in the same class.

Let us illustrate the above by an example that is a special case of what will be considered in more detail in Section 3. Let S be the set of integers and define the relation ρ by

$$a \, \rho \, b \quad \text{means that} \quad a - b \quad \text{is even}.$$

You should show that this is an equivalence relation. The corresponding classification is: S_0 is the set of all even integers and S_1 is the set of all odd integers. Every integer is in exactly one of S_0 and S_1. On the other hand, if we had started with the sets S_0 and S_1, the relation of being in the same subset is defined as above by $a - b$ being even. We shall have various other illustrations of the ideas of classifications and equivalence relations in succeeding sections.

3.   Other Examples of Isomorphisms

In Section 2 the examples of isomorphisms were very special ones. We now consider a fundamental one.

EXAMPLE 1

Theorem 3.1.   Let G be the multiplicative group of positive real numbers and G' the additive group of all real numbers. Let b be any positive number except 1. Then either of the two following equivalent correspondences is an isomorphism between G and G':

$$g \to g' = \log_b g \quad \text{or} \quad g' \to g = b^{g'},$$

where the operation for G is multiplication and that for G' is addition.

Proof: The equivalent correspondences show that g has a unique image g' and g' is the image of a unique g. Second,

$$g_1 \to \log_b g_1 \quad \text{and} \quad g_2 \to \log_b g_2$$

implies that

$$g_1 g_2 \to \log_b g_1 + \log_b g_2 = \log_b(g_1 g_2).$$

Or we could express the same correspondence in terms of exponents. Actually, this proof is merely a statement of properties of logarithms or exponents.

It is this isomorphism that makes computation with logarithms effective, since the isomorphism substitutes addition for multiplication. This goes hand in hand with the fact that the exponential function is the inverse of the logarithmic function.

EXAMPLE 2. The Argand diagram (see Section 3 of Chapter I) gives a familiar isomorphism between pairs of real numbers and complex numbers by means of the correspondence

(3.1) $$(x, y) \leftrightarrow x + iy, \qquad i^2 = -1.$$

Here adding vectors corresponds to adding the corresponding complex numbers, and both groups are additive. As a result of this correspondence we can designate a point in the plane either by a pair of real numbers or by a single complex number $x + iy$.

EXAMPLE 3. Here is another example of an isomorphism between a multiplicative group and an additive one. We know from de Moivre's theorem in trigonometry that

(3.2) $$(\cos \theta + i \sin \theta)^t = \cos(t\theta) + i \sin(t\theta)$$

for all rational numbers t. If we choose $\theta = 2\pi/n$, (3.2) becomes

(3.3) $$\left[\cos \frac{2\pi}{n} + i \sin \frac{2\pi}{n}\right]^t = \cos \frac{2\pi t}{n} + i \sin \frac{2\pi t}{n}.$$

Hence

(3.4) $$r = \cos \frac{2\pi}{n} + i \sin \frac{2\pi}{n}$$

is a root of $x^n = 1$, as may be seen by taking $t = n$ in (3.3). Now $r^n = 1$ implies that

(3.5) $$1, r, r^2, \ldots, r^{n-1}$$

are all roots of $x^n = 1$, that is, are nth roots of unity. They are distinct, since the right side of (3.3) is 1 if and only if t is a multiple of n. Since $x^n = 1$ has exactly n roots (shown in college algebra and also in Chapter III of this book), the numbers (3.5) are *the* nth roots of unity. The powers of r form a multiplicative cyclic group of order n. (Why?) Using the Argand diagram shows that the points corresponding to the powers of r lie on the *unit circle* (the circle of radius 1 with center at the origin) at the vertices of a regular n-sided polygon.

As in Example 1, we can set up an isomorphism between the powers of r and the exponents, but here there is a difference. For one thing, there are only n different powers of r. In fact,

$$r^s = r^u \quad \text{if and only if } s - u \text{ is a multiple of } n,$$

for suppose that $r^s = r^u$ with $s > u$. Then $r^{s-u} = 1$, and we showed immediately above that $r^t = 1$ if and only if t is a multiple of n.

For instance, when $n = 5$, the correspondence between powers of r and exponents is

$$r^0 = 1 \leftrightarrow 0, \qquad r^1 \leftrightarrow 1, \qquad r^2 \leftrightarrow 2, \qquad r^3 \leftrightarrow 3, \qquad r^4 \leftrightarrow 4.$$

So the correspondence between powers of r and these five integers is about as simple as it could be. But if we are to have an isomorphism, we must have operations. For powers of r, the operation is of course multiplication. But since $r^3 r^4 = r^7 = r^2$, then it must be understood that in adding exponents $3 + 4 = 2$; that is, after adding exponents just as we add numbers, we divide by 5 and let the remainder be the number that corresponds to the power of r.

In this connection, a notation of Gauss is useful.

Definition. *If s, t, and n are integers and n is positive, then $s \equiv t$ (mod n) (read "s is congruent to t modulo n") means: $s - t$ is a multiple of n.*

Since this notation is very useful in proving the isomorphism between the powers of r, an nth root of unity, and integers modulo n, we should first prove some properties of congruence and its connection with the previous section. The first set of properties is described in the following theorem.

Theorem 3.2. Given a positive integer n, congruence modulo n is an equivalence relation for the set $\mathbf{Z}$ of integers; that is, for integers a, b, and c, the following hold:

1. $a \equiv a$ (mod n).
2. If $a \equiv b$ (mod n), then $b \equiv a$ (mod n).
3. If $a \equiv b$ (mod n) and $b \equiv c$ (mod n), then $a \equiv c$ (mod n).

Partial proof: We leave the testing of the first two properties as exercises and prove the third. That is, we show that if $a - b$ and $b - c$ are multiples of m, then so is $a - c$. This follows, since $a - b = rm$ and $b - c = sm$ for integers r and s implies that $a - c = (r + s)m$ for the integer $r + s$.

Not only is congruence (mod n) similar to equality in that it is an equivalence relation, but one can also add a number to both sides of a congruence or multiply both sides of a congruence by a number without altering the truth of the congruence. We express this more precisely in the following theorem and leave the proof as Exercise 29.

Theorem 3.3. If $a \equiv b$ (mod n), then, for every integer c,

1. $a + c \equiv b + c$ (mod n).
2. $ac \equiv bc$ (mod n).

Corollary. If $a \equiv a'$ (mod n) and $b \equiv b'$ (mod n), then $a + b \equiv a' + b'$ mod n) and $ab \equiv a'b'$ (mod n).

We noted in Section 2 that whenever one has an equivalence relation among the elements of a set S, there is a partition of S into subsets or classes. So any positive integer n partitions **Z** into classes:

$$[0], [1], \ldots, [n - 1],$$

where two integers are in the same class if their difference is a multiple of n. Any member of a class serves to define it; for instance,

$$[1], [1 + n], [1 + 2n], \ldots, [1 - n], [1 - 2n], \ldots$$

are all the same class. Now we can define sum and products of classes as follows:

$$[a] + [b] = [a + b] \quad \text{and} \quad [a][b] = [ab].$$

This is possible since the corollary shows that the sum and product of classes are independent of the representatives used to define the class.

We carry these ideas further in later sections, but return now to the proof of the isomorphism for Example 3. We set up the following correspondence:

$$(3.6) \qquad\qquad r^k \leftrightarrow [k],$$

$$\text{multiplication} \leftrightarrow \text{addition of classes } [k] \text{ (mod } n).$$

Theorem 3.4. The correspondence (3.6) between the multiplicative group G of powers of r defined in (3.4) and the set of classes of integers modulo n under addition is an isomorphism.

Proof: First, we must show that (3.6) is a one-to-one correspondence. Each power of r certainly determines the class $[k]$. Since $r^s = r^t$ if and only if s and t are in the same class modulo n, it follows that the class determines the power of r.

It remains to show that the correspondence preserves the operations, that is, that $r^k \leftrightarrow [k]$ and $r^{k'} \leftrightarrow [k']$ imply that $r^{k+k'} \leftrightarrow [k + k']$. This follows from

$$r^{k+k'} = (r^k)(r^{k'}) \leftrightarrow [k] + [k'] = [k + k'].$$

This completes the proof.

We should not be so preoccupied with classes that we neglect another way of looking at the above. We could also consider the following set of integers:

$$(3.7) \qquad\qquad 0, 1, 2, \ldots, (n - 1)$$

and instead of ordinary addition take the operation to be addition modulo n. This means that to find the sum of two numbers of (3.7), we first go outside this set to get the usual sum and then return to the set by replacing the ordinary sum by its remainder when divided by n. For instance, if $n = 5$, $3 + 3 = 1$, since 1 is the remainder when 6 is divided by 5. To help fix these ideas, you should check the addition table modulo 5 shown in Table II.2.

Table II.2

	0	1	2	3	4
0	0	1	2	3	4
1	1	2	3	4	0
2	2	3	4	0	1
3	3	4	0	1	2
4	4	0	1	2	3

A companion theorem to Theorem 3.4 is the following.

Theorem 3.4a.   The correspondence $r^k \leftrightarrow$ (remainder when k is divided by n) is an isomorphism between the multiplicative group of powers of r, as defined in (3.4), and the numbers (3.7) under addition modulo n.

Exercises

1. Let $G = \{1, 3, 5, 7\}$ (mod 8) with multiplication modulo 8 as the operation. Prove that this is a group. To what group with four elements is it isomorphic?

2. Find a multiplicative group of numbers (mod 12) that is isomorphic to the group in Exercise 1.

3. Is there an additive group (mod m) that is isomorphic to the groups in Exercises 1 and 2?

4. Write the multiplication table for the Klein four-group.

5. Define $a \equiv b$ (mod π) to mean that $a - b$ is an integral multiple of π. Show that the real numbers modulo π form an additive group.

6. Explain why it is that if two groups are isomorphic to a third, they are isomorphic to each other.

7. Show that if two finite groups are isomorphic, they must have the same order.

8. Show why there is an isomorphism (3.1) between the group of vectors (x, y) and the group of complex numbers $x + iy$, where the operation is addition in both cases.

9. Show that if an Abelian group with six elements contains an element of order three, it is cyclic.

10. Let $S = \{x + y\sqrt{2}\}$, where x and y are rational numbers not both zero. Prove that S is a multiplicative group and that $x + y\sqrt{2} \leftrightarrow x - y\sqrt{2}$ is an isomorphism of S onto itself.

11. If in Exercise 10, S were an additive group, would the same correspondence be an isomorphism?

12. Let G be the group consisting of the two functions x and $1/x$; let H be the group consisting of the two functions x and $1 - x$ (see Section 5 of Chapter I), where the operation is composition. Prove that if the domain of the functions is the set of all real numbers different from 1 and 0, then G and H are isomorphic.

13. Prove that every group with four elements is either a cyclic group or isomorphic to the Klein four-group. Which is isomorphic to the group of Section 11 in Chapter I?

14. Let G be the additive group of all integers and G' the additive group of all even integers. Are the two groups isomorphic? Explain.

15. Prove that every group of order five is isomorphic to the cyclic group of order five.

16. Point out one place in Section 6 of Chapter I where there is an example of an isomorphism.

17. Show that in (1.3), b^2 must be e.

18. Show that the six elements in (1.3) are distinct.

19. Show that a group with six elements cannot have an element $g \neq e$ such that $g^5 = e$.

20. Prove that the powers of r defined by (3.4) form a multiplicative group.

21. Let $r = \cos(\pi/6) + i \sin(\pi/6)$ and $s = \cos(5\pi/6) + i \sin(5\pi/6)$. Find an isomorphism between the multiplicative groups R and S generated by r and s, respectively.

22. Write out the multiplication and addition tables of the numbers (3.7) for $n = 7$ and $n = 11$.

23. Prove Theorem 3.2.

24. Prove Theorem 1.2.

25. Prove the Corollary of Theorem 1.1.

26. Complete the proof of Theorem 2.1.

27. Complete the proof of Theorem 1.4.

28. Prove that two cyclic groups of the same order are isomorphic.

29. Prove Theorem 3.3 and the Corollary.

30. Is the multiplicative group **R*** (real numbers without zero) isomorphic to the additive group of all real numbers?

31. Let S be a subgroup of a group G, and for a and b two elements of G define $a \equiv b \pmod{S}$ to mean: $a^{-1}b$ is in S. Show that this is an equivalence relation. Does $a \equiv b \pmod{S}$ in this sense imply that either or both of $ac \equiv bc \pmod{S}$ and $ca \equiv cb \pmod{S}$?

32. If $a^2 \neq 1$, do the roots of the equation $x^n = a$, for n an integer greater than 2, form a multiplicative group? Why or why not?

33. Prove that a group G is Abelian if and only if the correspondence $g \to g^{-1}$ is an isomorphism.

34. Does the cancellation property hold for addition modulo m, that is, does $a + c \equiv b + c \pmod{m}$ imply that $a \equiv b \pmod{m}$? Prove the property or give an example in which it does not hold.

★35. (This exercise looks forward to later material.) Does the cancellation property hold for multiplication modulo m; that is, if $c \not\equiv 0 \pmod{m}$, does $ac \equiv bc \pmod{m}$ imply that $a \equiv b \pmod{m}$?

★36. If in the definition of an equivalence relation in Section 2, we choose $c = a$, it would seem that the third property implies the first. If this were so, the first property would be redundant; that is, it could be omitted from the list. Why is this not so?

★37. Let S be the set of ordered pairs of integers (x, y) in which $y \neq 0$. Define a relation ρ

$$(x, y) \, \rho \, (x', y') \qquad \text{whenever } xy' = yx'.$$

Prove that this is an equivalence relation. What is the corresponding partition of the set of such ordered number pairs into classes? What is a connection between this and fractions?

4.   *Cosets*

There are two important lines of thought which stem from the last example in Section 3 and from Section 2. We explore these in this section and the next.

As in Section 3 let $r = \cos(2\pi/n) + i \sin(2\pi/n)$ for n a positive integer. As we saw there, the powers of r can be represented as points on the unit circle that are the vertices of a regular polygon with n sides. With each r^u we associated the set $[u]$, which consisted of all the integers congruent to u modulo n. If H is the set of all multiples of n, we can write $[u]$ as $u + H$, which we use to denote the set of all integers of the form $u + h$, where u is fixed and h is any element of H. With this notation, each vertex of the regular polygon with n sides can be associated with one of the following sets of numbers:

(4.1) $H, 1 + H, 2 + H, \ldots, (n - 1) + H.$

We call each of these sets in (4.1) a coset of H. First we define this important concept for Abelian groups (using $+$ to denote the operation).

Definition.   *Let G be an Abelian group with the operation $+$ and H a subgroup of G. Then for each element g of G, the set $g + H$, that is, the set of elements $\{g + h\}$, where h is in H, is called a* coset *of H (in G).*

Since the idea of a coset is very useful here and in other parts of mathematics, it pays us to spend a little time looking into its properties. Now, by means of (4.1) the set of multiples of n, that is, H, partitions $\mathbf{Z}$ into n cosets. Each coset is determined by any one of its elements. That is, $(1 + kn) + H$ is the coset $1 + H$ for all integers k. Furthermore, two integers are in the same coset if and only if their difference is divisible by n, that is, is in H. Another way of showing this is to consider the condition that $g + H = g' + H$. This equality means that for each h in H, there is an h' in H such that $g + h = g' + h'$; that is, $g - g' = h' - h$. Since H is a group, this means that $g - g'$ is in H.

We shall more often be concerned with cosets, using the multiplicative notation so as not to imply commutativity. The corresponding definition is the following.

Definition.   *Let G be a group whose operation is written as if it were multiplication and let H be a subgroup of G. Then for each element g of G, the set gH, that is, $\{gh\}$ for all h in H, is called a* left *coset, and, similarly, Hg is called a* right *coset of H (in G).*

Here since multiplication need not be commutative, we must distinguish between left and right cosets. Before dealing with more properties of cosets, let us consider a multiplicative example.

EXAMPLE 1.   Let G be the octic group defined in Section 9 of Chapter I. There we showed that the elements of G are

$$\alpha^i \rho^j, \quad i = 0, 1 \quad \text{and} \quad j = 0, 1, 2, 3,$$

where

(4.2) $\alpha = \begin{pmatrix} 1 & 2 & 3 & 4 \\ 2 & 1 & 4 & 3 \end{pmatrix} \quad \text{and} \quad \rho = \begin{pmatrix} 1 & 2 & 3 & 4 \\ 2 & 3 & 4 & 1 \end{pmatrix}.$

In order to conform with the notation used for additive cosets, write the elements of G as $\rho^j\alpha^i$, take H to be the subgroup $\{\epsilon, \alpha\}$, and compute the following left cosets:

$$H = \{\epsilon, \alpha\}, \qquad \rho H = \{\rho, \rho\alpha\}, \qquad \rho^2 H = \{\rho^2, \rho^2\alpha\}, \qquad \rho^3 H = \{\rho^3, \rho^3\alpha\}.$$

In this case there are four cosets of H; G is partitioned into four classes. Note two things: First, H is the only coset that is a group (for instance, ρ is in ρH but ρ^2 is not); second, $H\rho = \{\rho, \alpha\rho\}$, which is not the same coset as ρH. We could equally well partition G into right cosets, but the partition would not be the same.

Now let us look at the idea of coset more abstractly, leaving additional examples until later.

Theorem 4.1.   Let G be a group, multiplicative in notation, and H a subgroup of G. Each g of G defines a left coset of H denoted by gH and consisting of all elements of the form gh, where h is in H. "Being in the same coset as" is an equivalence relation. The left cosets of G partition G into disjoint subsets. If H is finite, each left coset has the same number of elements as H. The same results hold for right cosets.

Proof: The equivalence relationship referred to is tied directly to the properties that H has by virtue of being a group:

1. Each element g of G is in the coset gH, since the identity element is in H.
2. If g is in the coset $g'H$, then g' is in the coset gH; for $g = g'h$ implies that $gh^{-1} = g'$ and h^{-1} is in H, since the inverse of every element of H is in H.
3. If g and g' are in the same coset and g' and g'' in the same coset, then g and g'' are in the same coset; for $g = g'h'$ and $g' = g''h''$ imply that $g = g''h''h'$ and $h''h'$ is in H, since the binary operation of G is closed over H.

Now we show that cosets are disjoint. Suppose that b is an element of both gH and $g'H$. Then, for some elements h and h' of H we have $b = gh = g'h'$. This implies that $g = g'h'h^{-1}$ and thus that gh'' is in $g'H$ for all h''. Hence gH is contained in $g'H$. Similarly, $g'H$ is contained in gH. This shows that gH and $g'H$ are the same if they have an element in common. That every element of G is in some coset follows from the construction of cosets. If H has a finite number of elements the cosets H and gH have the same number of elements, for we can set up the one-to-one correspondence $h \leftrightarrow gh$ for each h in H. This completes the proof.

The following fundamental theorem is almost a corollary of Theorem 4.1.

Theorem 4.2.   (Lagrange.)   If G is a finite group and H is any subgroup of G, then the order of H is a factor of the order of G.

Proof: If $H \neq G$, let g_2 be an element of G not in H. Then g_2 defines the coset g_2H. If $G = H \cup g_2H$, we are done. If not, let g_3 be in G but in neither H nor g_2H. Then g_3 defines g_3H. So we can continue until $H, g_2H, g_3H, \ldots,$ g_tH exhaust the elements of G. Furthermore, since no two different cosets have an element in common and since from Theorem 4.1 any two cosets have the same number of elements, it follows that if s is the order of H, then $st = n$, the order of G.

Corollary 1.   If the order of a group G is a prime number, the only subgroup of G other than itself is that consisting of the identity element alone.

Definition.   *If H is a subgroup of G, if s is the order of H, and n is the order of G, then $n/s = t$ is called the* index *of H in G.*

We use the notation $[G:H] = t$; that is, $t = o(G)/o(H)$.

Corollary 2.   The index of H in G is the number of left cosets of H. It is also the number of right cosets. (See Exercise 19.)

We can carry the above further, to the following natural result.

Theorem 4.3.   Let H be a subgroup of a finite group G and K a subgroup of H. Then $[G:K] = [G:H][H:K]$.

Proof: Let the cosets of K in G be

$$K, g_2K, g_3K, \ldots, g_vK.$$

By Corollary 2, with K in place of H, $v = [G:K]$. Now K is in H, by hypothesis. If g_2 is in H, then every element of the coset g_2K is in H, since then every element in g_2K is a product of two elements of H. Now we renumber the g's so that the first t are in H and the remaining ones are not. Every element of H is in some coset of K; hence

$$H = \{K, g_2K, g_3K, \ldots, g_tK\}$$

and we have that $t = [H:K]$. Now let $H, b_2H, \ldots, b_uH$ be the cosets of H in G. Then

$$b_iH = \{b_iK, b_ig_2K, \ldots, b_ig_tK\}, \qquad 2 \leq i \leq u.$$

Now for each i, the cosets b_ig_jK and b_ig_sK are disjoint if $j \neq s$ for if $b_ig_jk = b_ig_sk'$ for k and k' in K, then $g_j = g_sk'k^{-1}$, which is impossible. So every coset of H consists of t cosets of K. This shows that $[G:H] = u$ and $[G:K] = v = ut$, where $t = [H:K]$. This completes the proof.

It should be noted that if v is a divisor of n, the order of G, it is not necessary that G have a subgroup of order v. We shall give an example to show this in Section 11.

There are many important consequences of Lagrange's theorem, but we postpone these in favor of giving some examples of cosets.

EXAMPLE 2.   Consider the multiplicative groups $\mathbf{R}^*$ and $\mathbf{Q}^*$, the nonzero real and rational numbers, respectively. The group $\mathbf{Q}^*$ is, of course, a subgroup of $\mathbf{R}^*$. If r is any nonzero real number, $r\mathbf{Q}^*$ is a coset of $\mathbf{Q}^*$ and consists of all products rq, where q is a rational number. If r is itself rational, the coset is $\mathbf{Q}^*$ itself. Two real numbers are in the same coset if their quotient is a rational number.

EXAMPLE 3.   This example has to do with pairs of linear equations:

$$a_1 x + b_1 y = c_1,$$
$$a_2 x + b_2 y = c_2,$$

where not all the coefficients of x and y are zero. They may be written in matrix form as follows:

(4.3) $$(x, y)\begin{bmatrix} a_1 & a_2 \\ b_1 & b_2 \end{bmatrix} = (c_1, c_2).$$

Suppose that (x', y') and (x'', y'') are two solutions of (4.3). Then, using the distributive property of matrices, we see that

$$[(x', y') - (x'', y'')]\begin{bmatrix} a_1 & a_2 \\ b_1 & b_2 \end{bmatrix} = (0, 0).$$

So, now, if we associate with Equation (4.3) the corresponding pair of homogeneous equations

(4.4) $$(x, y)\begin{bmatrix} a_1 & a_2 \\ b_1 & b_2 \end{bmatrix} = (0, 0),$$

we see that if (x_0, y_0) is one solution of (4.3), all solutions will be of the form

$$(x, y) = (x_0, y_0) + (z, t),$$

where (z, t) is a solution of (4.4). If we let G be the additive group of ordered pairs (that is, vectors) of real numbers and H the set of solutions of (4.4), we see first that H is an additive group, since the sum and difference of two solutions of (4.4) is a solution. So the set of solutions of (4.3) is a coset of H:

$$(x_0, y_0) + H.$$

Note that if c_1 and c_2 are not both zero, the coset is not a group since then, for instance, the sum of two solutions of (4.3) is not a solution.

Now there may be only the trivial solution $(0, 0)$ of the pair of homogeneous equations (4.4), in which case there is only one solution of (4.3), namely, (x_0, y_0); this happens when the determinant $a_1 b_2 - b_1 a_2$ is not zero. However, if the determinant is zero, H is an additive group consisting of all multiples of a vector. In fact, if the determinant is zero, $H = \{x(r, s)\}$, where $a_1 r + b_1 s = 0$, provided that a_1 and b_1 are not both zero. You will be asked to show in Exercise 8 that the points of the coset $(x_0, y_0) + H$ are the points of a line. For three dimensions, the cosets are lines or planes.

EXAMPLE 4.   This example has to do with differential equations. Let

(4.5) $$a_2 y'' + a_1 y' + a_0 y = f(x)$$

be a differential equation where a_i are real numbers, y is a twice-differentiable function of x, and y' and y'' are the first two derivatives of y with respect to x. Equation (4.5) is called a *linear* differential equation. Equation (4.5) is called *homogeneous* if $f(x)$ is zero. It is known from the theory of differential equations that if y_1 and y_2 are two solutions of (4.5), then $y_1 - y_2$ is a solution of the homogeneous equation obtained from (4.5) by replacing $f(x)$ by 0. Thus if y_0 is any solution, all solutions are given by

$$y = y_0 + H,$$

where H is the additive group of all solutions of the homogeneous equation. So the set of solutions of (4.5) form a coset of H in the additive group G of twice-differentiable functions.

EXAMPLE 5.   Let G be the set of all transformations $\sigma_{a,b}$ defined as follows over the real numbers: $x\sigma_{a,b} = ax + b$, where a and b are real numbers and $a \neq 0$. Let the operation be composition. That is,

$$x\sigma_{a,b}\sigma_{c,d} = (ax + b)\sigma_{c,d} = c(ax + b) + d,$$
$$x\sigma_{c,d}\sigma_{a,b} = a(cx + d) + b.$$

This shows that composition of these transformations is not commutative. You will be asked to show in Exercise 18 that G is a group and that

$$H = \{\sigma_{1,b}\} \qquad \text{and} \qquad H' = \{\sigma_{a,0}\}$$

over all real b and all real nonzero a, respectively, are Abelian subgroups of G. You will also be asked to find their cosets.

Exercises

1. Using the pattern of Example 3, express all solutions of the following equations as a coset:

$$3x_1 + 2x_2 = 4,$$
$$6x_1 + 4x_2 = 8.$$

2. Repeat Exercise 1 for the following pair of equations:

$$x_1 + 3x_2 = 2,$$
$$-3x_1 - 9x_2 = -6.$$

3. Let G be the group of symmetries of an equilateral triangle and H the set of rotations, including the identity. How many left cosets of H are there? How many right cosets? Are the right cosets equal to the left cosets? Find them.

4. Let H be the set of symmetries of a rectangle not a square and G the octic group. Find the right and left cosets of H.

5. Let H be the subgroup $\{e, (1\ 2)(3\ 4)\}$ of the octic group. Find the left and right cosets of H.

6. Let H be the set of rotations of the octic group and G the octic group. Find the left and right cosets of H in G. Are the right cosets equal to the left cosets?

7. Let H be a subgroup of G. Prove that the only coset of H which is a group is H itself.

8. Let $A = (a_1, a_2)$ and $B = (b_1, b_2)$ be two linearly independent vectors with real components and $H = \{x(b_1, b_2)\}$ for x a real number. What geometrically is the coset $A + H$ in the additive group of all vectors in two dimensions? Let C be the vector $(a_1 + b_1, a_2 + b_2)$ and show that every element of $A + H$ is of the form

$$D = r(a_1, a_2) + s(a_1 + b_1, a_2 + b_2),$$

with $r + s = 1$. Interpret A, B, C, and D as points and show that D is between A and C if and only if r and s are positive and less than 1.

9. Suppose that G is a group and H is any subgroup of G whose index is 2. Prove that if g is any element of G not in H, then the cosets gH and Hg are equal.

10. Why are the subgroups of the octic group found in Exercise 9 of Section 9 in Chapter I the only subgroups except (e) and the group itself?

11. Show that the correspondence $gH \leftrightarrow Hg^{-1}$ is a one-to-one correspondence between the right and left cosets H in a group G.

12. Let S be the set of lines in the plane. Call two lines equivalent if they have the same slope. Show that this is an equivalence relation.

13. Solve each of the following differential equations:
   (a) $y'' - 4y = \cos x,$
   (b) $y'' + 3y' + 2y = x^2 + 6x + 12,$
      where in each case y is a twice-differentiable function of x.

14. Prove Corollary 2 of Theorem 4.2.

15. In the light of results in this section, improve the proof, given in Section 1, that any two non-Abelian groups of order six are isomorphic.

16. Let g be some element of G and N_g the set of all elements of G that are commutative with g. Show that N_g is a subgroup of G. Describe the cosets of N_g. (N_g is called the *normalizer* or *centralizer* of g; see Section 12.)

17. Let G be a finite group of order n. Prove that $g^n = e$ for every g in G.

18. Complete the discussion of Example 5.

19. Prove that if G is a finite group and if H is a subgroup such that the number of left cosets gH is finite, then there are just as many right cosets of H as left cosets.

5. *Cyclic Groups*

Cyclic groups, that is, groups that consist of powers of a single generator, are the simplest of all groups. We can without much difficulty find out all we want to know about such groups and perhaps more. First, we show that every subgroup of a cyclic group G is also a cyclic group. To this end let g be a generator of G and S a subgroup. All the elements of S must be powers of g. Let g^u be the least positive power of g in S. We shall show that g^u generates S.

Now let g^v be any element of S. Since S is closed under multiplication, it must contain

$$(5.1) \qquad\qquad g^{ux + vy}$$

for all integers x and y. That is, any power of g whose exponent is a linear combination of u and v with integer coefficients is also in S. But Corollary 1 of Theorem 10.2 of Chapter I informs us that d, the greatest common divisor of u and v, is in the set $\{ux + vy\}$. Hence g^d is in S. But since u is the least positive power of g in S and d divides u, it follows that $d = u$ and u is a factor of v. That is, g^u is in S and if g^v is in S, it follows that u divides v. Hence we have shown the first part of the following result:

Theorem 5.1. Let G be a cyclic group, g a generator of G, and S a subgroup of G. If u is the least positive integral power of g that is in S, then S consists of all powers of g^u; that is, S is a cyclic group generated by g^u. If G is a finite group of order n, then u is a divisor of n and the order of S is n/u.

Proof: We have proved all of the theorem except for computing the order of S when G is a finite group of order n. This can be done by listing the elements of S:

$$(5.2) \qquad\qquad g^u, g^{2u}, g^{3u}, \ldots, g^{(n/u)u}.$$

No two of (5.2) can be equal, since $g^{ru} = g^{su}$ (with $n/u \geq r > s > 0$) would imply that $g^{(r-s)u} = e$. But the exponent $(r-s)u$ is positive and less than n. This is impossible because the first n powers of g are distinct. The proof is complete.

Corollary.   Every subgroup of a cyclic group is cyclic.

We also know from Lagrange's theorem that every subgroup of G has an order that divides n, the order of G. For cyclic groups, contrary to the case in general, for every divisor of the order of G, there is a unique subgroup having that order. This we state and show in the following theorem.

Theorem 5.2.   If m is any divisor of the order of a finite cyclic group G, then G has a cyclic subgroup G_m of order m. Furthermore, if g is a generator of G, then $g^{n/m}$ generates G_m. For each m, G_m is unique.

Proof: There are m distinct powers of $g^{n/m}$ (see the proof of Theorem 5.1) and hence $g^{n/m}$ generates a cyclic subgroup of order m. If G_m is any cyclic subgroup of order m, let g^u be the least positive power of g in G_m and see from Theorem 5.1 that $n/u = m$, that is, $u = n/m$. Hence g^u generates G_m. So G_m is unique and the theorem is proved.

It should be noted that this theorem does not hold for groups that are not cyclic, even if they are Abelian. For instance, the Klein four-group K is of order 4, 4 is a divisor of 4, but K has no *cyclic* subgroup of order 4.

Cyclic groups are important in the study of any group, for suppose that g is any element of a group G. Then g generates a cyclic subgroup of G. Recall that the order of this subgroup is called the *order* of g. If G is finite, the order of each of its elements must be finite, whereas if G is infinite, it *may* have some elements of infinite order.

This is chiefly what we want to know about cyclic groups, but it is hard to refrain from asking other questions, since we can answer them without much difficulty. Given one generator of a group G, what can one say about all the generators? For example, consider the multiplicative group G that consists of 1, 2, 3, 4, 5, 6 (mod 7). The following table shows that 3 generates G:

k	1	2	3	4	5	6
3^k (mod 7)	3	2	6	4	5	1

Every element of the group is a power of 3. But 2 does not generate G, since $2^3 \equiv 1 \pmod 7$. As a matter of fact, we should not expect $2 \equiv 3^2 \pmod 7$ to generate G, since $(3^2)^3 \equiv 1 \pmod 7$. It can be shown that the only powers of 3 that generate G are 3 and $3^5 \equiv 5 \pmod 7$. It might appear from this limited example that the power of 3 for a generator can have no factor in common with 6, the order of the group. Let us explore this further.

So now suppose that each of g and g^u generate G. Then since g^u generates G, $g = g^{ux}$ for some integer x. Thus $g^{ux-1} = e$. If G is infinite, this can happen only if $ux - 1 = 0$; that is, $u = x = \pm 1$. If G is finite of order n, it must follow that $ux - 1$ is divisible by n. We prove the following lemma.

Lemma 1. If u and n are integers, there is an x such that $ux - 1$ is divisible by n if and only if u and n are relatively prime; that is, 1 is their g.c.d.

Proof: Suppose u and n had a factor d in common. Then $ux - 1 = yn$ for some integers x and y implies that d is a factor of 1. Thus $d = \pm 1$. On the other hand, suppose that 1 is the g.c.d. of u and n. Then by Corollary 2 of Theorem 10.2 in Chapter I we know that there exist integers x and y such that $ux + ny = 1$; that is, $ux - 1 = -ny$. This completes the proof.

So Lemma 1 and what we have shown prove the following result.

Theorem 5.3. If g is a generator of an infinite group G, the only other generator is g^{-1}. If g is a generator of a group G of order n, g^u is also a generator if and only if u and n are relatively prime.

The above has an immediate consequence for roots of unity, that is, solutions of $x^n = 1$. We saw in Section 3 that $r = \cos(2\pi/n) + i\sin(2\pi/n)$ has the property that its powers are the nth roots of unity; that is, r is a generator of G, the cyclic group of nth roots of unity. In this context the lemma implies the following corollary.

Corollary 1. If $r = \cos(2\pi/n) + i\sin(2\pi/n)$, then r^u is a generator of the multiplicative group of nth roots of unity if and only if u and n are relatively prime.

Definition. *A root of* $x^n = 1$ *is called a* primitive nth *root of unity if its powers are the* nth *roots of unity, that is, if its first* n *powers are distinct.*

Corollary 2. If b is a primitive nth root of unity, the primitive nth roots of unity are b^j for j and n relatively prime.

Recall that $r = \cos(2\pi/n) + i\sin(2\pi/n)$ is a primitive nth root of unity. Corollary 2 gives us a means of finding all the primitive nth roots of unity.

For reference we should define formally a term we have just used. (See Section 10 of Chapter I.)

Definition.   *Two integers are said to be* relatively prime *if* 1 *is their g.c.d. In particular,* 1 *is* relatively prime *to every integer.*

There is a convenient function called the Euler phi function, $\phi(n)$, for the number of positive integers less than n that have no factors greater than 1 in common with n, that is, the positive integers less than n that are relatively prime to n. For instance $\phi(7) = 6$, $\phi(6) = 2$, and $\phi(12) = 4$. We define $\phi(1)$ to be 1. Therefore, in this notation we have the following corollary.

Corollary 3.   The number of distinct elements that generate a cyclic group of order n is $\phi(n)$.

Now we apply Lagrange's theorem, leading to two theorems in the theory of numbers connected with Euler's phi function.

Theorem 5.4.   Let m be any positive integer and S the set of numbers between 1 and m inclusive that are relatively prime to m. Then S is a group under multiplication modulo m.

Proof: Recalling the definition of a group, we see that first we must show closure under multiplication; that is, if a and b are in S, so is ab. In other words, we wish to show that if a and b are relatively prime to m, then so is ab. Since this is quite obvious intuitively, we assume it for the moment and will prove it in Lemma 2.

The associative property holds, since $(ab)c = a(bc)$ implies that the remainders of the numbers on both sides are the same modulo m. Now 1 is in S because 1 is relatively prime to m. To show the existence of an inverse, let b be an element of S. From the first paragraph of this proof, all powers of b are in S. Hence S contains the cyclic group generated by b. If c is the order of b, then $b \cdot b^{c-1} - 1$ shows that b^{c-1} is the inverse of b. This completes the proof.

Corollary 1.   If $m = p$, a prime number, the numbers $1, 2, \ldots, (p - 1)$ modulo p form a multiplicative group.

If $m = ab$, with both a and b positive integers less than m, then there are two numbers in the sequence $1, 2, \ldots, m$ whose product is 0 modulo m. We have the following corollary.

Corollary 2.   The numbers $1, 2, \ldots, (m - 1)$ modulo m form a multiplicative group if and only if m is a prime number.

Since S in Theorem 5.4 is a group of order $\phi(m)$, then, by Lagrange's theorem, the order of every subgroup is a factor of $\phi(m)$; in particular, the order c of the cyclic subgroup generated by b any element of S is a factor of $\phi(m)$. Writing $cd = \phi(m)$, we see that $b^{\phi(m)} = (b^c)^d \equiv 1 \pmod{m}$. So we have proved the following.

Theorem 5.5. (Euler's theorem.) If b is a number relatively prime to m, then $b^{\phi(m)} \equiv 1 \pmod{m}$.

In case $m = p$, a prime number, the theorem implies the following.

Corollary. (Fermat's theorem.) If p is a prime number that does not divide b, then $b^{p-1} \equiv 1 \pmod{p}$.

The corollary follows because if p is a prime number, $1, 2, 3, \ldots, (p - 1)$ are the positive integers less than p and relatively prime to p. Thus $\phi(p) = p - 1$.

We must delay no longer proving the following result, which is assumed in the proof of Theorem 5.4.

Lemma 2. If a and b are two integers relatively prime to m, then their product is also.

Proof: By Corollary 2 of Theorem 10.2 of Chapter I we know the existence of integers x, y, z, and t such that $ax + my = 1$ and $bz + mt = 1$. Thus $bz(ax + my) = bz + 1 - mt$. This implies that

$$ab(xz) + m(bzy + t) = 1.$$

So any common divisor of ab and m must be a divisor of 1. This completes the proof of the lemma.

There are three results which we now state and prove. At this point they are curiosities, but we shall use two of them later (in Section 12 of Chapter III) to prove that if m is a prime number in Theorem 5.4, the multiplicative group is cyclic. Also, we shall prove a generalization of this result.

Theorem 5.6. If n is an integer, then n is the sum of $\phi(d)$ over all divisors d of n, including 1 and n.

Proof: Let G be a cyclic group of order n. If d is a divisor of n, we know by Theorem 5.2 that there is a unique cyclic subgroup G_d of order d. Conversely, the order of each cyclic subgroup of G is a factor of n. So, we have

1. There is a one-to-one correspondence between the divisors d of n and the cyclic subgroups G_d (of order d) of G.

Now, by Corollary 3 of Theorem 5.3, G_d has $\phi(d)$ generators. Furthermore, each element of G is a generator of some cyclic subgroup and the order of the element is the order of the subgroup that it generates. Hence

2. For each divisor d of n, there are $\phi(d)$ elements of G that are of order d.

But every element of G is of some order that is a divisor of n. If we classify the elements of G according to their order, the class of all elements of order d will have $\phi(d)$ elements. The total number of elements in all these classes is n, the order of G. Hence n is the sum of $\phi(d)$ over all the divisors of n. The theorem is proved.

Let us illustrate this theorem for the case $n = 15$. Then the divisors of 15 are 1, 3, 5, and 15. There is one generator of the subgroup 1, namely, 1. There are $\phi(3) = 2$ generators of G_3, $\phi(5) = 4$ generators of G_5, and $\phi(15) = 8$ generators of G_{15}. (For instance, the elements of G_{15} are g^u for $u = 1, 2, 4, 7, 8, 11, 13, 14$.) And, as is to be expected, $1 + 2 + 4 + 8 = 15$.

Theorem 5.7. If G is a finite Abelian group of order n with the property that $x^k = e$ has at most k solutions in G for all positive integers k, then G is a cyclic group.

Proof: Let G_d be some cyclic subgroup of G of order d. Now d is a divisor of n. Then, from Corollary 3 of Theorem 5.3, there are $\phi(d)$ generators of G_d. Furthermore, given any d, there is at most one cyclic subgroup of G of order d, since if h is a generator of G_d, $h, h^2, \ldots, h^d = e$ are solutions of $x^d = e$, and by the condition of this theorem there are no more. Since every element of G is a generator of some subgroup, it follows that n is the sum of $\phi(d)$ over those divisors d of n for which there is a cyclic subgroup G_d. But by Theorem 5.6, n is the sum of $\phi(d)$ over *all* divisors d of n. Thus for each divisor of n there is a cyclic subgroup of that order. In particular, this is true if $d = n$. This proves the theorem.

It should be noted that without the condition on the number of solutions $x^k = e$, the theorem is false. For instance, for the Klein four-group, $x^2 = e$ has four solutions, since each element is of order 2 or 1. And the Klein four-group is not a cyclic group. It is, in fact, the result of the condition on the solutions of $x^k = e$ which assures us that there is only one cyclic subgroup of order d for each divisor of n.

Theorem 5.8. If $g^u = e$ for g in a group G with identity e and if u is the least positive integer for which this is so, then $g^m = e$ for a positive integer m implies that u is a factor of m.

Proof: This is almost a corollary of Theorem 5.1, for consider the group H generated by g^u. Then $g^m = e$ is in H; hence g^m is a power of g^u, which implies that m is a multiple of u.

All the results of this section carry over immediately to nth roots of unity, that is, roots of the equation $x^n = 1$. For instance, for each divisor d of n, there are $\phi(d)$ primitive dth roots of unity. Furthermore, if r is a primitive nth root of unity, then r^u is also a primitive nth root of unity if and only if 1 is the g.c.d. of n and u, as in Corollary 2 of Theorem 5.3.

6.   Permutation Groups

In Section 7 we shall show that, in a certain sense, every group can be represented as a permutation group. To prepare for this, we find in this section a more compact method of representing permutations and derive some important properties of groups of permutations.

Recall that in Section 9 of Chapter I we defined the symmetric group S_n to be the group of all permutations of n symbols. Its order is $n!$. First consider the permutation

(6.1)
$$\begin{pmatrix} 1 & 2 & 3 & 4 & 5 \\ 2 & 3 & 4 & 5 & 1 \end{pmatrix}.$$

This is called a *cyclic permutation*, since if we write the five numbers around a circle in, say, a counterclockwise direction, the permutation above amounts to moving around the circle from 1 to 2, 2 to 3, and so forth, until finally we move from 5 to 1. A shorter way to write (6.1) is

(6.2)
$$(1\ 2\ 3\ 4\ 5),$$

which we call a *cycle*. In such a representation it is understood that each number is taken into the one on its right and the last is taken into the first. Notice that two other ways to write (6.2) are

$$(2\ 3\ 4\ 5\ 1) \quad \text{and} \quad (3\ 4\ 5\ 1\ 2).$$

Furthermore, if p is a cycle with r symbols, p^r is the identity and for no smaller positive exponent is this true. We call r the *order* or *length* of the cycle. A cycle of length 2 is called a *transposition*.

Now an important advantage of the cyclic notation is that every permutation can be written as a product of disjoint cycles, that is, cycles without common elements. To see how this goes, consider first the following example. Suppose that

$$\sigma = \begin{pmatrix} 1 & 2 & 3 & 4 & 5 & 6 & 7 \\ 2 & 5 & 3 & 1 & 4 & 7 & 6 \end{pmatrix}.$$

Then $1\sigma = 2$, $2\sigma = 5$, $5\sigma = 4$, $4\sigma = 1$ is equivalent to the cycle (1 2 5 4). Furthermore, $3\sigma = 3$, allowing us to omit 3; also, $6\sigma = 7$ with $7\sigma = 6$ gives us the cycle of two elements (6 7). Hence the permutation σ can be written

$$(1\ 2\ 5\ 4)(6\ 7).$$

Any symbol that is not changed is omitted from the product of cycles.

To write any permutation as a product of cycles, start with any number a_1; if it is unchanged by the permutation, omit it and go on to another. If $a_1\sigma = a_2 \neq a_1$, the cycle begins $(a_1 a_2 \cdots)$. Then compute $a_2\sigma$, and so on, to get the chain of images:

$$a_1\sigma = a_2,\ a_2\sigma = a_3,\ \ldots,\ a_r\sigma = a_{r+1},\ \ldots.$$

Since each permutation in S_n has a finite number of elements, there must be one point at which an element occurs which occurred before; call it $a_i = a_j$, with $i < j$. From our method of construction it follows that $a_{i-1}\sigma = a_{j-1}\sigma$. Since each permutation has an inverse, this implies that $a_{i-1} = a_{j-1}$ if $i > 1$. If $i > 2$, this implies in the same manner that $a_{i-2} = a_{j-2}$, and so on to the point where $a_1 = a_{s+1}$ for some s. If we assume that the least such positive subscript is $s + 1$, then we have a cycle $(a_1 a_2 a_3 \cdots a_s)$. If $s \neq n$, we can carry through the same process for another symbol until all are dealt with. Thus we have proved part 1 of the following theorem.

Theorem 6.1

1. Every permutation can be written as a product of disjoint cycles.
2. Every permutation can be written as a product of transpositions.
3. A permutation can be written as a product of disjoint cycles in only one way except for the order of the cycles. [Recall that, for instance, (1 2 3) and (2 3 1) are the *same* cycle.]

Proof: To prove part 2, we need merely show that any cycle can be written as a product of transpositions. To see this, we merely compute

$$(a_1 a_2)(a_1 a_3) \cdots (a_1 a_r) = (a_1 a_2 a_3 \cdots a_r).$$

Part 3 follows from the method of construction, since the permutation determines the image of each symbol. This completes the proof.

Note first that the transpositions in part 2 are not disjoint; second, that the product of two disjoint cycles is commutative. Also, in part 3 the condition that the cycles be disjoint is important. For instance,

$$(1\ 2)(1\ 3)(1\ 4) = (1\ 2\ 3\ 4) \qquad \text{and} \qquad (2\ 3)(2\ 4)(2\ 1) = (2\ 3\ 4\ 1),$$

whereas $(1\ 2\ 3\ 4) = (2\ 3\ 4\ 1)$. But it turns out to be true that if any permutation is written in two ways as a product of transpositions, the number of transpositions in the products are either both even or both odd. Before proving this, assume for the moment that this is true and call a permutation on n letters *even* if it can be expressed as a product of an even number of transpositions and *odd* if it is the product of an odd number of transpositions. For example, let us list the elements of S_3 and S_4 expressed as products of disjoint cycles:

The group S_3

Even permutations: e, (1 2 3), (1 3 2).
Odd permutations: (1 2), (1 3), (2 3).

The group S_4

Even permutations: e, (1 2 3), (1 3 2), (1 2 4), (1 4 2), (1 3 4), (1 4 3),
(2 3 4), (2 4 3), (1 2)(3 4), (1 3)(2 4), (1 4)(2 3).

Odd permutations: (1 2), (1 3), (1 4), (2 3), (2 4), (3 4), (1 2 3 4),
(1 2 4 3), (1 3 2 4), (1 3 4 2), (1 4 2 3), (1 4 3 2).

Notice that any cycle of length 3 can be expressed as a product of two transpositions and any cycle of length 4 as a product of three transpositions. However, it is important to note that we have not yet shown that (2 3 4), for instance, cannot be expressed in some strange way as a product of an odd number of transpositions.

In the case of these two groups this classification as odd or even has an interesting intuitive connection. For the group S_3, the symmetries of an equilateral triangle, the even permutations are just those symmetries which can be performed without going out of the plane and they also preserve orientation, described in Section 9 of Chapter I.

The group S_4 is that of the regular tetrahedron (see Exercise 23 of Section 9 in Chapter I), and with a model you can easily convince yourself that the even permutations are those which can be performed with a solid tetrahedron. For an odd permutation one would have to turn the solid inside out. Your attention should be called to the last three even permutations listed above. It is hard to see from the figure what they mean geometrically until you find by computation that each is a product of two rotations in "opposite directions."

However, this connection between motions in the plane and evenness or oddness does not always hold. For instance, the rotations for the square are odd permutations.

Now we state a basic theorem and prove it. One trouble with the proof is that we use a rather artificial device.

Theorem 6.2. If a permutation is written in two ways as a product of transpositions, the number of transpositions in the products are both odd or both even.

Proof: To prove this result, we use an expression whose sign depends on the permutation but which is altered by a transposition. Consider first the polynomial

$$P(4) = (x_1 - x_2)(x_1 - x_3)(x_1 - x_4)(x_2 - x_3)(x_2 - x_4)(x_3 - x_4).$$

Any permutation of the subscripts in the x's either leaves $P(4)$ unchanged or changes its sign. Whether it changes the sign or not depends on the permutation, not on the way it is expressed. So if we can show that a transposition changes the sign of $P(4)$, two results will follow: (1) If the permutation is a product of an even number of transpositions, it does not change the sign of $P(4)$; (2) if the permutation is a product of an odd number of transpositions, it does change the sign of $P(4)$; it cannot do both. You should try this out for $P(4)$ in some special cases before looking at the rest of the proof below. Let

$$P(n) = \prod_{i<j} (x_i - x_j),$$

where the product is over all i and j between 1 and n with $i < j$.

What is the effect upon P of the transposition $(r\ s)$, where $r < s$? There are three types of differences to deal with:

1. $(x_r - x_s)$.
2. Those containing one but not both r and s.
3. Those containing neither r nor s.

Interchanging r and s in the first changes the sign, and interchanging r and s in the third does not affect the sign. For the second possibility choose some t different from r and s and pair two differences as follows:

(6.3) $$\pm (x_r - x_t)(x_s - x_t),$$

where the sign is minus if t is between r and s and plus otherwise. Now, interchanging r and s in (6.3) merely interchanges the two factors and hence does not change the sign of the product. Thus the only change in sign of $P(n)$ comes from the factor $(x_r - x_s)$. Hence interchanging r and s changes the sign of $P(n)$.

To repeat, if the permutation does not change the sign of $P(n)$, it must be a product of an even number of transpositions; if it does change the sign, it must be a product of an odd number of transpositions. This completes the proof and gives us the right to speak of odd and even permutations.

Corollary. A cycle of order r is an even permutation if r is odd and an odd permutation if r is even.

Definition. A_n, *the set of even permutations of* S_n, *is called the* alternating group. (*We show later that it is a group.*)

Note that for the groups S_3 and S_4 the number of even permutations is half the order of S. This result is contained in the following theorem.

Theorem 6.3.   The set of even permutations of S_n is a group of order $(n!)/2$.

Proof: Let A_n be the set of even permutations of S_n. Since it is a subset of the group S_n, we need, by Theorem 7.1 of Chapter I, to prove only that A_n is closed and contains the inverse of each of its elements. Certainly, the product of two even permutations is even. Also, if t_i are transpositions,

$$[t_1 t_2 \cdots t_{s-1} t_s]^{-1} = t_s t_{s-1} \cdots t_2 t_1.$$

Thus the inverse of an even permutation is even. So we have shown that A_n is a group.

Suppose that A_n has s elements: $a_1, a_2, \ldots, a_s$. Each of these is a product of an even number of transpositions. Then look at the set

(6.4) $(1\ 2)a_1, (1\ 2)a_2, \ldots, (1\ 2)a_s.$

These are all odd permutations of S_n. They are all distinct, since $(1\ 2)a_i = (1\ 2)a_j$ implies that $a_i = a_j$. Furthermore, (6.4) includes all the odd permutations, for suppose that b is an odd permutation. Then $(1\ 2)b$ is an even permutation, and hence $(1\ 2)b = a_i$ for some i. This implies that $b = (1\ 2)a_i$ and hence is in (6.4). So (6.4) includes all the odd permutations. Since any permutation is either even or odd, it follows that $2s = n!$, the order of S_n. This completes the proof.

A somewhat more general theorem can be proved in the same way that we proved Theorem 6.3. We leave the proof as Exercise 11.

Theorem 6.3a.   Let G be a subgroup of S_n and let the order of G be r. If G contains an odd permutation, then the set of even permutations of G forms a subgroup of order $r/2$.

There is one caution that should be observed with regard to even or odd permutations. When we speak of two permutation groups being isomorphic, we refer to the binary operation—not to the evenness or oddness of their permutations. For instance, the permutation $(1\ 2)$ is odd and $(1\ 2)(3\ 4)$ is even, yet the following subgroups of S_4 are isomorphic:

$$H_1 = \{e, (1\ 2)\}, \qquad H_2 = \{e, (1\ 2)(3\ 4)\}.$$

Exercises

1. Compute the following products and find the order of each.

   (a) (1 2 4)(3 5). (b) (1 3 5 6)(2 4 5 3).
   (c) (1 2 3)(2 3 4)(5 6 7). (d) (1 2 3)(1 4)(1 3 2).
   (e) (1 2)(3 4)(1 2 3)(3 4)(1 2). (f) (1 2 3)(3 4 5)(5 6 7).

2. Show that the product (1 2)(3 4)(1 2 3) is a cycle of length 3. Use this fact or other means to prove that every even permutation can be expressed as a product of cycles of length 3.

3. For $r = 5, 6,$ and 7, express the following as a product of disjoint cycles:

$$(a_1 a_2 a_3)(a_2 a_3 a_4) \cdots (a_{r-2} a_{r-1} a_r).$$

4. Show that a cycle of length r is an even permutation if r is odd and an odd permutation if r is even.

5. Suppose that a permutation is written as a product of disjoint cycles. How from this representation can one determine if the permutation is even or odd?

6. Suppose that in Exercise 5 the word "disjoint" is omitted. Will this make any difference in the result?

7. Show that if p is an even permutation, then qpq^{-1} is also an even permutation for any permutation q.

8. List the even permutations of the octic group. This set of permutations is composed of the symmetries of what figure less specialized than the square?

9. Show that the multiplicative groups of numbers relatively prime to m (mod m) for $m = 8$ and $m = 12$ are isomorphic. To which group of four elements are they isomorphic?

10. If p^m is the order of a group for m a positive integer and p a prime number, prove that the group contains a subgroup of order p.

11. Prove Theorem 6.3a.

12. Is the set of all odd permutations of some subgroup of S_n a group?

13. Show that the product of two different transpositions is commutative if and only if they are disjoint.

14. Is it true that the product of two distinct cycles of order three is commutative if and only if they are disjoint?

15. Let p be a prime number different from 2 and 5. Show that the order of 10 (mod p) is a factor of $p - 1$. Deduce from this that the number of

digits in the periodic part of the decimal expansion of $1/p$ is a factor of $p - 1$.

16. Prove that if p is a prime number, $\phi(p^k) = p^k - p^{k-1} = p^k(1 - 1/p)$.

17. Assume that $\phi(ab) = \phi(a)\phi(b)$ for a and b relatively prime and prove that

$$\phi(n) = n \prod (1 - 1/p),$$

where the indicated product is over all prime divisors p of n.

18. Is the symmetric group on three symbols isomorphic to any additive or multiplicative group (mod m) for any integer m? Give reasons for your answer.

19. Let C and C' be two isomorphic cyclic groups of order n. How many isomorphisms are there between C and C'?

20. Let a and b be two elements of a group G. Prove that ab and ba are of the same order.

21. Show that every Abelian group of order 6 is cyclic.

22. Let G and G' be two isomorphic groups. Do all the isomorphisms of G onto G' form a group?

23. Let $s^p = 1$ and $s \neq 1$, where p is a prime number. Show that s is a primitive pth root of unity and hence if $s^r = 1$ for some positive integer r, then p divides r.

24. Let q be a prime factor of $2^p - 1$, where p is a prime number. Show, by use of Fermat's theorem or otherwise, that p is a factor of $q - 1$.

25. Prove that if a finite group G has no subgroups except e and itself, then the order of G is a prime number.

26. Show that $(1\ 2), (1\ 3), \ldots, (1\ n)$ generate S_n.

27. Prove that the number of primitive nth roots of unity is $\phi(n)$.

28. Prove that every infinite cyclic group is isomorphic to the additive group of integers.

29. Let $2^s + 1 = p$, a prime number, and show that in the multiplicative group $\{1, 2, \ldots, p - 1\}$ (mod p), the order of 2 is $2s$. Use Fermat's theorem and note that $2s$ is a factor of $p - 1 = 2^s$. Show that s must be a power of 2.

*30. Let b and c be two elements of an Abelian group G and suppose that the orders of b and c are r and s, respectively, with r and s relatively prime. Prove that the order of bc is rs.

*31. Show that (1 2 3), (1 2 4),..., (1 2 n) generate A_n, the alternating group.

*32. Prove that (1 2 3$\cdots$($n-1$)) and ([$n-1$]n) generate S_n.

*33. Let σ be an isomorphism between two permutation groups G and G' and let H be all those permutations of G that leave some number fixed (assuming that the permutations are those of numbers). Let H' be those elements of G' that correspond to H. Is H' necessarily a subgroup of G'? Must the elements of H' leave some number fixed? Explain your answers.

*34. Prove that every permutation of order 14 in S_{10} is odd.

*35. Express the following as a product of disjoint cycles (see Exercise 3):

$$(a_1a_2a_3)(a_2a_3a_4)(a_3a_4a_5) \cdots (a_{r-2}a_{r-1}a_r).$$

7.   Cayley's Theorem

In this section we shall prove, as we promised to do in the beginning of Section 6, that every group can, in a sense, be represented by a permutation group. We first state and prove the theorem for finite groups.

Theorem 7.1. (Cayley.)   If G is any group of order n, it is isomorphic to some subgroup of S_n, the symmetric group on n symbols.

Proof: Let $e, g_2, g_3, \ldots, g_n$ be the (distinct) elements of G. Then, for g any element of G, consider the set

(7.1) $$eg, g_2g, g_3g, \ldots, g_ng.$$

There are n elements in (7.1) and no two are equal, since $g_ig = g_jg$ would imply by the cancellation property that $g_i = g_j$. Thus the set (7.1) is a permutation or reordering of the elements of G. (Consider it a permutation of $1, 2, \ldots, n$, according to the subscripts, if you like.) So with each element of G there is associated a unique permutation of S_n. Here we are using the word "permutation" in the dynamic sense, that is, as a transformation or mapping of the elements of G onto the elements of G. Call σ_g the permutation associated with g, that is, the transformation σ_g defined by

(7.2) $$g_i\sigma_g = g_ig \quad \text{for } i = 1, 2, \ldots, n.$$

Let P be the set of all such transformations for the various elements of G. P is a subset of elements of S_n from its definition. We want to show that it is isomorphic to G. Once this is done, we know from Theorem 1.4 that P must also be a group.

We have shown that every element of g determines an element σ_g of P. No two elements of G can determine the same element of P, since

$$g_i g = g_i \sigma_g = g_i \sigma_{g'} = g_i g'$$

implies that $g = g'$. So the correspondence between G and P is one-to-one (onto). To show that the correspondence is an isomorphism, we need to verify that $\sigma_{gg'} = \sigma_g \sigma_{g'}$. This is done by the following calculations:

$$g_i \sigma_{gg'} = (g_i)gg' \quad \text{and} \quad g_i \sigma_g \sigma_{g'} = (g_i g)\sigma_{g'} = (g_i g)g' \qquad \text{for all } i.$$

Hence $\sigma_{gg'} = \sigma_g \sigma_{g'}$ follows from the associative property of elements of G. This completes the proof.

Definition. *The transformation σ_g of the theorem is called the* regular representation *of g.*

The same argument with minor changes shows a corresponding result for infinite groups. Here one must rely more heavily on the idea of a transforma-
tion of the group onto itself, for it may not be possible even to number the elements of G. So it is worthwhile to go through the proof again from the different point of view. In place of (7.2) we define σ_g by

(7.2a) $$x\sigma_g = xg \qquad \text{for all } x \text{ in } G.$$

Under the transformation σ_g each x in G has an image in G, since G is closed under multiplication. Furthermore, if h is any element of G, $xg = h$ is solvable for x in G and hence h is the image of an element of G under the transformation σ_g. Therefore, σ_g is a one-to-one transformation of G onto itself—a permutation of the elements of G. Furthermore, G is isomorphic to the set of transformations σ_g by the same calculation as that used in the finite case if g_i is replaced by x. Thus we have the following theorem.

Theorem 7.1a.   If G is any group, it is isomorphic to a subgroup of the group of permutations of G.

To acquire some understanding of what regular representations are, let us compute two for the symmetric group on three symbols and two for the octic group.

Regular Representations in S_3

Computation is a little easier if we choose as generators of S_3 the elements $\rho = (1\ 2\ 3)$ and $\alpha = (1\ 2)$. Then note that $\rho^3 = \epsilon = \alpha^2$ and $\alpha\rho = \rho^2\alpha$. We number the six elements a_i as follows:

ϵ	ρ	ρ^2	α	$\alpha\rho$	$\alpha\rho^2$
a_1	a_2	a_3	a_4	a_5	a_6

$a_i\sigma_{a_4} = a_i\sigma_\alpha$ $a_1 a_4 = a_4$ | $a_2 a_4 = a_6$ | $a_3 a_4 = a_5$ | $a_4 a_4 = a_1$ | $a_5 a_4 = a_3$ | $a_6 a_4 = a_2$

Hence the regular representation of a_4 is (1 4)(2 6)(3 5).

a_1	a_2	a_3	a_4	a_5	a_6
$a_i \sigma_{a_2} = a_i \sigma_\rho$ $a_1 a_2 = a_2$	$a_2 a_2 = a_3$	$a_3 a_2 = a_1$	$a_4 a_2 = a_5$	$a_5 a_2 = a_6$	$a_6 a_2 = a_4$

Hence the regular representation of a_2 is (1 2 3)(4 6 5).

Regular Representations in the Octic Group

Using the same notation as in Section 9 of Chapter I, we number the elements as follows:

ϵ	ρ	ρ^2	ρ^3	α	$\alpha\rho$	$\alpha\rho^2$	$\alpha\rho^3$
a_1	a_2	a_3	a_4	a_5	a_6	a_7	a_8
$a_i \sigma_{a_2} = a_i \sigma_\rho$ a_2	a_3	a_4	a_1	a_6	a_7	a_8	a_5
$a_i \sigma_{a_5} = a_i \sigma_\alpha$ a_5	a_8	a_7	a_6	a_1	a_4	a_3	a_2

Hence the regular representation of a_2 is (1 2 3 4)(5 6 7 8) and that of a_5 is (1 5)(2 8)(3 7)(4 6).

From these two examples one can observe that regular representations seem to have the following two properties. We leave the proof of the first as Exercise 8.

1. In a regular representation no permutation except the identity permutation (corresponding to the identity element e of G) leaves any symbol unaltered.
2. If g is an element of order r, its regular representation consists of n/r disjoint cycles of length r.

We state the second property as the following theorem.

Theorem 7.2.   Let g be an element of order r in a group G whose order is n. The regular representation of g as an element of G has the property that g corresponds to a product of n/r disjoint cycles of order r.

Proof: To simplify notation, renumber the elements of G so that the first r elements are the powers of g. That is, let

$$a_i = g^{i-1} \qquad \text{for } 1 \le i \le r,$$

where $g^0 = e$. Then

$$a_i g = a_{i+1}, \qquad \text{for } 1 \le i \le r-1 \quad \text{and} \quad a_r g = a_1.$$

Thus the first part of the representation of g consists of the cycle (1 2 3 $\cdots$ r). Call H the subgroup of powers of g. Then if $H \ne G$, let g_2 be an element of G not in H and form the left coset $g_2 H$. Let

$$a_{r+i} = g_2 g^{i-1} \qquad \text{for } 1 \le i \le r$$

and see that the a_{r+i} form another cycle of order r. We may continue in this way for a complete set of left cosets of H and thus show that the regular representation of g is a product of cycles of order r. The number of such cycles is the number of cosets, that is, n/r. This completes the proof.

We should make two comments about regular representations. The first is a reemphasis of those made at the close of Section 6. In the example above for the octic group, the element $a_2 = \rho$ considered as an element of S_4 is an odd permutation, since it is a cycle of order 4. But its representation in S_8 is even, being a product of two cycles of order 4. The group generated by a_2 in S_4 is isomorphic to that generated by its representation in S_8; but the isomorphism refers only to the operation of the group, not to the way in which it can be expressed within a given S_n.

Second, if one looks at the examples of this section, it is apparent that the regular representation of a subgroup of S_n is usually not as simple as that of the subgroup with which we started. For instance, for each element of S_3 we replace a permutation of three symbols by one with six. This would lead us to wonder what, in any case, the advantage of the regular representation is. Certainly, at this stage there is no advantage. But the fact is that this theorem of Cayley leads into a whole field of inquiry about groups which is very useful to physicists and others who apply mathematics. We can outline the process up to a point. Suppose that we have a group G of order n. Each of its elements has a regular representation that is a permutation on n symbols. By means of this representation we can associate with each element of the group a matrix of order n that is obtained from the identity matrix by applying the given permutation in S_n to its rows. By means of certain techniques one can often replace this matrix by one of smaller size and at the same time find certain simpler ways of representing the given group. So the regular representation is an important link between the group itself and its representation by matrices.

Exercises

1. Let A_4 be the alternating group on four symbols. Find the regular representation of (1 2)(3 4) in this group.

2. Find the regular representations of the elements of the Klein four-group.

3. Find the regular representation of σ_{ρ^2} in S_3, using the notation of this section.

4. Find the regular representation of $\sigma_{\alpha\rho}$ in the octic group, using the notation of this section.

5. Let G be the additive group of integers. What is the regular representation

of the positive integer n? This requires an extension of the idea of a regular representation.

6. If g is a generator of a cyclic group G, is the regular representation of g cyclic? If not, under what conditions will it be cyclic?

7. Suppose that instead of setting up the correspondence $g_i \sigma_g = g_i g$ in the proof of Theorem 7.1 we had written the product in the other order, namely, $g_i \sigma_g = g g_i$. Could one still prove the isomorphism required?

8. Show that in a regular representation of a group element different from the identity, no symbol is left unaltered.

9. Let p be a cycle of length r in S_n and p' its regular representation in S_m, where $m = n!$. Show that p' is odd only if r is even and $n \leq 3$.

10. Let g be a fixed element of a group G and, as in this section, define σ_g by $x\sigma_g = xg$ for all x in G. As we showed, σ_g is a one-to-one representation of G onto itself. Is σ_g an isomorphism? Explain.

8. Automorphisms

When we considered the symmetries of geometric figures, we noted that the nature of the symmetries gave us insight into the properties of the figures themselves. The symmetries are isomorphisms, in the sense that they leave distance unaltered. An analogous idea for groups instead of figures is that of an automorphism. An automorphism of a group is by definition an isomorphism of a group onto itself, that is, one for which the group and its image are the same. We can define this concept without reference to an isomorphism.

Definition.   *An* automorphism *of a group G is a one-to-one transformation of the group onto itself that preserves structure: that is, if σ is a one-to-one transformation of G onto itself, then σ is an automorphism if*

$$(8.1) \qquad (ab)\sigma = (a\sigma)(b\sigma) \qquad \text{for all elements } a \text{ and } b \text{ of } G.$$

(Note the meaning of "onto." See Section 5 of Chapter I.)

Since an automorphism is an isomorphism, Theorem 1.1 about isomorphisms applies equally well for automorphisms.

Theorem 8.1.   If σ is an automorphism of G, then the identiy of G is its own image and $(x\sigma)^n = x^n\sigma$ for all integers n (positive, negative, or zero).

Now a group can have more than one automorphism. For instance, suppose that G is the multiplicative group $\{1, r, r^2\}$, where $r^3 = 1$. Two automorphisms of G are the following:

$$1\sigma_1 = 1, \qquad r\sigma_1 = r, \qquad r^2\sigma_1 = r^2$$

and

$$1\sigma_2 = 1, \qquad r\sigma_2 = r^2, \qquad r^2\sigma_2 = r.$$

The first is the identity automorphism. The only product that needs to be tested for the second is

$$(r\sigma)(r\sigma) = (r^2)(r^2) = r = r^2\sigma.$$

In both cases, of course, the identity corresponds to itself. For this group there can be only two automorphisms, since there are only two possible images for r, namely, r and r^2; once the image of r is determined, then that of r^2 is determined as well.

Now the set of automorphisms itself forms a group, a kind of "second-generation group" (richly endowed with structure). We show this by proving the following theorem.

Theorem 8.2. The set of automorphisms of a group G is itself a group, where the operation is composition.

Proof: We must check in turn the properties of a group. If σ and σ' are two automorphisms of G, then $\sigma\sigma'$ is an automorphism for, first, the "product" is one-to-one. Second,

$$(ab)\sigma\sigma' = [(ab)\sigma]\sigma' = [(a\sigma)(b\sigma)]\sigma'$$
$$= [(a\sigma)\sigma'][(b\sigma)\sigma'] = [a\sigma\sigma'][b\sigma\sigma'].$$

Hence composition is a binary operation. Second, composition of automorphisms is associative, since an automorphism is a transformation. Third, $g\sigma = g$ for all g in G is the identity automorphism. Finally, if $g\sigma = g'$, the fact that σ is one-to-one implies that σ^{-1} exists and is defined by $g = g'\sigma^{-1}$. To show that σ^{-1} is an automorphism, note that $a\sigma = a'$ and $b\sigma = b'$ are equivalent to $a = a'\sigma^{-1}$ and $b = b'\sigma^{-1}$. Then, on the one hand, $(a'\sigma^{-1})(b'\sigma^{-1}) = ab$ and, on the other, since σ is an automorphism $a'b' = (a\sigma)(b\sigma) = (ab)\sigma$. Putting the two results together, we have

$$(a'\sigma^{-1})(b'\sigma^{-1}) = ab = (a'b')\sigma^{-1},$$

which shows that the inverse of an automorphism is an automorphism and completes the proof.

Now let us explore how we would go about finding the automorphisms of a group. One way to do this is to use the generators. Suppose that a group G has two generators g and h. What are some restrictions on possible automorphisms? Certainly, any automorphism must take the identity element into itself, and it must take g into some element of the same order, and similarly for h. But that may well not be enough, for there are interrelationships between the products of powers of g and h that must be preserved. Starting to find the automorphisms can be fairly straightforward, but it can become complex before we progress very far. Let us first find the group of automorphisms for three groups. The first example is more complex than the other two but in some respects it is more enlightening.

EXAMPLE 1.   We wish to find the group of automorphisms of S_3, the symmetric group on three symbols. (See Section 9 of Chapter I.) We shall do this from first principles even though a little later we shall find a much more efficient way of dealing with this problem. To avoid too many Greek letters, let us change our notation for the generators of S_3 to $r = (1\ 2\ 3)$, $a = (1\ 2)$, and e the identity element. Then

$$(8.2) \qquad\qquad r^3 = e = a^2 \qquad \text{and} \qquad ar = r^2a.$$

Now conditions (8.2) determine the multiplication table of S_3, and since the multiplication table determines a group up to an isomorphism, it follows that a one-to-one transformation of S_3 onto itself will be an automorphism if and only if the images of r and a satisfy (8.2). Furthermore, each element of S_3 is expressible uniquely in the following form:

$$(8.3) \qquad\qquad r^ia^j, \qquad i = 0, 1, 2 \quad \text{and} \quad j = 0 \quad \text{or} \quad 1.$$

Hence the image of every element of S_3 will be determined by the images of r and a in accord with

$$(r^ia^j)\sigma = (r\sigma)^i(a\sigma)^j.$$

What are the possibilities for the images of r and a? By Theorem 8.1, if σ is an automorphism, $r\sigma$ must be of order 3, and hence either r or r^2; while $a\sigma$ must be of order 2 and thus one of a, ra, r^2a. Hence if σ is to be an automorphism,

$$(8.4) \quad e\sigma = e, \quad r\sigma = r^u, \quad \text{and} \quad a\sigma = r^ta, \qquad \text{where } u = 0 \text{ or } 1$$
$$\text{and} \quad t = 0, 1, \text{ or } 2.$$

To show that any such σ is an automorphism, we must show that conditions (8.2) hold for the images. We chose σ so that the first two equalities hold. To test the third perform the following computations:

$$(a\sigma)(r\sigma) = r^tar^u,$$
$$(r^2\sigma)(a\sigma) = r^{2u}r^ta = r^{t+2u}a.$$

Now $ar = r^2a$ implies that $ar^u = r^{2u}a$ for $u = 0$ and 1. Hence $r^tar^u = r^{t+2u}a$, and we have shown that $(a\sigma)(r\sigma) = (r\sigma)^2(a\sigma)$. Thus all the conditions (8.2) hold for the images of r and a under any transformation σ of (8.4). Therefore, any σ defined by (8.4) is an automorphism. Hence we have shown that there are exactly six automorphisms of S_3.

If we can show that the group of automorphisms of S_3 is not Abelian, then, by Theorem 1.3, it must be isomorphic to S_3. To show it to be non-Abelian, look at (8.4) and define two automorphisms as follows:

$$r\sigma_1 = r, \quad a\sigma_1 = ar; \qquad r\sigma_2 = r^2, \quad a\sigma_2 = a.$$

Then

$$ar\sigma_1\sigma_2 = (ar^2)\sigma_2 = ar^4 = ar.$$

and

$$ar\sigma_2\sigma_1 = (ar^2)\sigma_1 = arr^2 = a.$$

This shows that $\sigma_1\sigma_2 \neq \sigma_2\sigma_1$, and hence the group of automorphisms of S_3 is isomorphic to S_3.

You will be asked to show in Exercise 11 that σ_1 and σ_2 are generators of the group of automorphisms. It is partially an accident that the two groups are isomorphic, as we shall see from the following examples.

EXAMPLE 2.   Let the group G be the additive group $\mathbf{Z}$. Then $\mathbf{Z}$ is a cyclic additive group with 1 as its generator. The image of 1 determines the automorphism. Let $(1)\sigma = x$, where x is in Z. Then $n\sigma$ must be nx for all integers n if σ is to be an automorphism. Now the transformation is onto. Hence every integer must be the image of some integer under the automorphism. But $n\sigma = nx$ implies that every image must be a multiple of x. This condition excludes some integers unless $x = 1$ or -1. So there are just two possible automorphisms: the identity and σ defined by $(n)\sigma = -n$ for all integers n. The latter is an automorphism, since it is one-to-one and

$$(a\sigma) + (b\sigma) = (-a) + (-b) = -(a + b) = (a + b)\sigma.$$

Note that in this case we have an infinite group with just two automorphisms. Furthermore, σ^2 is the identity automorphism.

EXAMPLE 3.   Let G be the multiplicative group of $\mathbf{Q}^*$, the set of nonzero rational numbers. Let σ be an automorphism of G. Of course, $(1)\sigma = 1$, and since -1 is the only integer of order 2, $(-1)\sigma = -1$. What are the images of the integers? Once we determine these images, the images of all rational numbers will be determined by multiplication. Now we assume, as you probably know and as we shall prove in Chapter III, that every integer greater than 1 can be expressed as a product of powers of prime numbers uniquely except for the order of the factors. This means that every rational number except 1 and -1 can be expressed uniquely in the form

$$\pm p_1^{a_1} p_2^{a_2} \cdots p_k^{a_k},$$

where each a_i is a positive or negative integer and the p's are distinct prime numbers. Now each p_i generates an infinite cyclic group so that a transformation σ will be an automorphism if and only if it satisfies the following conditions:

$$(1)\sigma = 1, \qquad (-1)\sigma = -1, \qquad p_i\sigma = p_{a_i},$$

where the infinite set a_i is some permutation of the positive integers. (We are also assuming that there is an infinite number of prime numbers.)

Although the above calculations are manageable for S_3 and the other groups, it is clear that many such calculations could quickly become tedious. Hence it is fortunate that there is one kind of automorphism which is easy to define and easy to deal with. Sometimes it is even true that all automorphisms for a given group are of this particular type—the so-called inner automorphisms.

Let b be some element of a group G and define the transformation σ_b by the correspondence

$$(8.5) \qquad x \xrightarrow{\sigma_b} b^{-1}xb, \qquad \text{that is, } x\sigma_b = b^{-1}xb \text{ for all } x \text{ in } G.$$

For instance, if $G = S_3$ and $b = a$ and we use the notation above, we have the following table:

x	e	r	r^2	a	ar	ar^2
$a^{-1}xa$	e	r^2	r	a	ar^2	ar

In fact, a little calculation will show that $\sigma_a = \sigma_2$, using the notation in Example 1. So we need the following theorem.

Theorem 8.3.   The transformation σ_b defined by (8.5) is an automorphism of the group G.

Proof: The correspondence (8.5) shows that each element of G has a unique image. To show that every element of G is the image of no more than one element of G, suppose that $x\sigma_b = y$; that is, $b^{-1}xb = y$. This implies that $x = byb^{-1}$; hence there can be no more than one such x. But, for this x,

$$x\sigma_b = b^{-1}(byb^{-1})b = y.$$

We have shown that the correspondence is one-to-one. To show that the correspondence preserves products note that

$$(b^{-1}xb)(b^{-1}yb) = b^{-1}(xy)b.$$

Thus (8.5) defines an automorphism of G and the theorem is proved.

This theorem gives us the right to the following definition.

Definition.   *The transformations $x \rightarrow b^{-1}xb$ of a group G onto itself, where b and x are in G, are called* inner automorphisms. *Any automorphism that is not an inner automorphism is called an* outer automorphism.

We also have the following theorem.

Theorem 8.4.   The set of inner automorphisms of a group G is itself a group.

Proof: Since the set of inner automorphisms is a subset of the group of all automorphisms, we need merely show closure and the existence of an inverse (Theorem 7.1 of Chapter I). Let σ_a be the transformation $x \rightarrow a^{-1}xa$. Then

$$(x\sigma_a)\sigma_b = b^{-1}(x\sigma_a)b = b^{-1}a^{-1}xab = (ab)^{-1}xab = x\sigma_{ab}.$$

Since $a(a^{-1}xa)a^{-1} = x$, we see that σ_c is the inverse of σ_a if $c = a^{-1}$. So the inverse of an inner automorphism is an inner automorphism. This completes the proof.

How many inner automorphisms are there for a given group? If the group is Abelian, the only inner automorphism is the identity, since then $a^{-1}xa = x$ for all x. Suppose, in general, that the inner automorphisms defined by elements a and b are the same; that is,

$$a^{-1}xa = b^{-1}xb \qquad \text{for all } x \text{ in } G.$$

This is equivalent to $ba^{-1}x = xba^{-1}$ for all x. Thus ba^{-1} must be commutative with every element of G. Therefore, the answer to the question at the beginning of this paragraph depends on knowledge of the set of elements C of G that are commutative with all elements of G. Now C must contain at least the identity element but may contain others. For instance, a look back will show you that C is the identity element for S_3 but has two elements for the octic group. The following is easily proved.

Theorem 8.5.   For a given group G, the set C of all elements commutative with all elements of G is a group.

Proof: If $ax = xa$ and $bx = xb$ for all elements x, then $bax = bxa = xba$ for all elements of x; the binary operation is closed for C. Also, $ax = xa$ implies that $a^{-1}x = xa^{-1}$, and thus the inverse of an element of C is also in C. This shows that C is a group.

Definition.   *The set of all elements commutative with all elements of a group G is called its* center.

Now, getting back to inner auotmorphisms and using the terminology just developed, we see that $\sigma_a = \sigma_b$ if and only if ba^{-1} is in C, the center of G.

That is, $\sigma_a = \sigma_b$ if and only if b is in Ca, the right coset of C defined by a. (See Section 4.) Since every element of C is commutative with a, then $Ca = aC$ and we may omit the qualification "right." Thus the number of distinct inner automorphisms of a finite group G is the number of cosets of its center (right cosets or left cosets). We state this in the following theorem.

Theorem 8.6.   The number of distinct inner automorphisms of a finite group G is equal to the number of cosets of C, its center.

Corollary 1.   The number of distinct inner automorphisms of a group G is equal to the index of C, its center, in G if the index is finite.

Corollary 2.   If a group G is Abelian, there is only one inner automorphism. (This is also easily shown without the theorem.)

For instance, for the octic group, the number of distinct inner automorphisms is 4, since its center is of order 2. There are six inner automorphisms of S_3, since its center is of order 1. If we look at Example 1, we see that there are only six automorphisms. Hence, in this case, all the automorphisms are inner.

Two somewhat similar examples will illustrate the effect of an inner automorphism.

EXAMPLE 4.   Let p be some element of S_n, the group of permutations on n symbols. Then it takes each number a in the set $1, 2, 3, \ldots, n$ into another number b of the set. That is, thinking of p as a transformation we can write $ap = b$. Now, if q is another permutation in S_n, we have $(ap)q = bq$; that is,

$$(aq)(q^{-1}pq) = bq.$$

This shows that if p takes a into b, then $(q^{-1}pq)$, the image of p under the inner automorphism σ_q, takes aq into bq. It is here that the contrast between two ways of looking at a transformation come into play. For $ap = b$ we can take the "alibi" point of view: p takes or "moves" a into b. For aq and bq we can take the "alias" point of view: multiplication by q gives a and b new names. If a and b take the new names aq and bq, then the former is taken into the latter by the transformation $q^{-1}pq$ instead of p.

For example, suppose that in S_3, p is the transposition (1 2) and q the cycle (1 2 3). Then, if $a = 1$, $ap = 2 = b$; if $a = 2$, $ap = 1 = b$; and if $a = 3$, $ap = 3 = b$. By similar calculations we get the following diagram:

$$
\begin{array}{ccc}
1,2,3 & \xrightarrow{\ p\ } & 2,1,3 \\
\downarrow{\scriptstyle q} & & \downarrow{\scriptstyle q} \\
2,3,1 & \xrightarrow{\ q^{-1}pq\ } & 3,2,1
\end{array}
$$

Now 2, 3, 1 are the new names for 1, 2, 3, respectively. The transformation p interchanges the first two numbers under the original names, and $q^{-1}pq$ interchanges the first two numbers under the second names.

EXAMPLE 5. The transformations discussed in Section 6 of Chapter I can be treated in a fashion similar to Example 4. Let T be the multiplicative (really composition) group of nonsingular transformations on vectors in two dimensions and π and σ two transformations of T. Then, as in Example 4, we can think of σ as changing the names of the vectors (a change of axes if you like) so that vector a has the new name $a\sigma = a'$ and b has the new name $b\sigma = b'$. Now let π be the transformation that "moves" a into b. Then, as above, we have the following diagram for vectors a, b, a', and b':

$$
\begin{array}{ccc}
a & \xrightarrow{\ \pi\ } & b \\
\downarrow{\scriptstyle\sigma} & & \downarrow{\scriptstyle\sigma} \\
a' & \xrightarrow{\ \sigma^{-1}\pi\sigma\ } & b'
\end{array}
$$

In particular, suppose that $(a, b)\sigma = (b, a)$ for every pair. Geometrically, this means that we interchange the two axes. Let π "transport" (a, b) into $(2a, b)$; that is, π doubles the first coordinate and leaves unchanged the second coordinate. Then computation shows that

$$(a, b)(\sigma^{-1}\pi\sigma) = (b, a)\pi\sigma = (2b, a)\sigma = (a, 2b).$$

This means that after the change of name, the transformation doubles the second coordinate instead of the first, which is only natural.

Exercises

1. Show that σ_1 and σ_2 defined in Example 1 are automorphisms of S_3.

2. Set up a one-to-one correspondence between S_3 and its automorphisms.

3. Prove that A_4 and S_4, the alternating group and symmetric group on four symbols, each have (e) as their center.

4. Which of the symmetries of a square are automorphisms of the multiplicative group $\{1, i, -1, -i\}$ considered as a set of powers of i, the square root of -1?

5. Find all the automorphisms of the octic group and point out which are inner automorphisms.

6. Given a group G and S the set of its outer automorphisms, is S a group?

7. How many automorphisms has a group of order p, where p is a prime number?

8. How many automorphisms has a cyclic group of order pq, where p and q are distinct prime numbers?

9. How many automorphisms are there of a cyclic group of order n?

10. Suppose that for a group G we define $x\sigma_a = axa^{-1}$ for each x in G. Is this an automorphism? Why or why not?

11. Show that σ_1 and σ_2 in Example 1 are generators of the group of automorphisms of G.

12. Carry through Example 4 for p the cycle (1 2 3) and q the transposition (1 2).

13. Carry through Example 5, interchanging the roles of π and σ; that is, π changes the name and σ "moves" the vectors.

14. Let G and G' be a pair of isomorphic groups. Prove that the number of different isomorphisms between G and G' is equal to the number of automorphisms of G.

9.   Homomorphisms

We have come close to the idea of a homomorphism several times. Now we have enough examples, including one in Section 8, to give meaning to the idea. A homomorphism can be thought of as a generalization of an isomorphism in which the correspondence is many to one instead of one to one. Although we shall apply it to groups in this chapter, we shall define it in terms of sets partly because such a definition simplifies a forthcoming proof.

Definition.   *Let S and S' be two sets with respective binary operations $\circ$ and $\circ'$. We call a transformation of S onto S' a* homomorphism *if it preserves the structure with respect to the binary operations; that is,*

$$(9.1) \qquad \textit{if } s_1 \rightarrow s_1' \quad \textit{and} \quad s_2 \rightarrow s_2', \quad \textit{then } (s_1 \circ s_2) \rightarrow (s_1' \circ' s_2').$$

If such a homomorphism exists, we say that S is homomorphic *onto S' and that S' is the* image *of S under the homomorphism. If σ is the homomorphism, then (9.1) may be written $(s_1 s_2)\sigma = (s_1 \sigma)(s_2 \sigma)$.*

An isomorphism is a homomorphism that is one to one, that is, if there is a homomorphism in both directions. One could also, as in the case of an isomorphism, speak of a homomorphism into instead of onto, in which not every element of S' is the image of an element of S. But for our purposes there is no special advantage in considering this, for it is the image set that is important for us.

Note some immediate consequences of this definition. Suppose that S is homomorphic onto S'. Then if S has an identity element, its image must be an identity element of S', for $e \to e'$ and $e \circ s = s$ for all elements of S implies that $e' \circ' s' = s'$ for all elements of S'. Let us write the same statement in more concise form, taking σ to be the homomorphism and writing the binary operations as if they were multiplication. Since σ is a homomorphism, $(ab)\sigma = (a\sigma)(b\sigma)$ for all elements of G. In particular, $s\sigma = (es)\sigma = (e\sigma)(s\sigma)$, for all s in S, so that $e\sigma$ is the identity in S'; thus the identity elements correspond. Also, if the inverse of s in S exists, then $(s\sigma)(s^{-1}\sigma) = e\sigma = e'$ shows that $(s^{-1}\sigma)$ is the inverse of $s\sigma$. In fact, $s^n\sigma = (s\sigma)^n$ for all integers n. (Compare the corresponding properties for an isomorphism in Section 1.) Similarly, if S is associative, so is S'. In fact, if S is a group that is homomorphic to a set S', then S' must also be a group. If S is Abelian, so is S'; if S is cyclic, so is S'. We leave the proofs of these statements as exercises.

Now let us explore what a homomorphism of a group G onto a group H is like. First, there will be a subset of elements of G that correspond to the identity element of H. Since this set is important, we give it a name—the *kernel* of the homomorphism. Thus if K is the kernel, then $K\sigma = e'$, the identity of H; that is, $k\sigma = e'$ for every k in K, and K includes every element of G whose image is e'. This set K is a group, as we now show.

Theorem 9.1.   Let σ be a homomorphism that takes a group G onto H. Then the kernel of σ is a subgroup of G.

Proof: Let G be a group and σ a homomorphism of G onto H. Since the kernel K of σ is a subset of G, we need to show closure and that K contains the inverse of every element of K. Now, $a\sigma = e'$ and $b\sigma = e'$, where e' is the identity element of H, implies that $e' = (a\sigma)(b\sigma) = (ab)\sigma$, which shows closure. Also $(a\sigma) = e'$ and $(a\sigma)(a^{-1}\sigma) = e\sigma = e'$ implies that $a^{-1}\sigma = e'$ and completes the proof.

Suppose that elements b and a of G have the same image under σ. Then $b\sigma = a\sigma$ implies that

(9.2) $$(a^{-1}\sigma)(b\sigma) = (a^{-1}\sigma)(a\sigma) = e\sigma = e'.$$

Thus $e' = (a^{-1}\sigma)(b\sigma) = (a^{-1}b)\sigma$. Hence $a^{-1}b = k$ for some element of K; that is, $b = ak$. This means that b is in the left coset aK. Conversely, if b is in aK, then $b\sigma = a\sigma$. Thus $b\sigma = a\sigma$ if and only if b and a are in the same left coset of K. We have proved the following theorem.

Theorem 9.2.   If σ is a homomorphism of a group G onto a group H, and if K is the kernel of the homomorphism, then two elements a and b of G have the same image in H if and only if they are in the same left coset of K.

Let us see what the above means in terms of two examples.

EXAMPLE 1.   Consider S_3 and the multiplicative group $H = \{1, -1\}$. Define the correspondence σ by

$$(1\ \ 2)\sigma = (1\ \ 3)\sigma = (2\ \ 3)\sigma = -1$$

and

$$e\sigma = (1\ \ 2\ \ 3)\sigma = (1\ \ 3\ \ 2)\sigma = 1,$$

where e is the identity permutation. This is a homomorphism, since each of e, $(1\ \ 2\ \ 3)$, and $(1\ \ 3\ \ 2)$ is a product of two transpositions. The kernel of the homomorphism is the subgroup $\{e, (1\ \ 2\ \ 3), (1\ \ 3\ \ 2)\}$.

EXAMPLE 2.   Let S be the multiplicative group M of nonsingular matrices of order 2 with rational elements. Define a transformation σ by $A\sigma = \det(A)$ for each A in M (see Section 4 of Chapter I); that is,

$$\begin{bmatrix} a_1 & a_2 \\ b_1 & b_2 \end{bmatrix} \sigma = a_1 b_2 - b_1 a_2.$$

You were asked to prove in Exercise 23 after Section 4 in Chapter I that $\det(AB) = (\det A)(\det B)$. This is what is needed to show that σ is a homomorphism of M onto the multiplicative group of nonzero rational numbers Q^*. The kernel of σ is M_0, the subgroup of M consisting of all matrices whose determinant is 1. From Theorem 9.2 two matrices of M have the same image in Q^* if and only if their determinants are the same.

Theorem 9.3.   A homomorphism of a group G onto H is an isomorphism if and only if its kernel is the identity element.

Proof: We need to show that every element of H is the image of a unique element of G. From Theorem 9.2 we know that a and b will have the same image if and only if they are in the same left coset of K. If K consists of the identity element only, then two different elements of G cannot be in the same left coset and cannot have the same image. If K contains more than the identity element of G, it has at least two different elements, and they will have the same image. This completes the proof.

Now let us see how these theorems apply to two examples considered previously.

EXAMPLE 3.   As in Section 8 let G be a group and A its set of inner automorphisms. To every element of G corresponds an inner automorphism

$$x\sigma_g = g^{-1}xg.$$

You will be asked to show in Exercise 20 after Section 11 that the correspondence $g \to \sigma_g$ is a homomorphism and its kernel is the center of C of G.

EXAMPLE 4. In Section 3 we defined $a \equiv b \pmod{m}$, for a positive integer m, to mean that $a - b$ is a multiple of m. We showed in Sections 3 and 4 that this amounted to partitioning the set of integers into cosets modulo m. Thus if G is the additive group of integers and H the subgroup of all multiples of m, we can set up the correspondence

$$g \rightarrow g + H.$$

This correspondence, as you will be asked to show in Exercise 21 after Section 11, is a homomorphism and its kernel is H.

EXAMPLE 5. A particular case of Example 4 is given by taking the point of view of Theorem 3.4a. Let m be a positive integer and S the set of integers: $0, 1, 2, \ldots, (m - 1)$ with addition modulo m. Denote by a_m the remainder when the integer a is divided by m. Then the correspondence

$$a \rightarrow a_m$$

is a homomorphism of the additive group $\mathbf{Z}$ onto S.

10.   *Normal Subgroups and Quotient Groups*

Looking at the proof of Theorem 9.2, we see that we could just as well have written in place of (9.2),

$$(b\sigma)(a^{-1}\sigma) = (a\sigma)(a^{-1}\sigma) = e\sigma = e'.$$

This would imply that ba^{-1} is in K, and thus b is in the right coset Ka. But, in general, the right coset Tg for a subgroup T of G is not the same as the left coset gT. (We shall give an example following Theorem 10.2.) It would be embarrassing if the cosets were not the same for the kernel K. Fortunately, for the kernel of a homomorphism, the cosets gK and Kg are the same. Since this property is an important one, we give it a name.

Definition.   *If T is a subgroup of a group G, it is called a* normal (*or* invariant) *subgroup if the cosets gT and Tg are the same for all elements g of G.*

Notice that every subgroup of an Abelian group is normal. But we shall see that non-Abelian groups also have normal subgroups. It is easier to show that the kernel of a homomorphism is a normal subgroup of G if we have an alternative way to characterize it. This is given by the following theorem.

Theorem 10.1.   If T is a subgroup of G, it is a normal subgroup if and only if $g^{-1}Tg = T$ for all elements g of G, where $g^{-1}Tg$ means the set of elements $\{g^{-1}tg\}$ for t in T. That is, if T is taken into itself by every inner automorphism of G, then T is a normal subgroup.

Proof: Suppose that $Tg = gT$. One is tempted merely to multiply on the left by g^{-1}, but we must bear in mind that T is a set. Thus $Tg = gT$ means that for each t in T, there is a t' in T such that $tg = gt'$. *Now* we can multiply on the left by g^{-1} and see that $g^{-1}tg = t'$ is in T. This means that $g^{-1}Tg \subseteq T$. Also, for every t' in T, there is a t such that $tg = gt'$; that is, $g^{-1}tg = t'$ and hence $g^{-1}Tg \supseteq T$. These two results together show that $g^{-1}Tg = T$.

Conversely, suppose that $g^{-1}Tg = T$. This means that for each t in T, there is a t' in T such that $g^{-1}tg = t'$ and hence $tg = gt'$. This shows that $Tg = gT$ and completes the proof.

In practice a slight modification of the above is more useful.

Theorem 10.1a.   In Theorem 10.1, T is a normal subgroup of G if (and only if) $g^{-1}Tg \subseteq T$ for all g.

Proof: Let g be an element of G. Then $g^{-1}Tg \subseteq T$ for all g implies that it is also true for g^{-1} in place of g. Thus $gtg^{-1} = t'$ is in T. But this is equivalent to $t = g^{-1}t'g$, which is in $g^{-1}Tg$. Hence $T \subseteq g^{-1}Tg$ and thus $T = g^{-1}Tg$. The "only if part" is easy, since $g^{-1}Tg = T$ implies that $g^{-1}Tg \subseteq T$.

With these alternative ways of showing a subgroup to be normal, we can prove the following theorem.

Theorem 10.2.   The kernel K of a homomorphism of a group G onto H is a normal subgroup of G.

Proof: Using Theorem 10.1a, we want to show that $g^{-1}kg$ is in K for all k in K. If σ is the homomorphism, then

$$(g^{-1}kg)\sigma = (g^{-1}\sigma)(k\sigma)(g\sigma) = (g\sigma)^{-1}e'(g\sigma) = e'.$$

This shows that $g^{-1}kg$ is in K and completes the proof.

To illustrate these ideas, let us look at two examples of subgroups of S_3.

EXAMPLE 1.   Let T be the subgroup $\{e, (1\ 2)\}$ of S_3. The cosets of T are as follows:

Left cosets	*Right cosets*
T	T
$(1\ 3)T = \{(1\ 3), (1\ 3\ 2)\}$	$T(1\ 3) = \{(1\ 3), (1\ 2\ 3)\}$
$(2\ 3)T = \{(2\ 3), (1\ 2\ 3)\}$	$T(2\ 3) = \{(2\ 3), (1\ 3\ 2)\}.$

Here the left cosets are not the same as the corresponding right cosets. This means that T cannot be the kernel of any homomorphism of G onto a group H.

EXAMPLE 2.   Let T' be the subgroup $\{e, (1\ 2\ 3), (1\ 3\ 2)\}$. Then the cosets of T' are

Left cosets	*Right cosets*
T'	T'
$(1\ 2)T' = \{(1\ 2), (1\ 3), (2\ 3)\}$	$T'(1\ 2) = \{(1\ 2), (2\ 3), (1\ 3)\}.$

Thus T' is a normal subgroup and can be the kernel of a homomorphism.

Actually, *any* normal subgroup is the kernel of a homomorphism. Before showing this in general, let us see how we can set up such a homomorphism for subgroup T' above. Now T' should be the kernel of the homomorphism. Taking 1 to be the identity element of the image group, we see that the homomorphism σ should be chosen so that $t\sigma = 1$ for all t in T'. If there is a homomorphism of G, it must be onto a group, and since all the elements of the coset $(1\ 2)T'$ have the same image, the image group must have order 2. Hence the other element of the image group can be taken to be -1. Then the homomorphism takes T' into 1 and the elements of the coset $(1\ 2)T'$ into -1. To see that it actually is a homomorphism, look at $(a\sigma)(b\sigma)$, where a and b are elements of G. If a and b are in the same coset of T', then their product is in T' [for example, $(1\ 2)(1\ 3)$ is in T']. Thus

$$(a\sigma)(b\sigma) = (\pm 1)(\pm 1) = 1 = (ab)\sigma;$$

if a and b are in different cosets, their product is in $(1\ 2)T'$, and

$$(a\sigma)(b\sigma) = (\pm 1)(\mp 1) = -1 = (ab)\sigma.$$

We have come back to Example 1 of Section 9 from another direction.

EXAMPLE 3.   Consider A_4, the alternating group on four symbols (the group of even permutations), and

$$K = \{e, (1\ 2)(3\ 4), (1\ 3)(2\ 4), (1\ 4)(2\ 3)\}.$$

Let S be the set of remaining elements of A_4, namely,

$$(1\ 2\ 3), (1\ 3\ 2), (1\ 2\ 4), (1\ 4\ 2), (1\ 3\ 4), (1\ 4\ 3), (2\ 3\ 4), (2\ 4\ 3).$$

(A geometrical connection with these elements is given in Exercise 23 in Section 9 of Chapter I.) To show that K *can be* the kernel of a homomorphism, we now show that K is a normal subgroup of A_4.

First, to show that K is a group, notice that the square of each element of K is the identity. Computation shows that

$$(1\ 2)(3\ 4)(1\ 3)(2\ 4) = (1\ 4)(2\ 3),$$

and since all products of two of the last three elements of K are obtained by permutations of the numbers 1, 2, 3, 4, we see that K is closed under multiplication and that each element has an inverse. Hence K is a group.

Second, to show that K is a normal subgroup of A_4, we could perform $3 \cdot 8$, or 24, calculations. But, fortunately, it is easy to avoid this. Let p be any element of S and k any element of K other than e. Then pkp^{-1} has the property that

$$(pkp^{-1})^2 = pkp^{-1}pkp^{-1} = e.$$

Hence pkp^{-1} is of order 2. But the elements of K, except e, are all of order 2, and every element of S is of order 3. Hence pkp^{-1} is in K for all k in K. This shows that K is a normal subgroup of A_4. Thus from Theorem 10.2, K could be the kernel of a homomorphism of A_4 onto some group H.

Suppose that there were a homomorphism σ with kernel K of A_4 onto some group H. Let us see how far we can go toward finding H. Since K is to be the kernel of σ, we have $k\sigma = e$ for all elements k of K. That is, four elements of A_4 are mapped onto the identity of the group H that we wish to find. Now if g is any element of A_4, we have

$$(gk)\sigma = (g\sigma)(k\sigma) = g.$$

This shows that for every g in A_4, four elements, including g, have the same image as g. Furthermore, if $g'\sigma = g\sigma$, it follows that

$$g^{-1}g'\sigma = (g^{-1}\sigma)(g'\sigma) = (g^{-1}\sigma)(g\sigma) = e\sigma.$$

Hence $g^{-1}g' = k$ and $g' = gk$ for some k in K. Also, two elements of A_4 have the same image under σ if and only if they are in the same coset of K. Since K is of order 4 and A_4 of order 12, the group H that we are seeking should be of order $\frac{12}{4} = 3$. It would therefore be very convenient if, in some sense, the cosets of K turned out to be a group. We see that the cosets of K are

$$K, \quad gK, \quad g'K,$$

where g is not in K and g' is in neither K nor gK.

Now we show that, for some binary operation, the cosets are a group. The natural choice for an operation would seem to be multiplication, defined as follows for every pair of elements a and b of G:

$$(aK)(bK) = \{ak_1bk_2\} \qquad \text{over all } k_i \text{ in } K.$$

First we should show that this is a binary operation, that is, the product of two cosets is a coset. Since K is a normal subgroup, $bK = Kb$, that is, $k_1b = bk_3$ for some k_3 in K. Then

$$ak_1bk_2 = abk_3k_2 \qquad \text{is in } abK.$$

Conversely, each element of abK is in $(aK)(bK)$, since $abk = (ae)(bk)$ and e is in K. So "product," as defined above, is a binary operation. We state that it is associative and has K as its identity, and each coset has an inverse. Since

we prove these in the course of the proof of the next theorem, it is not necessary to do it for this case. The long and short of it is that the homomorphism
is onto H, the group of cosets. This example should give us an idea of what is
to happen.

On the basis of the examples, we should prove that if K is a normal subgroup of G, then, first, the cosets of K form a group H and, second, there is a
homomorphism of G onto H whose kernel is K.

Theorem 10.3.   Let K be a normal subgroup of a group G. The cosets of K
form a group H under the operation of product defined as follows:

(10.1) $(aK)(bK) = \{akbk'\}$ over all k and k' in K.

Also, $(aK)(bK) = abK$. (Note that we have stated this in terms of left
cosets, but since K is normal, it could equally well be stated in terms of right
cosets.)

Proof: First, we must show that the product (10.1) is a binary operation.
This is done just as in the discussion of Example 3. To show that such an
operation is associative, we note that

$$[(aK)(bK)](cK) = [abK][cK] = [(ab)c]K,$$

and

$$(aK)[(bK)(cK)] = [aK][bcK] = a(bc)K.$$

Then the product of cosets is associative, since the product of elements of the
group G is associative; that is, $(ab)c = a(bc)$. Furthermore, K is the identity
element of H, since $(aK)K = aK$. The inverse of aK is $(a^{-1})K$. This completes
the proof.

Definition.   *If K is a normal subgroup of a group G, then the group consisting
of the cosets of K under the operation* (10.1) *is called a* quotient group *and is
written G/K.*

From Corollary 2 of Theorem 4.2 we know that if G is a finite group, the
number of cosets of K is the index of K in G. Thus we have the following
corollary.

Corollary.   If K is a normal subgroup of a finite group G, then

$$o(G/K) = [G:K].$$

Theorem 10.4.   If K is a normal subgroup of a group G, there is a homomorphism σ of G onto G/K. The kernel of σ is K.

Proof: We set up a correspondence

(10.2) $g\sigma = gK$ for all g in G.

Now (10.2) means that we are letting each element g of G correspond to a single entity gK, which is a "bunch" of elements of G. This is a many-to-one correspondence if K is of order greater than 1, for if a and b are any two elements of the coset gK, then $a\sigma = b\sigma$. But the image under σ of each element of G is a unique coset. It remains to show that the correspondence (10.2) preserves products. The following shows it:

$$ab\sigma = abK = (aK)(bK) = (a\sigma)(b\sigma)$$

for every a and b in G. Thus the theorem is proved.

Let us digress for a moment to look in Example 2 at the subgroup $T' = \{e, (1\ 2\ 3), (1\ 3\ 2)\}$ of S_3. This is the alternating group on three symbols. Since the alternating group is of index 2 in S_3, there can be only two right cosets and two left cosets. A short argument shows that corresponding cosets must be equal. We state a more general result as follows.

Theorem 10.5.   If G is a group and K a subgroup of index 2 in G, then K is a normal subgroup of G.

Proof: Since K is of index 2, there are just two left cosets, K and gK, and just two right cosets, K and Kg'. Now the coset gK consists of all elements of G not in K, and the same can be said for the coset Kg'. Hence $Kg' = gK$ if g and g' are not in K, and the proof is complete.

Now where are we? Theorem 10.4 affirms that if K is a normal subgroup of G, there is a homomorphism σ of G onto G/K and the kernel of σ is K. Theorem 10.2 states that if σ is a homomorphism of G onto a group H, then the kernel K of σ is a normal subgroup. In the first case we started with a normal subgroup and found a homomorphism; in the second case we started with a homomorphism and found a normal subgroup. It would seem reasonable that the group H in the second case should be isomorphic to G/K of the first case. We thus tie things together by proving that this is so.

Theorem 10.6.   Let G be a group and σ a homomorphism of G onto some group H. Then the kernel K of σ is a normal subgroup of G and H is isomorphic to G/K.

Proof: Since Theorem 10.2 affirms that K is a normal subgroup, we have only the isomorphism to prove. It follows from Theorem 9.2 that for two elements a and b of G, $a\sigma = b\sigma$ if and only if a and b are in the same (left) coset of K. That is, σ takes every element of the coset gK into the same element of H. Thus associated with the homomorphism σ of G onto H is the transformation $\bar{\sigma}$ of the cosets of G onto H. That is, the transformation $\bar{\sigma}$ of

G/K onto H is defined by $(gK)\bar{\sigma} = (gk)\sigma = g\sigma$. Furthermore, $\bar{\sigma}$ is a homomorphism because

$$(gK)\bar{\sigma} = g\sigma \qquad \text{and} \qquad (g'K)\bar{\sigma} = g'\sigma$$

imply that

$$(gg'K)\bar{\sigma} = gg'\sigma = (g\sigma)(g'\sigma) = [(gK)\bar{\sigma}][(g'K)\bar{\sigma}].$$

Thus we have shown that $\bar{\sigma}$ is a homomorphism of G/K onto H. To show that it is an isomorphism, we need by Theorem 9.3 to show that *its* kernel is the identity coset, that is, K. Thus we need to show that

$$(gK)\bar{\sigma} = e' \quad \text{implies that } gK = K.$$

But $e' = (gk)\sigma = g\sigma$ implies that g is in K and hence that $gK = K$. This shows that $\bar{\sigma}$ is an isomorphism and completes the proof of the theorem.

Since Theorems 10.4 and 10.6 are two of the most important theorems in this chapter, let us recapitulate the information they give us. In both cases one starts with a group G. Then one may go in either of two directions. First, one can have a normal subgroup of G, call it K. From this one can define a homomorphism ϕ of G onto the quotient group G/K with the elements of K as its kernel. Second, one can start with a homomorphism of G onto a group H; this has a kernel K, and the quotient group G/K is isomorphic to H. Hence we come out with the same relationships whichever way we start: a group G, a normal subgroup K, and a homomorphism of G onto a group H that is isomorphic to G/K.

One can exhibit these interrelationships by the following diagram:

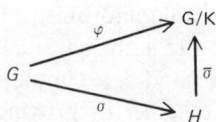

where ϕ, σ, and $\bar{\sigma}$ are as defined previously. The homomorphism ϕ is sometimes called "the natural homomorphism" of G onto G/K. We could write $H = G\sigma = G\phi\bar{\sigma}$ and $G/K = G\phi = G\sigma\bar{\sigma}$. One can "go" directly from G to H or by way of G/K.

In a way, what is most important about the results of Theorems 10.4 and 10.6 is that there *is* such a relation between the kernel of the homomorphism of G onto H and G/K. But we can use it, too. From a "practical" point of view, one of K and H is apt to be simpler than the other, and we can use the simpler one to get information about the other. For instance, H may be Abelian along with G/K, which tells us something about K.

11.   Examples

EXAMPLE 1.   To illustrate the last point of Section 10, let G be some finite group and K a subgroup of index 2. We know from Theorem 10.5 that K is normal and from the Corollary of Theorem 10.3 that G/K is a group of order 2. Since every group of order 2 is isomorphic to the multiplicative group $H = \{1, -1\}$, we know that G/K is isomorphic to H. In this isomorphism, K corresponds to 1 and Kg, for any element g not in K, corresponds to -1. That is, using the notation of the proof of Theorem 10.6, $K\bar{\sigma} = 1$ and $(Kg)\bar{\sigma} = -1$. This means that $g\sigma = 1$ if g is in K and $g\sigma = -1$ if g is not in K. Thus since σ is a homomorphism, $(g)^2\sigma = (g^2)\sigma = 1$ implies that the square of every element in G is in K. We state this result as follows.

Theorem 11.1.   If G is a finite group and K a subgroup of index 2, then K contains the squares of all elements of G.

As a consequence of this we can prove that A_4 has no subgroup of index 2. Suppose that K were a subgroup of index 2 in A_4. Then from Theorem 11.1 with $G = A_4$, K must contain the squares of all elements of A_4. But A_4 contains eight cycles of 3 and each is a square; for example, (1 2 3) = (1 3 2)^2. This is impossible, since K has only $\frac{12}{2} = 6$ elements. This contradiction proves the result. This is an example promised earlier of a group G and a number k (6 in this case), which is a divisor of the order of G such that G has no subgroup of order k.

The same method can be used to show that A_5 has no subgroup of index 2. This is left as Exercise 31.

EXAMPLE 2.   We showed in Example 3 of Section 10 that

$$K = \{e, (1\ 2)(3\ 4), (1\ 3)(2\ 4), (1\ 4)(2\ 3)\}$$

is a normal subgroup of A_4. Then A_4/K is of order 3 and hence isomorphic to the multiplicative group $\{1, r, r^2\}$ with $r^3 = 1$. If we use the same argument used to prove Theorem 11.1, it follows that the cube of every element of A_4 is in K. This is evident from first principles, since the cube of every element of A_4 not in K is the identity.

The same argument used for A_4 in Example 3 of Section 10 could be used to show that K is also a normal subgroup of S_4. This is left as Exercise 12. But it should be noted that K's being a normal subgroup of A_4 does not automatically imply that it is a normal subgroup of S_4. For instance, if G is the octic group, there is a chain of subgroups $G_1 \subset G_2 \subset G$, where each is of index 2 and hence normal in its successor, but G_1 is not a normal subgroup of G.

Notice that Theorem 11.1 does *not* affirm that if K is a subgroup of index 2 in a finite group G, then every element of K is the square of an element of G. For instance, let G be the Klein four-group

$$\{e, a, b, ab\},$$

where $ab = ba$, $a^2 = b^2 = e$, and $K = \{e, a\}$. Then K contains the squares of all elements of G, since it contains e. But the element a is not the square of any element of G. A theorem whose proof we leave as Exercise 29 is the following.

Theorem 11.2.   If K is a subgroup of G of index 2 and g is an element of G not in K, then K consists of the products xg, where x ranges over the elements of the coset gK.

Exercises

1. Which of the following correspondences are homomorphisms of the multiplicative group G of nonzero rational numbers $\mathbf{Q}^*$ onto a subset S of $\mathbf{Q}^*$? If the correspondence is a homomorphism, identify S and the kernel of the homomorphism. Also identify G/K.

   (a) $r \rightarrow |r|$.   (b) $r \rightarrow 2r$.   (c) $r \rightarrow 1/r$.
   (d) $r \rightarrow -r$.   (e) $r \rightarrow r^3$.   (f) $r \rightarrow \sqrt{r}$.

2. Are any of the homomorphisms of Exercise 1 automorphisms? If so, what ones?

3. Let G be the multiplicative group of all complex numbers $x + iy$, where x and y are real and not both zero. Define the transformation σ by

$$(x + iy)\sigma = \sqrt{x^2 + y^2}.$$

   Show that this is a homomorphism. Find the image group and the kernel.

4. Let G be the set of pairs of real numbers (a, b) with a and b not both zero. Define $(a, b) \circ (c, d) = (ac - bd, ad + bc)$ and prove that G is a group under this operation. Define a transformation σ by

$$(a, b)\sigma = \sqrt{a^2 + b^2}$$

   and show that it is a homomorphism. What is the kernel of σ and what is the image group? What relation is there between this exercise and Exercise 3?

5. Let G be the octic group. Find a chain of subgroups $G_1 \subset G_2 \subset G$ such that each is of index 2 in its successor and yet G_1 is not a normal subgroup of G.

6. Is a homomorphism an equivalence relation? If not, what properties does it lack?

7. Suppose that H is a normal subgroup of G. If $H \subset K \subset G$ for some subgroup K, is H normal in K?

8. Suppose that H is a subgroup of G of index p, where p is a prime number. Must H be a normal subgroup of G?

9. Show that in Example 1 of Section 10 the product of two left cosets is not a coset.

10. Show that if G is Abelian, every subgroup is normal.

11. Verify Theorem 11.1 for the alternating group on three symbols.

12. In Example 3 of Section 10 show that K is a normal subgroup of S_4.

13. Show that the intersection of two normal subgroups of a group G is a normal subgroup of G.

14. If the operation of a group G is addition and T is a subgroup of G, what is the meaning of the quotient group G/T?

15. Let G be the additive group of integers and K the subgroup of multiples of m, for some integer m greater than 1. What does the quotient group G/K become in this case?

16. Let K be a normal subgroup of G of index p, where p is a prime number. Prove that if g is in G, then g^p is in K.

17. Let S be a subgroup of G and, as in Exercise 31 of Section 3, define $a \equiv b \pmod{S}$ to mean that $a^{-1}b$ is in S. Prove that $a \equiv b \pmod{S}$ implies that both $ac \equiv bc \pmod{S}$ and $ca \equiv cb \pmod{S}$ if and only if S is a normal subgroup of G.

18. The dihedral group of order n, that is, the set of symmetries of a regular n-gon, is generated by two elements g and h, where

$$g^n - e - h^2 \quad \text{and} \quad hgh - g^{n-1}.$$

Find its center.

19. Show that if H and K are normal subgroups of a group G, then $hkh^{-1}k^{-1}$ is in $H \cap K$ for h in H and k in K. Show that if $H \cap K = \{e\}$, then $hk = kh$ for all h in H and k in K.

20. Prove that the correspondence in Example 3 of Section 9 is a homomorphism. Prove that its kernel is C.

21. Prove that the correspondence in Example 4 of Section 9 is a homomorphism. Prove that its kernel is H.

22. Let K be the subset of all elements of A_4 of order 2 or 1. Prove that not only is K a group, but also it is a normal subgroup of S_4.

23. Let G be a group and H the set of all elements of G of order 2 or 1. Show by an example that H is not necessarily a group. Prove that if H is a group, then it is a normal subgroup.

24. Let K be a subgroup of a group G. Prove that if every product of two left cosets of K is a left coset, then K is a normal subgroup of G.

25. For Example 3 of Section 10 show that if s is any 3-cycle of the set S, then $a^{-1}sa$ is of order 3 for every a in A_4. Does this prove that S together with the identity element is a normal subgroup of A_4? Explain.

26. Suppose that G is a group which is homomorphic onto a set S' with a binary operation. Prove that S' is associative.

27. Under the conditions of Exercise 26 prove that S' is a group. Prove that if G is cyclic, so is S'.

28. Let N be a normal subgroup of G such that G/N is Abelian. Prove that N contains the *commutator subgroup* of G defined as follows: the group generated by all elements of the form $ghg^{-1}h^{-1}$ for all elements g and h of G. (Note that not every element of the commutator group need be of the form $ghg^{-1}h^{-1}$.)

★29. Prove Theorem 11.2.

★30. Let C_G denote the commutator subgroup of G as defined in Exercise 28. Prove that C_G is a normal subgroup of G and that G/C_G is Abelian.

★31. Prove that A_5, the alternating group of five symbols, has no subgroup of index 2.

★32. Let $g^{-1}hg = h^i$. Prove that $g^{-k}hg^k = h^t$, where $t = i^k$. Show that if g is of order m, then

$$h^{i^m} = h$$

and hence that $i^m \equiv 1 \pmod{q}$, where q is the order of h. If m and q are prime numbers, show that m is a factor of $q - 1$ unless $gh = hg$.

★33. Let H and K be subgroups of a group G and K a normal subgroup of G. Define HK to be the set of all products hk, where h is in H and k in K. Prove:
(a) HK is a subgroup of G.
(b) K is a normal subgroup of HK.
(c) The transformation $H \overset{\sigma}{\to} (HK)/K$ defined by $h\sigma = hK$, with h in H, is a homomorphism of H onto HK/K. What is its kernel?
(d) The group $H(H \cap K)$ is isomorphic to the group HK/K.

12.   *Automorphisms and Conjugate Elements*

Now let us apply the above results to get further information about inner automorphisms discussed in Section 8. In Theorem 8.6 we proved that the number of inner automorphisms of a finite group G is equal to the number of cosets of C, its center. This was done by showing that two elements of G, a_1 and a_2, determine the same inner automorphism $a_i^{-1}xa_i$ if and only if a_1 and a_2 are in the same coset of C. Also for any element of C, $c^{-1}xc = x$ for all x; that is, the elements of C correspond to the identity inner automorphism. So in the language of Section 11 we have a homomorphism of the elements of G onto the group of inner automorphisms, and the kernel of this homomorphism is the center of G. Hence we have shown the following theorem.

Theorem 12.1.   There is a homomorphism of a group G onto the group of inner automorphisms of G, and the kernel of this homomorphism is C, the center of G.

Corollary.   (See Theorem 9.3.)   If the center of a group G consists of the identity element alone, then G is isomorphic to the group of its inner automorphisms.

One should note, however, that even if a group G is isomorphic to the group of its inner automorphisms, it still may have some outer automorphisms. This is not true for S_3, since 6 is the number of automorphisms and there are six inner automorphisms. But when we consider A_4 we have a different situation. Since its center is (e) (see Exercise 3 after Section 8), the corollary above shows that A_4 has exactly 12 inner automorphisms. On the other hand, S_4, whose center is also (e), has 24 inner automorphisms. Now every inner automorphism of S_4 is an automorphism of A_4, for let $x\sigma_g = g^{-1}xg$ for all x in S_4. Then if x is in A_4, so is $g^{-1}xg$, since A_4 is a normal subgroup of S_4. Furthermore, no two distinct inner automorphisms of S_4 are the same automorphism of A_4, for suppose that $g_1^{-1}xg_1 = g_2^{-1}xg_2$ for all x in A_4. Then $g_2g_1^{-1}x = xg_2g_1^{-1}$ and $g_2g_1^{-1}$ is commutative with all x in A_4. But from the answer to Exercise 3 after Section 8, e is the only element of S_4 that is commutative with all elements of A_4. Hence $g_1^{-1}xg_1 = g_2^{-1}xg_2$ implies that $g_1 = g_2$, and the automorphisms of S_4 are not distinct. Thus we have shown that the 24 inner automorphisms of S_4 are distinct automorphisms of A_4. Hence A_4 has at least 24 automorphisms, only 12 of which are inner automorphisms.

We also have the following theorem.

Theorem 12.2.   Let A be the group of automorphisms of a given group G and N the subgroup of inner automorphisms. Then N is a normal subgroup of A.

Proof.   Let σ_a be the inner automorphism $x\sigma_a = a^{-1}xa$ and let τ be any automorphism. We want to show that $\tau^{-1}\sigma_a\tau$ is an inner automorphism. Now

$$x\tau^{-1}\sigma_a\tau = [a^{-1}(x\tau^{-1})a]\tau = (a^{-1}\tau)x(a\tau) = (a\tau)^{-1}x(a\tau) = x\sigma_{a\tau}$$

shows that if σ_a is an inner automorphism, so is $\tau^{-1}\sigma_a\tau$ for every automorphism τ. This proves the theorem.

Inner automorphisms can be used to classify the elements of a group in a fruitful way. Let g be any element of a group G. We call the elements $x^{-1}gx$, for x in G, *conjugate elements* of g. If G happens to be Abelian, there is only one conjugate of g, namely, g itself. In fact, the same is true if g is in the center of G. But if there are elements of G not commutative with g, the element g will have conjugates different from itself. Now consider the set G_g of elements of G that are conjugates of g. What can we discover about G_g? It certainly is not necessarily a group, for suppose that p is a transposition. Then any conjugate $x^{-1}px$ is of order 2 and is an odd permutation. But $x^{-1}pxy^{-1}py$ is a product of two conjugates of p and is certainly not a conjugate of p, since it is an even permutation. But we can count the number of elements conjugate to an element g in the same way that we counted cosets, by asking under what conditions two conjugates $x^{-1}gx$ and $y^{-1}gy$ are equal. The answer is as before: if and only if yx^{-1} is commutative with g. So let N_g be the set of elements of G commutative with g and call N_g the *normalizer* (or centralizer) of g. You were asked to show in Exercise 16 of Section 4 that N_g is a group. Thus we have proved the following theorem.

Theorem 12.3.   Let g be an element of a group G and N_g its normalizer. The number of distinct elements of G conjugate to g is equal to the index of N_g in G.

Now we can use the above as a means of partitioning a finite group G into conjugate classes. To do this, start with C, the center of G. Each element of C is in a conjugate class of one element. Then if G is not Abelian, pick an element g not in C and put into a class all its conjugates. [The number of elements in this class is some divisor of $o(G)$. Why?] If this does not exhaust G, choose an element g' not a conjugate of g and not in C and consider a class composed of its conjugates. Now the set of conjugates of g is disjoint from the set of conjugates of g' because being a conjugate is an equivalence relation. (Why?) In this fashion we can continue until every element of G is in some conjugate class. Notice that different classes can have different numbers of elements, but in each case the number of elements is a factor of $o(G)$. The conjugate classes are disjoint. So we have

(12.1) $o(G) = o(C) + \sum dn_d,$

where n_d is the number of distinct conjugate classes with d elements and the sum is over all divisors greater than 1 of $o(G)$.

From this, one can prove (left as Exercise 8) the following curious result.

Theorem 12.4.   If G is a group whose order is p^s for p a prime number and s a positive integer, then the center of G has at least p elements.

Equation (12.1) can also be used to prove the following theorem (see comments in Section 13), but we shall give a slight modification of the very ingenious proof of J. H. McKay in *The American Mathematical Monthly*, vol. 66 (1959), p. 119. The only place in this book where we require this theorem is in Section 15 of Chapter VI.

Theorem 12.5.   (Cauchy.)   If p is a prime number that divides the order of a group G, then G has a subgroup of order p.

Proof: The proof consists in showing that there is an element of G different from e, the identity, which is a solution of $x^p = e$. Once this is shown, we know that such a solution must generate a subgroup of G of order p, from Lagrange's theorem.

The key idea may be seen by looking at the case when $p = 2$. Consider the products $ab = e$, where a and b are in G. If $a = b$, then $x = a$ is a solution of $x^2 = e$. If $a \neq b$, then two different products, ab and ba, are equal to e. So, if there are s pairs of distinct elements whose product is e, this accounts for $2s$ elements of S. Now if n is the order of G and is divisible by $p = 2$, then $n - 2s$ is an even integer, and hence there must be an even number of solutions of $x^2 = e$. One such solution is $x = e$. There must be another one and that is what we are looking for.

If when reading a detective story you prefer to stop before the end to see if you can guess how it comes out, you might like to stop here and see if you can complete the proof, or read a little further and then stop.

Let S be the set of ordered p-tuples of elements of G

$$(a_1, a_2, \ldots, a_p)$$

with the restriction that the product $a_1 a_2 a_3 \cdots a_p$ is equal to e. Now if n is the order of G, S has n^{p-1} elements (p-tuples), since the first $p - 1$ of the a_i may be chosen arbitrarily and then a_p chosen to be the inverse of the product of the previous a_i.

Let σ be the permutation $(1\ 2\ 3 \cdots p)$ and consider it a transformation on the p-tuples. That is,

$$(a_1, a_2, \ldots, a_p)\sigma = (a_2, a_3, \ldots, a_p, a_1).$$

In other words, σ permutes cyclically the elements of a p-tuple. We call two p-tuples α and β *equivalent* if $\alpha\sigma^i = \beta$ for i one of $0, 1, \ldots, p - 1$. Such equivalence is an equivalence relation. (Why?) Now if α is in S, then $\alpha\sigma$ is also and hence all p-tuples equivalent to it, for

$$a_1(a_2a_3a_4 \cdots a_p) = e = (a_2a_3a_4 \cdots a_p)a_1.$$

(The product in parentheses is the inverse of a_1.) By means of this relation, S may be partitioned into classes.

Now we show that such a class of p-tuples either contains exactly one p-tuple or exactly p different p-tuples. Since σ generates a cyclic group H of order p, no class can contain more than p different p-tuples. If a class contains less than p different p-tuples, then $\alpha\sigma^i = \alpha\sigma^j$ for α a p-tuple of the class and i and j in the range from 0 to $p - 1$ inclusive with, say, $i > j$. Then we can apply σ^{-j} on the right and have $\alpha\sigma^k = \alpha$ for some k between 1 and $p - 1$ inclusive. But the order of σ^k must divide the order of H; hence the order of σ^k is p and σ^k generates H. Thus every power of σ takes α into itself; that is, all cyclic permutations of the a_i in α take α into itself. But that can happen only if all the a_i are equal, in which case the class contains only one p-tuple. So we have shown that each class of p-tuples contains exactly one p-tuple or exactly p different p-tuples.

Let r be the number of classes of p-tuples of S that contain only one p-tuple and s the number of classes that contain p different p-tuples. Then counting the elements of S in two different ways, we have

$$n^{p-1} = r + sp.$$

Since, by hypothesis, p is a factor of n, the above equation shows that p must be a factor of r.

Now each class of one p-tuple looks like this: $(a, a, \ldots, a)$. Since the p-tuple is in S, $a^p = e$. Hence each class of one p-tuple gives a solution of $x^p = e$ in G, and each solution gives such a class so that r is the number of solutions in G of $x^p = e$. We showed in the previous paragraph that r is divisible by p. But r cannot be zero, since one solution of $x^p = e$ is e. Hence r is at least p. We have shown that $x^p = e$ has a solution not e, and the theorem is proved.

Exercises

1. Partition S_3 into conjugate classes and verify (12.1).

2. Partition the octic group into conjugate classes and verify (12.1).

3. Partition A_4 into conjugate classes and verify (12.1).

4. Show that if g is an element of G, the number of distinct elements that are conjugate to g is equal to the index in G of the normalizer N_g of g.

5. Is the normalizer N_g of g, an element of a group G, necessarily a normal subgroup of G?

6. Prove that if g is an element of a finite group G, the number of elements (including g) that are conjugate to g is a divisor of the order of G.

7. Show that the relationship "g is a conjugate element of h" is an equivalence relationship of all elements g and h of a group.

8. Prove Theorem 12.4.

9. Prove that the equivalence of classes defined in the proof of Theorem 12.5 is an equivalence relation.

⋆10. Prove that if $o(G) = p^2$ for p a prime number, then G is Abelian. (*Hint:* Use Exercise 32 after Section 11.)

⋆11. We know that if a group is Abelian, every subgroup is normal. Is the converse true? That is, if every subgroup of a group G is normal, must G be Abelian? Prove it or give an example to show that it is not so.

⋆12. Let G be a group whose order is pq, where p and q are distinct prime numbers. Prove the statements in parts (a) through (d) and answer, with reasons, the questions in part (e).
  (a) G contains a subgroup H whose order is p and one whose order is q.
  (b) If the order of H is p, then either H is a normal subgroup or it has exactly q conjugate subgroups (including itself).
  (c) If H is of order p but is not a normal subgroup, then the $pq - q(p-1) - 1 = q - 1$ elements of G not in any conjugate subgroup of H, together with e, form a normal subgroup K of order q.
  (d) If G is not Abelian and has a normal subgroup H of order p, then each element g in G but not in H generates a subgroup of order q that is not a normal subgroup of G. Furthermore, q is a factor of $p - 1$. (*Hint:* Exercise 32 after Section 11 can be useful here.)
  (e) Is there a non-Abelian group of order 15? Is there one of order 21?

13. Further Recommended Reading

We shall mention here just two extensions of the theory developed in this chapter. First, one can define conjugate subgroups of a given group as we defined conjugate elements and obtain some fruitful results. Second, Cauchy's theorem (Theorem 12.5) is a special case of what is sometimes called a theorem of Sylow—that if p is a prime number and $p^u \mid o(G)$, then G has a subgroup of order p^u. A proof of the latter may be found, among other places, in Herstein's *Topics in Algebra* (see the Bibliography). Many of the topics that we treated here are dealt with there from a more sophisticated viewpoint, as can also be said of the books by van der Waerden and Birkhoff and Mac Lane listed in the Bibliography. There is a kind of special charm to Herstein's book, for he takes pains to communicate to the reader his enthusiasm for the subject.

For more advanced theory of groups see Marshall Hall's *The Theory of Groups*. This goes into detail, for example, on group representations, that is, representing groups by means of matrices, as mentioned in Section 6. A detailed bibliography appears in Hall's book. Two books on representations that should be mentioned are Bryan Higman's *Applied Group Theoretic and Matrix Methods* (New York: Oxford University Press, 1955) and *Representation Theory of Finite Groups and Associative Algebras* by C. W. Curtis and Irving Reiner (New York: Wiley-Interscience, 1962).

14.   Biographical Notes

Augustin-Louis Cauchy (1789–1857) was surpassed only by Euler and Cayley in volume of publication, and certainly he was a greater mathematician than Cayley. He was a devout royalist and Catholic without change in changing times, which sometimes furthered his career, sometimes inhibited it, but never stopped it. He made important contributions not only to mathematics, but also to astronomy, optics, and mechanics. Although Cauchy is largely responsible for the abstract definition of a group and other results, including the last theorem of Section 11, his contributions to the field of analysis and infinite series are much more extensive. To most mathematicians "Cauchy's theorem" refers to a fundamental result in the theory of complex variables.

Pierre de Fermat (1601–1665) was a jurist by profession and the king's parliamentary counselor in Toulouse, France, but his hobby was mathematics. He made many discoveries but apparently did not feel it necessary to record the proofs. In some cases the methods he indicated could be used to give proofs; in others there is great doubt that he had any proof at all. For instance, the numbers $2^{2^k} + 1$ bear his name (see Chapter VI) because he claimed he had a proof that all such numbers are prime numbers. Euler disproved this by showing that if $k = 5$, the number is the product of 641 and 6,700,417. No larger prime Fermat number has been found, and some have been proved not prime. He claimed to have a proof that $x^n + y^n = z^n$ is not solvable in positive integers for $n > 2$, his "Last Theorem." So far no one has proved or disproved this result. With regard to his theorem mentioned in Section 5, it is interesting to note that apparently the Chinese knew in the year 500 B.C. that $2^p - 2$ is divisible by p if p is a prime number. Fermat said he had a proof. The first written proof was that of Leibniz, who used the expansion of $(1 + 1 + \cdots + 1)^p$. But, although Fermat gave few proofs, his contribution to mathematics is of the highest order.

Leonard Euler (1707–1785) was a native of Basel, Switzerland, and a student of the famous mathematician Johann Bernoulli, whom he soon surpassed. Most of his academic life was spent in the Academy of Science at Berlin under Frederick the Great and in St. Petersburg (now Leningrad) under Catherine the Great. He was a man of broad culture and in the mathematical world has not been surpassed in the volume and depth of publication. He made fundamental contributions as well to the presentation of mathematics, and much of the approach to calculus in textbooks can be traced back to him. The use of e, π, and i began with him, and the fundamental formula $e^{i\theta} = \cos\theta + i\sin\theta$ is due to him. He made notable contributions to mathematical physics. George Simmons called him "the Shakespeare of mathematics: universal, richly detailed, and inexhaustible."

Joseph Louis Lagrange (1736–1813) was born in Turin and at the age of 19 became professor of mathematics at the Royal Artillery School. When Euler left Berlin and the court of Frederick the Great to go to St. Petersburg in 1766 at Euler's recommendation, Lagrange succeeded him. He was in a number of respects Euler's successor, for he worked in isoperimetric problems (one such is to find the curve with given perimeter having greatest area) in the calculus of variations and he developed techniques for solving related problems that had eluded Euler. The two collaborated on a number of problems in this field. Later he became interested in algebra and number theory. Perhaps his greatest work was the unification of general mechanics in his *Mécanique analytique*.

Karl Friedrick Gauss (1777–1855), usually considered the greatest mathematician of modern times, was born in Brunswick, Germany, of poor parents. At the age of 18 he solved the problem of determining which regular polygons can be constructed with ruler and compass (see Chapter VI). He set out to put mathematics on a rigorous foundation. In fact he took such pains with rigor that he carefully covered the steps by which he found his proofs. Abel is said to have remarked about Gauss: "He is like the fox, who effaces his tracks in the sand with his tail." He had little enthusiasm for teaching but taught exceedingly well. His work ranged throughout mathematics, physics, and optics. The many terms that bear his name are only a sign of the universality of his contributions to mathematics and science.

Arthur Cayley (1821–1895) spent the first eight years of his life in St. Petersburg, and then his family moved back to England. At the age of 17 he entered Cambridge University. Law was his profession until he could devote full time to mathematics on the occasion of becoming Sadlerian Professor of Mathematics at Cambridge University in 1863. He was very prolific, producing between the ages of 25 and 29 between 200 and 300 papers. His chief fields were elliptic functions, analytic geometry of n dimensions, determinants, matrices, and abstract groups.

Christian Felix Klein (1849–1926) was born in Dusseldorf and entered the University of Bonn at the age of $16\frac{1}{2}$. Between 1872 and 1886 he was professor of mathematics at Erlangen, and it was there that he promulgated his "Erlangen program," which cast new light on the study of geometry before and after his time. His point of view is that *a* geometry is the study of properties of figures that remain invariant under certain groups of transformations. That is, Euclidean geometry is the study of properties of points that remain unchanged under the group of rigid motions, that is, Euclidean transformations. If it is a topological group, then the geometry becomes topology. If it is the group of nonsingular linear transformations, we have projective geometry. During his second period in Göttingen (1892–1911) he was chiefly concerned with organizing mathematics and mathematics instruction. His ideas were felt not only in Europe but in the United States as well. His influence even extended to the reform of the mathematics curriculum in the United States shortly after the turn of the century.

III

Fields, Integral Domains, and Rings

1. Definition of a Field

In the second section of Chapter I we defined in passing what is meant by a field. It is time we dealt with this topic in detail. The simplest examples of fields are $\mathbf{Q}$, the set of rational numbers; $\mathbf{R}$, the set of real numbers; and $\mathbf{C}$, the set of complex numbers. In each of these we can add, subtract, multiply, and, except for zero, divide. First, we give a definition of a field without reference to ideas of groups and then show how this definition may be abbreviated by stating it in terms of groups. We shall phrase it in somewhat abstract terms, since we want to include fields beside those mentioned above, but you would find it useful to see how the properties listed jibe with those of the three sets mentioned.

Definition. *A field is a set S of two or more elements a, b, c, ... with two binary operations: multiplication, indicated by $a \cdot b$ or ab, and addition, indicated by $a + b$, with the following properties:*

1. *Associativity: $a + (b + c) = (a + b) + c$ and $a(bc) = (ab)c$.*
2. *Existence of identity elements: for addition 0 such that $a + 0 = 0 + a = a$ for all a; for multiplication 1 such that $1 \cdot a = a \cdot 1 = a$ for all a.*
   [As for groups (see below), the identities are unique.]

3. *Existence of inverse elements: For each a in S, there is an additive
   inverse that we denote by (−a) such that a + (−a) = (−a) + a = 0;
   if a ≠ 0, there is a multiplicative inverse that we call a⁻¹ such that
   aa⁻¹ = a⁻¹a = 1. [As for groups (see below), the inverses are
   unique.]*
4. *Commutativity: a + b = b + a and ab = ba.*
5. *Distributivity: a(b + c) = ab + ac and (b + c)a = ba + bc.*

It should be noted in view of property 4 that we do not need the second
equation in the distributive property. Also, we observe the usual convention
that $ab + ac$, for instance, means $(ab) + (ac)$. Some authors choose to omit
the commutative property for multiplication in the definition of a field. If *we*
have occasion to omit this property, we shall use one of the terms "skew
field" or "noncommutative field."

In the definition of a field we could have omitted the word "binary" and
added the condition of *closure*: If a and b are in S, then $a + b$ and ab are
unique elements of S. This is a rather trivial remark, but we make it because
it has a bearing on the following alternative definition of a field in terms of
groups.

Definition.   *A* field *is a set S of two or more elements a, b, c, . . . with two binary
operations: addition, indicated by a + b, and multiplication, indicated by
a · b or ab, with the following properties:*

1g. *S is an Abelian group under addition.*
2g. *S with 0 omitted is an Abelian group under multiplication.*
5. *The distributive property holds.*

Now, property 2g refers to the set S with zero omitted, which we denote by
S^*. If the operation of multiplication is to be a binary operation for S^*, it
must be closed for S^*. That property is stated as number 13 below. The proof
of that property depends only on the distributive property, the properties of
addition, *and* that $0 \cdot a$ is in S. Therefore, multiplication must be a binary
operation for the whole set S as well as for the subset S^*. The two definitions
are equivalent. One is tempted to leave the proof of this statement as an
exercise, but we resist the temptation because the proof seems merely busy
work.

Describing a field in terms of groups not only has the advantage of brevity,
but also allows us to make certain deductions from known properties of
groups. Referring back to additional properties of a group in Section 1 of
Chapter I, we see that the five properties which define a field imply the
following:

6. The additive identity 0 and the multiplicative identity 1 are unique;
   that is, no other elements have the property that describes them.

7. The additive and multiplicative inverses of each element of a field are unique.
8. The equations $a + x = x + a = b$ and, for $c \neq 0$, $cx = xc = b$ are solvable for unique elements x.
9. The cancellation properties hold: (a) if $a + b = a + c$, then $b = c$, and (b) if $ab = ac$ with $a \neq 0$, then $b = c$.
10. The additive inverse of $a + b$ is $(-a) + (-b)$ and the multiplicative inverse of ab is $b^{-1}a^{-1}$ when the individual inverses exist.

There are three other properties which follow and which combine the two operations in some manner.

11. $0 \cdot a = a \cdot 0 = 0$ for all elements a.
12. $a(-b) = -(ab)$ and $(-a)(-b) = ab$.
13. If $ab = 0$, then at least one of a and b is zero.

We prove properties 11 and 13 and leave the proof of 12 as Exercise 10 after Section 2. For property 11 we use the identity for multiplication and the distributive property to compute $0 \cdot a + a = 0 \cdot a + 1 \cdot a = (0 + 1)a$. But $0 + 1 = 1$ and hence $0 \cdot a + a = 1 \cdot a = a$. If we add $(-a)$ to both sides of $0 \cdot a + a = a$, we get $0 \cdot a = 0$, by use of the associative property.

To show property 13, suppose that $a \neq 0$. Then a^{-1} exists; we can multiply $ab = 0$ on the left by a^{-1} and use the associative property to get $(a^{-1}a)b = a^{-1} \cdot 0 = 0$, by property 11. Thus $1 \cdot b = b = 0$, and we have shown that if $ab = 0$ and $a \neq 0$, then $b = 0$. As we noted above, one effect of this is that the operation of multiplication is not only closed over a field F but also over F^*, the elements of F with zero omitted. (We often use the asterisk to denote the omission of zero.)

This is quite a long list of properties. Fortunately, you are familiar with these, since they form the basis for manipulations with numbers and in algebra. Hence it is usually not necessary to refer to them specifically.

Now the elements of a field need not be numbers. For instance, consider the set of all rational functions of x, that is, all functions expressible in the form $f(x)/g(x)$, where f and g are polynomials in x with, say, real coefficients and $g(x)$ is not the number 0. Denote this set by $\mathbf{R}(x)$. It is not necessary to check all the properties in detail, but we should note that 1 and 0 are the multiplicative and additive identities just as for numbers, there is closure for addition, subtraction, and multiplication. The multiplicative inverse of $f(x)/g(x)$ is $g(x)/f(x)$ if $f(x) \neq 0$.

The identities of a field can be designated by symbols other than 0 or 1. But we shall see shortly that every field contains a subfield that is isomorphic to a field in which the identity elements are 0 and 1. Hence we might just as well use these symbols from the beginning.

For future use, just as for groups, we should find conditions under which a subset of the elements of a field is itself a field. So let S be a subset of elements of a field F. The associative and commutative properties for multiplication and addition hold for S, since they hold for F and F contains S. The same is true for the distributive property. We need to show that the elements of S form an additive subgroup of elements of F and that the nonzero elements of S form a subgroup of the multiplicative group of elements of F^*. Looking back (Theorem 7.1 of Chapter I) at the conditions that a subset of a group be a group, we have the following.

Theorem 1.1. A subset S of the elements of a field F is itself a field if and only if

1. When s and s' are two elements of S, then $s + s'$ and ss' are in S.
2. When s is in S, then $-s$ is in S and, for $s \neq 0$, s^{-1} is in S.

We can express the conditions of the theorem by saying that S is a subfield of F if S is closed under addition, subtraction, multiplication, and (except for zero) division. A subfield of F that is not F is called a *proper* subfield.

Before looking at the construction of fields in general, let us consider a specific one not only because it gives us a field different from the familiar ones but because the group properties come into play. Let S be the set of numbers $\{a + b\sqrt{2}\}$, where a and b are rational numbers. Since S is a subset of the real numbers, we may apply Theorem 1.1. To show closure under addition and multiplication, see that

$$(a + b\sqrt{2}) + (c + d\sqrt{2}) = (a + c) + (b + d)\sqrt{2},$$

and

$$(a + b\sqrt{2})(c + d\sqrt{2}) = (ac + 2bd) + (bc + ad)\sqrt{2}.$$

Closure under subtraction is shown by $-(a + b\sqrt{2}) = (-a) + (-b)\sqrt{2}$. It remains to show closure under division. You have probably already proved that $\sqrt{2}$ is irrational. (This follows from Theorem 13.5 of this chapter.) Since $\sqrt{2}$ is irrational and a and b are rational, $a + b\sqrt{2} = 0$ implies that both a and b are zero. Thus we want to show that $a + b\sqrt{2}$, with not both a and b zero, has a multiplicative inverse. For this we compute

$$\frac{1}{a + b\sqrt{2}} = \frac{a - b\sqrt{2}}{(a + b\sqrt{2})(a - b\sqrt{2})} = \frac{a - b\sqrt{2}}{a^2 - 2b^2} = r + s\sqrt{2},$$

where

$$r = \frac{a}{a^2 - 2b^2} \quad \text{and} \quad s = \frac{-b}{a^2 - 2b^2}.$$

Since $a^2 - 2b^2 = 0$ implies that $a = b = 0$, this shows that the set of numbers $a + b\sqrt{2}$ with not both a and b zero is closed under division. Notice that without the restriction that a and b be rational, this conclusion would be false.

2.   *The Structure of Fields*

Now let us look into the possibilities for a field F. First, F must contain an identity for addition and one for multiplication, which we have agreed to designate by 0 and 1, respectively. (We shall at this point proceed as if these *were* the numbers 0 and 1. Then later in this section we review the whole process from a more abstract standpoint.) There are two possible outcomes of this addition: Either some repetition occurs or it does not. If there is no repetition, the field must contain an infinite number of elements and hence all the positive integers. Then, since a field contains additive inverses and, except for zero, multiplicative inverses, it must contain the reciprocals of the positive integers and their negatives—hence all the rational numbers. That is, if there are no repetitions in the sequence of sums of 1's, the field F must have the field of rational numbers as a subfield.

The other possibility is that repetition occurs, that is, that the sum of r ones is equal to the sum of s ones for some $r < s$. This *must* occur if the field has a finite number of elements but *may* occur when the field has an infinite number of elements. In this case, using the cancellation property for addition, we see that for some positive integer n, the sum of n ones is zero; that is, $n = 0$. If n is the least positive integer for which this is so, we call n the *characteristic* of the field. Then the sum of $n + 1$ ones is 1, the sum of $n + 2$ ones is 2, and so on. In fact, to find what the sum of r ones is in the field we divide r by n, and the remainder after this division is what r is in the field. So we have shown that if a field F has characteristic n, instead of containing all the integers, it contains the numbers $0, 1, \ldots, (n - 1)$, where addition is modulo n. (See Section 3 of Chapter II, in particular Theorem 3.4a.) Now from Corollary 2 of Theorem 5.4 in Chapter II we know that the numbers $1, 2, \ldots, (n - 1)$ modulo n form a multiplicative group if and only if n is a prime number. Now if n is not a prime number, it is equal to the product rs for r and s both different from n, and hence $rs \equiv 0 \pmod{n}$, which denies property 13 for a field.

Therefore, we have shown most of the following theorem.

Theorem 2.1.   If F is a field, it either contains (see the comments at the end of this section) the field of rational numbers as a subfield or the field $\mathbf{Z}_p$ of numbers $0, 1, 2, \ldots, (p - 1)$ modulo p for some prime number p.

Proof: All that remains to be shown is that the numbers modulo p form a field. That the numbers $1, 2, \ldots, (p - 1) \pmod{p}$ form a multiplicative group follows, as we noted above, from Corollary 2 of Theorem 5.4 in Chapter II. For addition, the numbers $0, 1, 2, \ldots, (p - 1)$ form a cyclic group $(\bmod\ p)$ with 1 as a generator—the set is isomorphic to the multiplicative group of pth roots of unity. The distributive property holds, since

$a(b + c) = ab + ac$ for integers, and hence the remainders when $a(b + c)$ and $ab + ac$ are divided by p are the same. This completes the proof.

Definition.   *If a field F contains the subfield of rational numbers, it is said to have* characteristic 0 *(or infinity), and if it contains the field of integers modulo p for a prime number p, F is said to have* characteristic p.

Let us explore further what may happen if a field is finite, that is, contains a finite number of elements. As shown above, it must contain the numbers $0, 1, \ldots, (p - 1)$ modulo p for some prime p. If there are no other elements, then F is that field. If there is another element, we can call it r_2 and see that F must also contain all numbers $x_1 + x_2 r_2$, where x_1 and x_2 are in $\mathbf{Z}_p$. If this exhausts the elements of F, fine. If not, there is another element r_3 not in the set $x_1 + x_2 r_2$. Then the numbers $x_1 + x_2 r_2 + x_3 r_3$ are all in F. So, if F is finite, we may continue until every element of F is expressed in the form

$$x_1 + x_2 r_2 + x_3 r_3 + \cdots + x_k r_k.$$

No two such numbers can be equal unless coefficients of corresponding r_i are all equal, for suppose that

(2.1) $x_1 + x_2 r_2 + x_3 r_3 + \cdots + x_k r_k = y_1 + y_2 r_2 + \cdots + y_k r_k.$

Then if $y_k \neq x_k$, we can express r_k as a linear combination of the previous r's with coefficients in $\mathbf{Z}_p$, which is contrary to the method of construction. So $y_k = x_k$. Then we can subtract $x_k r_k$ from both sides of (2.1) and get a similar expression with r_k omitted. Repeating the process shows that $x_i = y_i$ for all i. So we have proved the following theorem.

Theorem 2.2.   If F is a finite field, it has characteristic p for some prime p. If $F \neq \mathbf{Z}_p$, it contains elements $r_2, r_3, \ldots, r_k$ such that every element of F can be expressed uniquely in the form

$$x_1 + x_2 r_2 + \cdots + x_k r_k,$$

where the x_i are in $\mathbf{Z}_p$. The field F contains then p^k elements.

Why does one speak of the field of rational numbers as being of characteristic 0? The key is to note what happens for characteristic p. Two integers in $\mathbf{Z}$ correspond to the same number in $\mathbf{Z}_p$ if their difference is an integral multiple of p. Two integers in $\mathbf{Z}$ correspond to the same number in $\mathbf{Q}$ if their difference is a multiple of zero, that is, if they are equal. From another point of view it is sometimes said that $\mathbf{Q}$ has characteristic infinity, since no positive integer corresponds to 0 in $\mathbf{Q}$.

Now let us review more carefully what we have just done. We denoted by 0 and 1 the identities for addition and multiplication. Suppose that we had designated them by z and u, respectively, instead. Then we would know that

$u, u + u, u + u + u, \ldots$ would all be in F, and we could carry through the same argument with u that we did with 1. Thus if there are no repetitions, we can set up a one-to-one correspondence between sums of the u's and sums of the 1's; that is, $nu \leftrightarrow n$ for every positive integer n. Closure under subtraction would imply that n could be a negative integer as well. By this time it should be clear that we have an isomorphism. So we define it formally.

Definition.   *Let F and F' be two fields. They are said to be* isomorphic *if there is a one-to-one correspondence between the elements of the two fields, that is, a one-to-one transformation σ of F onto F', such that*

$$(a + b)\sigma = a\sigma + b\sigma \quad and \quad (ab)\sigma = (a\sigma)(b\sigma).$$

This means that for two fields to be isomorphic their additive and multiplicative groups must be isomorphic. Actually, corresponding to what we saw for groups, we would not need to assume that F' is a field but only that addition and multiplication are binary operations in F'; then the isomorphism with F would show that F' is also a field.

One could carefully continue the above process with u and z, but it would be tedious and of little value. We shall therefore merely give an example and restate the two theorems in terms of an isomorphism.

For example, let F be the field $\{a + b\sqrt{2}\}$ with a and b rational. Then the transformation σ defined by $(a + b\sqrt{2})\sigma = a - b\sqrt{2}$ is an isomorphism of F onto itself, that is, an *automorphism*. You will be asked to show this in Exercise 18.

Theorem 2.1a.   If F is a field, it contains a subfield that is isomorphic to the field of rational numbers or one that is isomorphic to the field $\mathbf{Z}_p$ of integers modulo p for some prime p.

Theorem 2.2a.   If F is a finite field, it is isomorphic to $\mathbf{Z}_p$ or contains elements $r_2, r_3, \ldots, r_k$ such that every element of F is expressible uniquely in the form

$$x_1 + x_2 r_2 + \cdots + x_k r_k,$$

where the x_i are in a field that is isomorphic to $\mathbf{Z}_p$.

We shall quite often consider two isomorphic fields to be the same and revert to the nomenclature in Theorems 2.1 and 2.2.

Notice that it is possible to have infinite fields whose characteristic is finite. For instance, in Section 1 we mentioned $R(x)$, the field of rational functions of x with real coefficients. If we choose $\mathbf{Z}_p$ to be the coefficients of the polynomials $f(x)$ and $g(x)$ in $f(x)/g(x)$, then the field would have characteristic p while still having an infinite number of elements.

In the past all finite fields were called Galois fields, but *they* have come to have a more specialized meaning. We deal extensively with Galois fields in the more modern sense in Chapter VI. At the end of that chapter is a short biographical sketch of Evariste Galois, whose name these groups bear. Before we can proceed with the discussion of fields, we shall need more information. This we develop in the following sections.

Exercises

1. Find a field with just two elements.

2. Show that the numbers $a + bi$ form a field, where a and b are real and $i^2 = -1$.

3. Show that the numbers of the form $a + b\sqrt{3}$ form a field, where a and b are rational numbers.

4. Do the numbers of the form $a + b\sqrt[3]{2}$ form a field, where a and b are rational? Why or why not?

5. Let $x^2 = 3$ and consider the numbers of the form $a + bx$, where a and b are in $\mathbf{Z}_5$. Show that these form a field.

6. Let $x^2 = 3$ and consider the numbers of the form $a + bx$, where a and b are in $\mathbf{Z}_{11}$. Do these numbers form a field? Why or why not?

7. Give one positive integer k less than 10 such that there is no field with k elements.

8. Let $x^2 + 1 = 0$ in $\mathbf{Z}_3$ and consider the elements of the form $a + bx$, where a and b are in $\mathbf{Z}_3$. Prove that this set of elements forms a field.

9. Let $x^2 + 1 = 0$ in $\mathbf{Z}_5$. Show that the set $a + bx$, where a and b are in $\mathbf{Z}_5$, does not form a field.

10. Prove property 12 of a field as given in Section 1.

11. Do the numbers of the form $a + b\sqrt{12}$, where a and b are rational numbers, form a field?

12. Do the numbers of the form $a + b\sqrt{3}$, where a and b are real numbers, form a field?

13. Find a field with just four elements.

14. Show that the set of real numbers being a field implies that the set of complex numbers forms a field.

15. Let S be the set (a, b), where a and b are real numbers. Define $(a, b) + (c, d)$ to be $(a + c, b + d)$ and $(a, b)(c, d)$ to be $(ac - bd, ad + bc)$.

Prove that S is a field. To which of the fields previously mentioned in these exercises is it isomorphic?

16. Show that the set $\{a + bw\}$ forms a field, where w is one of the roots of $x^2 + x + 1 = 0$ and a and b are rational numbers.

17. Show that the set $\{a + bi + \sqrt{2}(c + di)\}$, for a, b, c, and d rational numbers, forms a field.

18. Let F be the field of elements $a + b\sqrt{2}$, where a and b are rational. Prove that the transformation σ defined by $(a + b\sqrt{2})\sigma = a - b\sqrt{2}$ is an isomorphism of F onto itself, that is, an automorphism.

19. Consider the fields $F = \{a + b\sqrt{2}\}$ and $F' = \{a + b\sqrt{3}\}$, where a and b are rational. Is the transformation σ defined by $(a + b\sqrt{2})\sigma = a + b\sqrt{3}$ an isomorphism of F onto F'?

20. Find all the automorphisms of the field $\mathbf{Z}_3$.

21. Find all the automorphisms of $\mathbf{Z}_p$ for an arbitrary prime p.

22. Consider the field F described in Exercise 17. Find an automorphism of F, not the identity, and justify your conclusion.

23. Let S and S' be two subfields of F. Show that their interesection is a subfield of F.

24. Is the union of two subfields of a field F necessarily a field? Give reasons for your answer.

★**25.** Show that the set $\{a + b\sqrt[3]{2} + c\sqrt[3]{4}\}$, for a, b, and c rational numbers, forms a field.

3. Integral Domains

The set of integers $\mathbf{Z}$ has most of the properties of a field. In fact, the only fundamental property that it lacks is the existence of multiplicative inverses for all nonzero elements. (For instance, the equation $3x = 2$ is not solvable in integers.) But $\mathbf{Z}$ does have a property which, for a field, is a consequence of the existence of an inverse; that is, in $\mathbf{Z}$ if $ab = 0$, then at least one of a and b is zero. This is property 13 in our list of properties of a field. This holds not only for integers, but also, for instance, for the set of all polynomials with real coefficients, since the set of polynomials is a subset of the field $\mathbf{R}(x)$ of rational functions of x. Also, this property holds for the set of Gaussian integers: $S = \{x + yi\}$, where x and y are integers, since S is a subset of the field of complex numbers. Since there are a number of examples of systems that have this property, we give them a name, as follows.

Definitions. *Let D be any set that satisfies all the properties of a field in its definition (our first five properties) except that the requirement that a multiplicative inverse exist is replaced by*

(3.1)   *If $ab = 0$ for a and b in D, then at least one of a and b is zero.*

Then D is called an integral domain. *[The property* (3.1) *is customarily described, somewhat oddly, by saying that D has no divisors of zero.] If a and b are in D and $ab = 1$, then a and b are called* units. *A subset of the elements of D with the same two operations is called a* subdomain *if it is an integral domain.*

In the light of this definition let us look at one example given just before we defined an integral domain. Since the set of polynomials in x with real coefficients is a subset of the rational functions, properties 1, (3.1), 4, and 5 of a field hold for polynomials. Closure for addition and multiplication hold from the usual rules for adding and multiplying polynomials. Also, $-f(x)$ is the additive inverse of $f(x)$. Finally, 1 and 0 are polynomials, showing that property 2 of a field holds. The same argument could be used to show that the Gaussian integers form an integral domain.

Theorem 3.1.   An integral domain has all 13 properties listed for a field in Section 1 except the existence of a multiplicative inverse and the solvability of $cx = b$ for $c \neq 0$.

Proof: If you look back at the proofs of the 13 properties of a field, you will see that we need to show only the cancellation property for multiplication, the uniqueness of the multiplicative identity and, when it exists, of the multiplicative inverse. To show the cancellation property, suppose that $ab = ac$ with $a \neq 0$. Then, by the distributive property, $a(b - c) = 0$, which by property (3.1) implies that $b - c = 0$; that is, $b = c$. This also implies the uniqueness of the multiplicative identity, since $a \cdot 1 = a \cdot e$ implies that $1 = e$. To complete the proof of the theorem, suppose that the inverse a^{-1} of a exists and $a^{-1}a = 1 = ba$. Then multiplication on the right by a^{-1} shows that $a^{-1} = b$. This completes the proof.

Immediately after the proof of Theorem 7.1 in Chapter I we noted that if a group is finite, closure implies the existence of an inverse. Therefore, in an integral domain with a finite number of elements, closure of the multiplicative group of nonzero elements implies the existence of an inverse, for let d be an element of the finite integral domain D. Then all the powers of d, just as in the note after Theorem 7.1 of Chapter I, are in D. Hence two powers must be equal, and thus, for some positive integer s, $d^s = 1$. Then d^{s-1} is the inverse of d. Hence we have proved the following theorem.

Theorem 3.2.   Any integral domain with a finite number of elements is a field.

Not only can we get an integral domain by changing one property of a field, but, starting with any integral domain, we can also recover or construct a field in which the domain lies. One way to do this is by the same process that we use to get the rational numbers from the integers. Let us see how this goes. Start with an integral domain D and consider the set of ordered pairs (a, b), where a and b are in D. Since this will be like the fraction a/b, we know that we must exclude $b = 0$. Then we translate into the notation of number pairs the known properties of fractions by replacing a/b by (a, b) and $=$ by $\simeq$; [$(a, b) \simeq (c, d)$ is read (a, b) is equivalent to (c, d).]

	Fractions	Number pairs
Definition:	The set a/b, where a and b are integers and $b \neq 0$.	The set (a, b), where a and b are in D and $b \neq 0$.
Equality:	$a/b = c/d$ if $ad = bc$.	$(a, b) \simeq (c, d)$ if $ad = bc$.
Sum:	$a/b + c/d = (ad + bc)/bd$.	$(a, b) + (c, d) \simeq (ad + bc, bd)$.
Product:	$(a/b)(c/d) = ac/bd$.	$(a, b)(c, d) \simeq (ac, bd)$.

Of course we could also use fractions for general integral domains, but the advantage in using number pairs is to avoid confusion with any preconceived ideas about fractions. If D were the set of integers, the above would give an isomorphism between the rational numbers and ordered pairs of integers.

In order to prove that under the above conditions ordered pairs of elements of an integral domain D form a field, we must show the following:

1. For $\simeq$, addition and multiplication are binary operations.
2. The relation $\simeq$ is an equivalence relation and $(a, b) \simeq (a', b')$ with $(c, d) \simeq (c', d')$ imply that
    (a) $(a, b) + (c, d) \simeq (a', b') + (c', d')$.
    (b) $(a, b)(c, d) \simeq (a', b')(c', d')$.
3. The set of ordered pairs (a, b), where $b \neq 0$, forms a group under addition, having identity element $(0, 1)$.
4. The set of ordered pairs (a, b), with $ab \neq 0$, forms a group under multiplication, having identity element $(1, 1)$.
5. The distributive property holds.

Now property 1 holds immediately from the definition of the two operations on ordered pairs. To show that $\simeq$ is an equivalence relation (see Section 2 of Chapter II), we test the three properties of such a relation: first, $(a, b) \simeq (a, b)$, since $ab = ba$; second, $(a, b) \simeq (c, d)$ implies that $(c, d) \simeq (a, b)$, since $ad = bc$ implies $cb = da$ (remember that we are in an integral domain); third, $(a, b) \simeq (c, d)$ and $(c, d) \simeq (e, f)$ imply that $ad = bc$ and

$cf = de$. Hence $adf = bcf = b(de) = bde$ and, by the cancellation property of multiplication in D, $af = be$, which implies that $(a, b) \simeq (e, f)$.

To prove property 2(a), we see that, by definition, $(a, b) + (c, d) \simeq (ad + bc, bd)$ and $(a', b') + (c', d') \simeq (a'd' + b'c', b'd')$. Hence, to show 2(a), it suffices to prove that $(ad + bc, bd) \simeq (a'd' + b'c', b'd')$; that is,

$$ab'dd' + bb'cd' = a'bdd' + bb'dc'.$$

This follows, since $(a, b) \simeq (a', b')$ implies that $ab' = ba'$ and $(c, d) \simeq (c', d')$ implies that $cd' = dc'$.

To prove property 3, note that the additive identity is $(0, 1)$, since $(a, b) + (0, 1) \simeq (a, b)$. In fact, $(0, 1) \simeq (0, r)$ for all r different from zero. So the additive identity is unique in the sense that if two ordered pairs are the identity, they are equivalent to each other. The additive inverse of (a, b) is $(-a, b)$. This inverse is unique for (a, b) in the same sense that the identity is unique. The associative property holds for number pairs because it holds in D.

We leave as Exercises 15 and 22 after Section 5 the proofs of 2(b), 4, and 5. Thus *we* have established the following theorem.

Theorem 3.3. Let D be an integral domain and S the set of ordered pairs (a, b), where a and b are in D and $b \neq 0$. If equivalence, sum, and product of pairs are defined as after Theorem 3.2, then S is a field.

Definition.   *The field derived from an integral domain D by the above process is called the* quotient field *of D.*

So we have shown (here the word "we" is justified, since much of the proof was left as an exercise) that starting with any integral domain, one can form in this fashion a field that "contains it."

At the end of the previous sentence we put "contains it" in quotation marks because if one is to be particular, the statement is not strictly true. The elements of the field $S = \{(a, b)\}$ are pairs of elements in D, and a pair of elements of D cannot be a single element of D. The actual situation is that there is a subset S_0 of S which "behaves like," that is, is isomorphic to D. Since we are using the word as defined in Section 2, we should be careful to relate it to that context. So, we let S_0 be the subset $\{(a, 1)\}$, where a is in D. We define the transformation σ by

$$(a, 1)\sigma = a \qquad \text{for all } a \text{ in } D.$$

We want to show that this is an isomorphism. First, it is one-to-one onto, since $(a, 1) \leftrightarrow a$. Then, since there are two operations involved, we must check the correspondence for each of addition and multiplication. So, we make the following calculations:

$$[(a, 1) + (a', 1)]\sigma = (a + a', 1)\sigma = a + a' = (a, 1)\sigma + (a', 1)\sigma,$$
$$[(a, 1)(a', 1)]\sigma = (aa', 1)\sigma = aa' = [(a, 1)\sigma][(a', 1)\sigma].$$

This shows that the correspondence is preserved under both addition and multiplication. Hence the isomorphism is established. Therefore, the field F consisting of ordered number pairs (a, b), with addition and multiplication as defined, contains a subset isomorphic to the integral domain D. In *that* sense F contains D. We also say that D is *embedded in F.*

4.   Polynomials

In Section 3 we defined an integral domain and showed how it could be embedded in a field. An integral domain can also be embedded in other integral domains. For example, $\mathbf{Z}$ is embedded in the integral domain $S = \{a + bi\}$, where a and b are in $\mathbf{Z}$ and $i^2 = -1$. Suppose that we generalize this a little and consider the set $S = \{a + bt\}$, where a and b are in an integral domain D. If t were already in D, we would have nothing not already in D. Hence we want t to be outside of D. There is no difficulty with addition —here the group properties are almost automatic. But we have to be careful about multiplication. If D is to be closed, then

$$(a + bt)(c + dt) = ac + (bc + ad)t + bdt^2$$

would also have to be in D. This means that t^2 must be in D; that is, $t^2 = r + st$ for elements r and s of D. One special case of this is when t^2 is in $\mathbf{Z}$, as when $t = i$. Of course, we would also have to look at property (3.1), but let us pass over this for the time being to concentrate on closure. If t^2 were not in D, we should have to consider the set $a + bt + ct^2$ and then the question would arise about closure in *this* set. Now it might happen that no matter how far we go we should not have closure. This can happen if $D = \mathbf{Z}$ and $t = \pi$, for example, although we shall not prove that this is so. In fact, it would not have to be any number at all if we have certain understandings.

Let us start over again with an integral domain D and a symbol t. Then consider all expressions of the form $a_1t + a_0$, where a_1 and a_0 are in D. Although t is just a symbol, we assume that it has the properties of elements of an integral domain, for instance, $at = ta$ for all a in D, $t + t - 2t$, $-t + t = 0$, $1 \cdot t = t$, and so on. Then suppose that t^2 is not in the set $a_1t + a_0$ and look at $a_2t^2 + a_1t + a_0$ with the same understandings. So continuing for all powers of t, we have a set of expressions

(4.1) $$f(t) = a_nt^n + a_{n-1}t^{n-1} + \cdots + a_1t + a_0,$$

where n is a nonnegative integer and the a_i are in some integral domain D. We are accustomed to call this a *polynomial* in t over D, that is, with coefficients in D. The number a_n is called its *leading coefficient* and n is the *degree* of $f(t)$ if a_n is not zero. We do not assign a degree to the zero polynomial $f(t) = 0$. Notice in the last case that we are not referring to an equation that is to be solved but to the polynomial which is, if you will, $0 \cdot t + 0$. Of course, a polynomial may not be in the form (4.1). For instance, $4t^2 + 3t^2 + 5t^3 - 3$ is a perfectly good polynomial, although it is not in the form (4.1). It is, however, by the definition of equality that we shall give, equal to one of the form (4.1). We want to define equality, sum, and product of polynomials so that the set of all polynomials with coefficients in D forms an integral domain. Intuitively, though t is not a number, we want it to behave as if it were except for division. What this boils down to is to define equality, sum, and product of polynomials as we are accustomed to do.

Since this is all familiar, we merely state the known results in general form. We take $f(t)$ as in (4.1) and represent $g(t)$ by

(4.2) $$g(t) = b_r t^r + b_{r-1} t^{r-1} + \cdots + b_1 t + b_0.$$

Then we define the sum and product as follows:

$$f(t) + g(t) = \sum_{i=1}^{n} (a_i + b_i) t^i,$$

$$f(t)g(t) = \sum_{k=0}^{n+r} c_k t^k, \qquad \text{where } c_k = \sum_{i+j=k} a_i b_j,$$

where, for the sums, it is assumed for simplicity of notation that $r \leq n$, and if $i > r$, then b_i is defined to be zero.

Then any sum of products of powers of t and elements of D can, using the above definitions, be written in the form (4.1). If we want to know whether two such sums are equal, we write them in the form of (4.1) and look at corresponding coefficients. If corresponding coefficients are equal, the sums are equal; if the corresponding coefficients are not all equal, the sums are not equal. We have chosen the definition of equality, sum, and product so that the following theorem is true.

Theorem 4.1.   The set of all polynomials in t with coefficients in an integral domain D is an integral domain. We call it $D[t]$.

Outline of the proof: Since proving this requires looking in turn at all the properties of an integral domain, it is very tempting to "leave the details to the reader." Rather than do this, on the one hand or, on the other hand, go through the demonstration in complete detail, it seems better to sketch the proof, dwelling only on the points that might cause difficulty. For addition we have closure, the associative property, the identity (the number 0 of D, which is a polynomial); the additive inverse of $f(t)$ is that in which each coefficient is replaced by its negative in D.

Closure for multiplication follows directly from the definition. The associativity property could be verified by writing products in full notation, but this scarcely seems necessary; the number 1 is the multiplicative identity. To check that there are no divisors of zero [property (3.1)], let $f(t)$ and $g(t)$ be written as in equations (4.1) and (4.2), with the assumption that the leading coefficients a_n and b_r are different from zero. Then the leading coefficient of the product is $a_n b_r$, which is different from zero, since the coefficients are in an integral domain. This shows that the product polynomial is not zero. In fact, we have shown that the degree of the product of two polynomials is the sum of the degrees of the members of the product. Checking the distributive property would complete the proof.

5.   Ordered Fields and Ordered Integral Domains

Some fields and integral domains have an additional property of being "ordered." Since every field is an integral domain *a fortiori*, we shall concentrate our attention in this section chiefly on ordered integral domains, realizing that everything here written about integral domains applies to fields as well. The most natural way to think of an ordered domain is to consider it to have elements that can be ordered according to a relationship of "greater than" or "less than." Fundamentally, to have such an ordering it is necessary that if a and b are two elements of D, then exactly one of the following three conditions holds:

1. $a = b$.
2. a is less than b, written $a < b$. (Another way to express the same
(5.1)   relationship is $b > a$ or "b greater than a.")
3. a is greater than b, written $a > b$. (Another way to express the same relationship is $b < a$ or "b less than a.")

But this is not all that we need. For instance, we should have such properties as : $a > b$ and $b > c$ imply that $a > c$; and if $c > 0$, then $a > b$ implies that $ac > bc$. It is rather more satisfactory if we begin with the set of positive numbers of D, set up certain desirable properties, and then derive properties of inequality from these.

Hence we start with the idea of positive elements and set up certain properties which we want them to have. Also, we want to make the list of requirements as short as possible for economy of effort. Here is one way to accomplish this.

Definition.   *An integral domain D is said to be* ordered *if it contains a subset D^+ called its* positive elements *such that*

1. *If b is in D, then exactly one of the following holds: $b = 0$, b is in D^+, or $(-b)$ is in D^+ (the trichotomy property).*
2. *The set D^+ is closed under addition and multiplication; that is, if b and c are in D^+, then $b + c$ and bc are in D^+.*

The nonzero elements of D that are not in D^+ are called its negative *elements.*

Four other properties are implied by the above two:

3. The product of two negative elements of D is positive.
4. The product of a negative and a positive element is negative.
5. The square of any element in D is either zero or positive.
6. The multiplicative identity of D is positive.

Property 3 follows from the fact that if b and c are negative, then $(-b)$ and $(-c)$ are positive and their product, $(-b)(-c) = bc$, is positive. For property 4, let b be negative and c positive. Then $(-b)$ is positive and $(-b)c = -(bc)$, which is positive. This shows from property 1 that bc is negative. Property 5 follows from properties 3 and 2.

To prove property 6, let a be any positive element of D and e be the identity element for multiplication. Then $ea = a$. If e were negative, then by property 4, $ea = a$ would be negative, contrary to fact. Since $e \neq 0$, property 1 implies that e is positive.

Notice that the set of integers, the field of rational numbers, and the field of real numbers are all ordered. The same is true of any subset of these. If an integral domain D is ordered, it is possible to order $D[t]$ by letting $(D[t])^+$ be those polynomials whose leading coefficient is in D^+. Showing that this is so is left as Exercise 17.

But the set of complex numbers is not an ordered field, since 1's being positive implies that -1 is negative and -1 is the square of a complex number, denying property 5. A field of characteristic p is not ordered, since $p - 1$ is the sum of $p - 1$ numbers 1 and hence must be positive, while $p - 1 = -1$, which is negative.

From the point of view of Section 9 of Chapter II, an ordered field F has the property that if F^* is the multiplicative group of nonzero elements, then there is a homomorphism of F^* onto the group consisting of 1 and -1 under multiplication, where the positive elements of F correspond to 1 and the negative ones to -1.

Now let us see what consequences the definition has for inequality. The connection between the properties of the positive elements of an ordered domain and inequality are as given in the following definition.

Definition.   *If D is an ordered integral domain or field, we define $a > b$ to mean that $a - b$ is positive. Alternative expressions of this relationship are $b < a$, b is less than a, and a is greater than b.*

Phrased in terms of inequality, properties 1 and 2 of an ordered integral domain are

i. If b and c are in D, then exactly one of the following holds: $b = c$, $b > c$, $b < c$. (*the trichotomy property*)
ii. If b and c are greater than zero, then so is their product and sum.

Other familiar properties follow immediately and their proofs are left as Exercise 20.

   iii. If $a > b$ and $b > c$, then $a > c$. (*the transitive property*)
   iv. If $a > b$ and $c > 0$, then $ac > bc$.
   v. If $a > b$ and $c < 0$, then $ac < bc$.
   vi. If $ac > bc$, then either $c > 0$ and $a > b$, or $c < 0$ and $a < b$.

Property vi can be seen to follow from the previous five without going back to the definition of inequality in terms of positive elements. To see this, suppose that $c > 0$; then $c^{-1} > 0$, for if $c^{-1} < 0$, property iv would imply that $cc^{-1} = 1 < 0$, contrary to fact. So $c^{-1} > 0$, and hence $(ac)c^{-1} > (bc)c^{-1}$; that is, $a > b$. If $c < 0$, a similar proof can be given.

Two symbols related to the above are $\geq$ meaning "greater than or equal to" or "not less than" and $\leq$ meaning "less than or equal to" or "not greater than." If you look carefully at properties ii through v of inequality, you will see that they hold equally well if throughout $>$ is replaced by $\geq$ and $<$ by $\leq$. There is one additional property that is useful in a surprising number of cases:

$$\text{If } a \leq b \text{ and } b \leq a, \text{ then } a = b.$$

Use of this property is often the simplest way to prove equality.

   Another concept in an ordered integral domain is the *absolute value*, written $|x|$. It is defined, as you know, as follows:

$$|x| = x \text{ if } x \geq 0 \qquad \text{and} \qquad |x| = -x \text{ if } x \leq 0.$$

The absolute value is never negative.

   In Section 3 we showed how an integral domain can be embedded in a field. If the integral domain is ordered, is the quotient field ordered? One would expect this to be true from experience with the integers and rational numbers.

Theorem 5.1.   Let D be an ordered integral domain and F its quotient field. Define $(a, b) > 0$ if $ab > 0$. Then F is an ordered field; that is, the positive elements of F have properties 1 and 2 of the definition of an ordered integral domain.

   We leave the proof of this as Exercise 11. Notice that part of the proof is to show that if $(a, b) > 0$ and $(a, b) = (c, d)$, then $(c, d) > 0$. Based on this theorem, inequality of number pairs can be defined as follows:

(5.2) $(a, b) > (c, d)$ when $(a, b) - (c, d) > 0$.

Note that from properties 1 and 2 of an ordered domain, properties 3, 4, 5, and 6 follow for F.

Exercises

1. Find the product $(3x^3 + 2x - 5)(x^2 + 5)$ in $\mathbf{Z}$, in $\mathbf{Z}_7$, and in $\mathbf{Z}_{11}$.

2. Is $x^5x^7 = x$ in $\mathbf{Z}_{11}$? Why or why not?

3. Show that the set $\{a + bi\}$, where a and b are in $\mathbf{Z}$ and $i^2 = -1$, is an integral domain. Find the units of this domain.

4. Show that the set $\{a + b\sqrt{2}\}$, where a and b are integers, is an integral domain.

5. Let S be the set $\{a + bw\}$, where $w^2 + w + 1 = 0$ and a and b are integers. Show that S is an integral domain.

6. If equivalence and product of number pairs are defined as in Section 3 but sum is defined by $(a, b) + (c, d) = (a + c, b + d)$, show that one does not have a field.

7. Let S be the set $\{(a, b)\}$, where a and b are integers. Define sum and product of number pairs by $(a, b) + (c, d) = (a + c, b + d)$ and $(a, b) - (c, d) = (ac - bd, ad + bc)$. Show that S is an integral domain.

8. Define $f(x) = |x| + x$ and $g(x) = |x| - x$, where x is any real number. Prove that the product $f(x)g(x) = 0$. Are $f(x)$ and $g(x)$ in an integral domain?

9. Define $g(x)$ and $f(x)$ as in Exercise 8 and calculate

$$f[g(x)] \quad \text{and} \quad g[f(x)].$$

10. Show that for number pairs as in Section 3, if $ab \neq 0$, then $(a, b)^{-1} \cong (b, a) \cong (b, 1)(a, 1)^{-1}$.

11. Prove Theorem 5.1.

12. Show that the units of the integral domain $a + b\sqrt{2}$, where a and b are integers, are the solutions in integers of $x^2 - 2y^2 = \pm 1$. Find three units.

13. Find the units of the integral domain $\{a + bw\}$, where a and b are integers and w is a root of $x^2 + x + 1 = 0$. (See Exercise 5.)

14. The set of rational numbers is an integral domain. What are its units?

15. Show that the multiplicative properties of a field are satisfied for the number pairs of integers whose equality, sum, and product are defined as in Section 3.

16. Is every subdomain of an ordered integral domain an ordered domain?

17. Let D be an ordered integral domain. Show that $D[t]$ is an ordered integral domain if we define $f(t)$ to be positive in $D[t]$ when its leading coefficient is positive.

18. Let a, b, c, and d be elements of an ordered integral domain. Prove:

    i. $|ab| = |a| \cdot |b|$.

    ii. $|a + b| \leq |a| + |b|$.

    iii. If $0 < a < b$, then $a^{-1} > b^{-1}$.

    iv. If $(a, b) \neq (c, d)$ with $bd > 0$, then one of the following holds: $(a, b) < (a + c, b + d) < (c, d)$ or $(a, b) > (a + c, b + d) > (c, d)$, if inequality of number pairs is defined by (5.2).

19. Prove that if a, b, c, and d are in an ordered field and inequality of number pairs is defined by (5.2), then $(a, b) > (c, d)$ if and only if $abd^2 > cdb^2$. Conclude that if $bd > 0$, then $(a, b) > (c, d)$ if and only if $ad > bc$.

20. Derive properties iii through vi listed for inequality.

21. Suppose that throughout the six properties of inequality one replaces $>$ by $\geq$ and $<$ by $\leq$. What changes need to be made?

22. Show that the distributive property holds for pairs of integers as defined in Section 3.

23. If the multiplicative group of nonzero elements of an integral domain D is homomorphic to the multiplicative group $\{1, -1\}$, must D be an ordered domain? Why or why not?

24. Rephrase Theorem 3.3 in terms of equivalence classes. (See Section 2 of Chapter II.)

25. Let F be a field. Suppose that we define F to be "semiordered" if the nonzero elements of D fall into one of two disjoint sets: (1) the set of squares of elements of F designated by F_S and (2) the set of nonsquares designated by F_N. If a semiordered field must be ordered, prove it. If not, indicate an additional condition that is necessary. Is an ordered field necessarily semiordered?

26. Is the field Z_7 semiordered in the sense of Exercise 25?

27. Suppose that for an integral domain D we assume a relationship "greater than" with properties i and ii. Then if we define D' to be the elements of D greater than zero, will properties 1 and 2 of an ordered integral domain be satisfied?

28. Find the quotient field of the integral domain of Exercise 3.

29. Find the quotient field of the integral domain of Exercise 7.

30. If an integral domain D is isomorphic to a field, must D be a field?

31. Suppose that S is the set of number pairs (a, b), $b \neq 0$, where a and b are rational numbers instead of merely integers. Define equivalence, addition, and multiplication of number pairs as in Section 3. What field is S?

32. Suppose that, in Exercise 31, a and b are numbers $r + s\sqrt{2}$, where r and s are integers. Show that S is isomorphic to the field $\{x + y\sqrt{2}\}$, where x and y are rational. Give an isomorphism.

33. Let the polynomial $a_n x^n + a_{n-1} x^{n-1} + \cdots + a_1 x + a_0$ correspond to the vector $(a_n, a_{n-1}, \ldots, a_1, a_0)$. Indicate how you would define addition and multiplication of vectors so that they will form an integral domain isomorphic to $F[x]$.

34. Is the transformation σ defined by $x\sigma = x + 1$ an automorphism of an integral domain D? Why or why not?

6.   *A Return to the Integers*

From one point of view, part of this chapter has been a narrowing-down process. Starting with an integral domain, we had a number of possibilities, including the integers, any field, numbers of the form $a + bi$, where a and b are integers, or polynomials. Then when the requirement of being ordered was imposed, some of these fell by the wayside, notably fields of characteristic different from zero and the complex numbers. In this section we deal with another property of the set of integers which, when imposed on an ordered integral domain, leaves us only with the integers or, if you please, with a system isomorphic to the integers.

Definition.   *A subset of S, an ordered integral domain, is called* well-ordered *if for every nonempty subset T of S it is true that T has a least element, that is, an element t_0 such that $t_0 \leq t$ for all elements of T.*

The property that characterizes the integers is the well-ordering postulate.

The Well-ordering Postulate.   The positive integers form a well-ordered set.

This is a property that we assume is true for the set of integers. We want to show that any ordered integral domain whose positive elements form a well-ordered set is isomorphic to the set of integers. First we need the following theorem.

Theorem 6.1.   If D is an ordered integral domain whose positive elements form a well-ordered set and if 0 and 1 denote the additive and multiplicative identities, respectively, then there is no element d of D such that $0 < d < 1$.

Proof: Let D^+ be the set of positive elements of D; D^+ is not empty, since it contains 1. Since D^+ is well-ordered, it must contain a least element:

call it d_0. If $0 < d_0 < 1$, then by properties of inequality we have $0 < d_0^2 < d_0 < 1$, which denies the assumption that d_0 is the least element of D^+. This completes the proof.

Now we can prove the principal result of this section.

Theorem 6.2.   Let D be an ordered integral domain whose positive elements form a well-ordered set, and let 0 and 1 denote its additive and multiplicative identities. Then D is the set of integers.

Proof: First, D must contain the set of integers, since $1, 1 + 1, 1 + 1 + 1$, and so on, are contained in D by the closure property and since D, being ordered, has characteristic zero. Let S be the set of positive elements of D that are *not* integers. To show that S is empty, we assume that it is not empty and reach a contradiction. Since S is a subset of D^+ and D^+ is well-ordered, then S has a least element: call it s_0. Now $s_0 - 1 \geq 0$, since 1 is the least element of D^+ and s_0 is in D^+. On the other hand, $s_0 - 1$ cannot be in S, since s_0 was chosen to be the least element of S. So $s_0 - 1 \geq 0$ and is not in S. But $s_0 - 1$ is in D. Hence $s_0 - 1$ is either 0 or a positive integer in D. In either case s_0 must be an integer, which contradicts our definition of S. Hence the contradiction shows that S is empty and that all elements of D are integers. This completes the proof.

The theorem thus asserts that if D is an ordered integral domain whose positive elements are well-ordered, then it must, up to a change in notation for the identities, be the set of integers. Hence we have the following corollary.

Corollary.   Let D be an ordered integral domain whose positive elements are well-ordered. Then D is isomorphic to **Z**.

7.   *Mathematical Induction*

There is a property of the positive integers $\mathbf{Z}^1$ that can be used in a method of proof which is called *mathematical* or *finite induction*.

Induction Property.   A set of integers N contains all the positive integers if it has the following two properties:

1. N contains the number 1.
2. For every n in N, it is true that $n + 1$ is in N.

When you look carefully at the induction property, it seems obvious intuitively—so obvious that nothing but a simple proof would make it more apparent. So if you are content merely to believe this property, you might better stop here. Nevertheless, there is some point in showing that it can be deduced from the well-ordering property, largely because we shall see that other important properties of the integers can also be deduced from the well-ordering property. The point is that if one can trace a number of properties back to a single one, he does not have to assume as much. In a sense we put all our assumptions in one basket.

So now we show how the induction property follows from the well-ordered postulate. (There is some similarity to the proof of Theorem 6.2.) The method is to suppose that there were a positive integer d not in N, and let S be the set of *all* positive integers *not* in N. We show that if d is an integer in S, then $d - 1$ is also. Then S would be a nonempty set with no minimum element, which contradicts the fact that S is well-ordered.

Hence to prove our result, we need to show that if d is in S, then $d - 1$ is also. Now $d - 1 \geq 0$, since d is a positive integer and 1 is the least positive integer. Furthermore, $d - 1 \neq 0$, since if $d - 1$ were 0, then d would be 1, which is in N—hence not in S. So $d - 1$ is a positive integer. Now $d - 1$ cannot be in N since, by property 2, $(d - 1) + 1 = d$ would then be in N, which is false. So $d - 1$ must be in S, which gives us the contradiction desired. We have shown that the induction property follows from well-ordering.

As we noted above, a rephrasing of this property gives a useful method of proof called "mathematical induction" or "finite induction." To see this, suppose that $Q(n)$ is a statement about the integer n which is either true or false. If we let N in the induction property be the set of integers n for which $Q(n)$ is true, we see that the following principle holds.

Induction Principle. A statement $Q(n)$ about the integer n is true for all positive integers n if

1. $Q(1)$ is true.
2. For each integer n, the truth of $Q(n)$ implies the truth of $Q(n + 1)$; that is, if $Q(n)$ is true, then $Q(n + 1)$ is true.

The truth of $Q(n)$ is often called the *induction hypothesis*, since it is the hypothesis of the implication in property 2. Now we give two illustrations of the use of this principle in proofs.

Theorem 7.1. If b and c are real numbers such that $0 < b < c$, then

(7.1) $$b^n < c^n \qquad \text{for all positive integers } n.$$

Proof: The statement $Q(n)$ is $b^n < c^n$. Now $Q(1)$ is true since, by hypothesis, $b < c$. To complete the proof, we need to show that for each integer n, $b^n < c^n$ implies that $b^{n+1} < c^{n+1}$. If we multiply $b^n < c^n$ on both sides by b and multiply both sides of $b < c$ by c^n, we get the inequalities

(7.2) $$b^{n+1} < bc^n \qquad \text{and} \qquad bc^n < c^{n+1}.$$

Using the transitive property of inequality, we can complete the proof.

Theorem 7.2.   Consider the sequence of integers $1, 1, 2, 3, 5, \ldots$, in which each after the first two is the sum of the two preceding ones (the *Fibonacci sequence*). Then any two consecutive terms are relatively prime and every third one is even.

Proof: Let the successive terms be u_k; that is, $u_1 = 1$, $u_2 = 1$, $u_3 = 2$, $u_4 = 3$, $u_5 = 5, \ldots$ and $u_{k+2} = u_{k+1} + u_k$. First, if $k = 1$, $u_k = u_1 = 1$ and $u_{k+1} = u_2 = 1$ are relatively prime. Hence the theorem is true for $n = 1$. Now we need to show that if u_k and u_{k+1} are relatively prime, so are u_{k+1} and u_{k+2}. Thus $u_{k+2} - u_{k+1} = u_k$ shows that if d is a common factor of u_{k+2} and u_{k+1}, then it is also a factor of u_k. That is, d is a factor of both u_{k+1} and u_k. Since, by the induction hypothesis, these two are relatively prime, then $d = 1$. This shows that u_{k+2} and u_{k+1} are relatively prime and completes the proof by mathematical induction.

Now we want to show that u_{3s} is even for all integers s. It is true for $s = 1$, since $u_3 = 2$. Here the induction is on s. Now,

$$u_{3s+3} = u_{3s+2} + u_{3s+1} = (u_{3s+1} + u_{3s}) + u_{3s+1} = 2u_{3s+1} + u_{3s}.$$

By the induction hypothesis, u_{3s} is even and hence u_{3s+3} is also. This is the proof.

The second part of the previous proof was an induction on multiples of 3. An induction does not really need to start with 1. For instance, if one replaces n by $n - 1$, this induction begins with 0 and, if successful, proves the theorem for all nonnegative values of n. Or one could replace n by $n - c$ for any integer c to prove the theorem for all integer values beginning with c, Sometimes, also, one wants to assume that the theorem holds not only for the previous value of n, but also for all previous values of n. This is an equivalent form of induction. (See Exercise 16.)

Unfortunately, many of the examples of induction proofs are somewhat artificial in that there are easier proofs by other methods. We shall try to avoid these. But there are occasions, as we shall see later in this chapter, when an induction proof is really the most satisfactory way to establish a result.

We mention the binomial theorem in this connection, since we shall be using it from time to time and since it is often proved by mathematical induction. (A simpler proof can be obtained using combinations and permutations.) We do not prove it, since a proof can be found in most elementary textbooks on algebra.

The Binomial Theorem.   If n is any positive integer,

$$(x + y)^n = x^n + \binom{n}{1}x^{n-1}y + \binom{n}{2}x^{n-2}y^2 + \cdots + \binom{n}{i}x^{n-i}y^i + \cdots$$

$$+ \binom{n}{1}xy^{n-1} + y^n,$$

where

$$\binom{n}{i} = \frac{n!}{(i!)(n-i)!} \qquad \text{for } 1 \le i \le n-1.$$

Although the usual proof is in a more special context, it holds for x and y in any integral domain. Actually, as will be pointed out in Section 17, we do not need to restrict x and y to be in an integral domain.

Exercises

1. Do the nonnegative integers form a well-ordered set? Give reasons for your answer.

2. Let s be an integer. Show that every nonempty subset of the integers that contains no element less than s is well-ordered.

3. Show that the set of positive rational numbers is not well-ordered.

4. Prove that $x^{2k+1} + y^{2k+1}$ is divisible by $x + y$ for all positive integers k.

5. Prove that $x^n - y^n$ is divisible by $x - y$ for all positive integers n.

6. In the Fibonacci sequence of Theorem 7.2, prove that u_{4s} is divisible by 3 for all positive integers s.

7. We showed above that in the Fibonacci sequence the elements u_{3s} are all divisible by 2 and u_{4s} by 3. Are all elements u_{5s} divisible by 4? Why or why not?

8. Prove that the sum of the first n odd integers is n^2.

9. Prove that if n is an integer greater than 1, then $(1 + 1/n)^n > 2$.

10. Prove that if n is an integer greater than 1, then $(1 + x)^n > 1 + nx$ for all real numbers $x > -1$.

11. Prove that if a and b are positive real numbers and n is a positive integer, then

$$\left(\frac{a+b}{2}\right)^n \le \frac{a^n + b^n}{2}.$$

12. Let S be the set of integers greater than the rational number k, where k may be negative. Is S a well-ordered set?

13. Show why the well-ordering postulate for integers implies that if S is any nonempty set of integers less than some integer k, then S has a maximum element.

14. Prove that in any ordered field the positive elements do not form a well-ordered set.

15. Let S be the set of rational numbers such that if s is in S, then ns is an integer for some integer $n \leq 7$. Is S a well-ordered set? Give reasons for your answer.

16. Prove that in the induction principle, one can replace statement 2 by: For each integer n, the truth of $Q(k)$ for all positive integers $k \leq n$ implies that $Q(n + 1)$ is true.

17. Let T be the set of positive rational numbers t such that if t is in T, then nt is an integer for some integer $n \leq 5$. Is T a well-ordered set? Give reasons for your answer.

18. Assume that the set of real numbers is ordered and that for every real number r, there is an integer greater than r. Prove that there is a greatest integer less than r.

8.   Divisibility in Integral Domains

The chief difference between a field and an integral domain that is not a field is in the properties of divisibility. For a field, division is always possible except by zero. This is not so for an integral domain that is not a field. In this section we shall be using much of the theory that has gone before to look into the properties of divisibility. It is not necessary to note each time that we are concerned with integral domains that are not fields, since most of the results are trivially true for fields. A distinct advantage in using the concept of an integral domain is that our results can then be proved simultaneously for the set of integers and the set of polynomials with coefficients in a field or integral domain, or indeed for a number of other systems. First, we deal with properties of any integral domain; later more restrictions are convenient. It is important to keep track of just what is assumed at each stage. Let us start with the following.

Definition.   *If b and c are two elements of an integral domain D, with $b \neq 0$, c is said to be divisible by b if there is an element d of D such that $bd = c$. The notation for this relation is $b \mid c$. The same relation can be described by writing "b is a factor or divisor of c" or "c is a multiple of b."*

Notice that there is a great difference between $b \mid c$ and b/c. The former is a relation between two elements of D and the latter is an element of the field in which D is embedded—a fraction if you will.

Definition.   *A divisor of the identity 1 is called a* unit.

The set of integers has just two units, 1 and -1. The units of the set $F[x]$ of all polynomials with coefficients in a field F are the nonzero elements of F, for suppose that $f(x)g(x) = 1$ for two polynomials over F. Then each must be of degree zero and hence is in F. Every nonzero element of F is a unit in $F[x]$, since if $a \neq 0$ is in F, so is a^{-1}, and $aa^{-1} = 1$. If the polynomials have coefficients in an integral domain D, then the units of $D[x]$ (the set of polynomials with coefficients in D) are the units of D.

We shall prove part of the following theorem, leaving the rest as Exercise 6 after Section 9. We should emphasize that by proving this about integral domains, we have simultaneous results about the set of integers and the set of polynomials with coefficients in any field or integral domain.

Theorem 8.1. If a, b, and c are any elements of an integral domain D, where $a \neq 0$; in part 3, $b \neq 0$; and in part 4, $bc \neq 0$, then

1. If $a \mid b$, then $a \mid bc$.
2. If $a \mid b$ and $a \mid c$, then $a \mid (b + c)$.
3. If $a \mid b$ and $b \mid c$, then $a \mid c$. (*The transitive property*)
4. If $b \mid c$ and $c \mid b$, then $b = uc$ for some unit u.

Proof of parts 3 and 4: For part 3, notice that $a \mid b$ implies that $ad = b$ and $b \mid c$ implies that $be = c$ for elements d and e of D. Then $ade = be = c$ shows that $a(de) = c$ and a divides c. For part 4, see that $b \mid c$ and $c \mid b$ implies that $bd = c$ and $ce = b$ for d and e in D. Then $bde = ce = b$, which, by the cancellation property, implies that $de = 1$; this shows that both d and e are factors of 1 and hence units.

Definition. *Let p be an element of an integral domain that is not a unit. If its only factors are units and unit multiples of itself, we call p a* prime *element. A nonzero element that is neither a unit nor a prime element is called* composite. (Thus an integral domain consists of four mutually exclusive kinds of elements: 0, units, prime elements, and composite elements.)

Note what this means in various kinds of domains. If $D = \mathbf{Z}$, then its prime elements are the prime numbers. (At this stage we allow prime numbers to be negative; later we may restrict them to be positive.) If $D = \mathbf{Q}[t]$, the prime elements, that is, the prime polynomials, are those polynomials of positive degree with rational coefficients that have no factors of lower positive degree. For instance, in $\mathbf{Q}[t]$ the polynomial $t^2 - 2$ is prime but $t^2 + \frac{11}{12}t + \frac{1}{6} = (t + \frac{1}{4})(t + \frac{2}{3})$ is not prime. However, in $\mathbf{R}[t]$ the polynomial $t^2 - 2$ is composite, since it is a product $(t - \sqrt{2})(t + \sqrt{2})$ of two polynomials of positive degree in $\mathbf{R}[t]$. Similarly, $t^2 + 1$ is prime in $\mathbf{R}[t]$ but not in $\mathbf{C}[t]$. Then whether or not a polynomial is prime depends on the field or integral domain in which its coefficients lie. To repeat, the polynomial $t^2 - 2$ is prime over the rational field or the set of integers but not over the field of real numbers. Prime polynomials are also called *irreducible* polynomials and composite polynomials are called *reducible*. Just as it is difficult except in special cases to determine whether a given integer is a prime, so it is in general hard to find whether or not a given polynomial is prime. An interesting question whose answer we postpone until Section 13 is: If a polynomial is prime in $\mathbf{Z}[t]$, is it also prime in $\mathbf{Q}[t]$?

We know that any integer can be expressed as a product of prime numbers and the same is true for polynomials. Instead of proving this separately for the two systems, we shall in Section 9 find a common ground that can serve as a basis for a common proof.

9. Euclidean Domains

In Section 10 of Chapter I and elsewhere we have assumed that one can divide an integer by another integer and get a quotient and a remainder that is nonnegative and less than the divisor. Since this is intuitively evident, one could well take this as a property or postulate for the set of integers, just as we could have taken the induction principle as a postulate. They are both perhaps more evident intuitively than the well-ordering property. But here, as for induction, there is an advantage in deducing the divisibility property from the well-ordering property, since by so doing we have one basic assumption instead of two. Thus we state formally the divisibility property and show that it follows from the postulate that the set of positive integers is a well-ordered set.

Theorem 9.1.   If a and b are two integers with b positive, then there exist unique integers q and r such that

$$(9.1) \qquad\qquad a = qb + r, \qquad 0 \le r < b.$$

Proof: Let us first explore how the intuitive meaning of (9.1) can lead us to a proof. The number q in (9.1) is the greatest multiple of b that is less than or equal to a. That is, q is the greatest integer such that $a - qb$ is nonnegative. To fit this to the well-ordering postulate, we need to state this in terms of a minimum instead of a maximum. An equivalent way to state the same property of q is to say that $b(q + 1)$ is the minimum multiple of b which is greater than a, that is, for which $a - (q + 1)b$ is negative. Now we can begin the proof.

Let S be the set of all integers x such that $a - bx < 0$. Now all such x are greater than a/b. From Exercise 12 (answered in the affirmative) of Section 7, S is a well-ordered set. It is not empty, since $x = |a| + 1$ is in S. So, it has a least element; call it $x = q + 1$. Thus

$$a - (q + 1)b < 0 \qquad \text{and} \qquad a - qb \ge 0.$$

Hence if $r = a - qb$, we have $r < b$ and $r \ge 0$.

To show uniqueness, suppose that $a = qb + r = q'b + r'$ with $0 \le r \le r' < b$. Then $(q' - q)b + r' - r = 0$ and $r' - r$ is a multiple of b. But $r' - r$ is less than b and nonnegative, which shows that $r' - r = 0$. Hence $r' = r$ and $q' = q$. This completes the proof.

Another form of (9.1) can be obtained by dividing by b to get

$$\frac{a}{b} = q + \frac{r}{b}, \qquad \text{that is,} \quad 0 \le \frac{a}{b} - q < 1.$$

This affirms that for any rational number a/b there is an integer q such that $a/b - q$ is nonnegative and less than 1. This result can be modified as you are asked to do in Exercise 8 after Section 9: If a/b is any rational number, there exists an integer q such that

$$\left| \frac{a}{b} - q \right| \le \frac{1}{2}.$$

It is clear intuitively that these results also hold if a/b is replaced by any real number.

The companion theorem for polynomials is as follows.

Theorem 9.1p.   Let $f(x)$ and $g(x)$ be two polynomials in $F[x]$ for some *field F* (a mere integral domain does not suffice) and if $g(x) \ne 0$, then there are unique polynomials $q(x)$ and $r(x)$ such that

(9.1p) $f(x) = q(x)g(x) + r(x), \qquad \text{where } r(x) = 0$

or

$$r(x) \text{ is of lesser degree than } g(x).$$

Before proving this, let us recall for a particular example how the division process works. Let $f(x) = x^4 + x^3 + x - 2$ and $g(x) = 2x^3 + 2x - 3$. Then

$$
\begin{array}{r|l|l}
2x^3 + 2x - 3 & x^4 + x^3 + 0 \ + \ x - 2 & \tfrac{1}{2}x + \tfrac{1}{2} \\
& \underline{x^4 \qquad\quad + x^2 - \tfrac{3}{2}x} & \\
& x^3 - x^2 + \tfrac{5}{2}x - 2 & \\
& \underline{x^3 \qquad + \ x - \tfrac{3}{2}} & \\
& -x^2 + \tfrac{3}{2}x - \tfrac{1}{2}. &
\end{array}
$$

So in this case $q(x)$ of (9.1p) is $\frac{1}{2}x + \frac{1}{2}$ and $r(x) = -x^2 + \frac{3}{2}x - \frac{1}{2}$.

We cannot base this proof on the well-ordering postulate for polynomials, since the set of polynomials with coefficients in a field need not even be ordered. But we can base our proof on the well-ordering postulate for the integers that are the degrees of polynomials. If the degree of f is less than that of g, we can merely take $r = f$ and $q = 0$. We assume that

(9.2)
$$f(x) = a_n x^n + a_{n-1} x^{n-1} + \cdots + a_1 x + a_0,$$
$$g(x) = b_s x^s + b_{s-1} x^{s-1} + \cdots + b_1 x + b_0,$$

where a_n and b_s are both different from zero and $n \geq s$. What we need to do is to divide f by g and get a remainder whose degree is less than that of g. Now we show in two ways how the proof may be completed; the second uses mathematical induction.

Method 1.   The first step in the process of division is to find some multiple of g whose leading term is the same as that of f, with the idea that we can then subtract to get a polynomial of lower degree than n. Now the leading term of $(a_n/b_s)x^{n-s}g(x)$ is $a_n x^n$ so that

(9.3)
$$f(x) - \left(\frac{a_n}{b_s}\right)x^{n-s}g(x) + h(x),$$

which is of degree $n - 1$ or less. Then if $n > 1$, we repeat the same process with $h(x)$ in place of $f(x)$ to get $h(x) - k(x)g(x) = h'(x)$, of degree lower than $h(x)$. That is,

$$f(x) - \left[\left(\frac{a_n}{b_s}\right)x^{n-s} + k(x)\right]g(x) = h'(x),$$

where $h'(x)$ has degree not greater than $n - 2$. Continuing this process completes the proof except for uniqueness.

Method 2.   If we are to use mathematical induction on n beginning with $n = 0$, we must first check the theorem for $n = 0$. In this case $f(x) = a_n$, $g(x) = b_s$, and $q(x) = a_n/b_s$ with $r(x) = 0$.

Now we suppose that the theorem holds for all polynomials of degree less than n; $h(x)$ in (9.3) is of degree less than n. Hence by the hypothesis of the induction, we may take $h(x) = q'(x)g(x) + r(x)$, where $r(x) = 0$ or $r(x)$ is of degree less than $g(x)$. When we replace $h(x)$ in (9.3) by this expression, we have

$$f(x) = \left(\frac{a_n}{b_s}\right)x^{n-s}g(x) + q'(x)g(x) + r(x).$$

Then equation (9.1p) holds with $q(x) = (a_n/b_s)x^{n-s} + q'(x)$, and the proof is complete except for uniqueness.

By either of these methods we have proved the theorem except for uniqueness. To prove this, assume that $f(x) = q'(x)g(x) + r'(x)$. Then, subtract this from the other expression for $f(x)$ in (9.1p); we have

(9.4)
$$0 = [q(x) - q'(x)]g(x) + r(x) - r'(x).$$

The degree of $r(x) - r'(x)$ is less than that of $g(x)$, while the rest of the right side of (9.4) is of degree at least that of $g(x)$, unless $q(x) - q'(x)$ is zero. So $q(x) = q'(x)$ and (9.4) implies that $r(x) = r'(x)$, which completes the proof.

If we look back at the two methods of proof, we see what is usually the situation for induction proofs. One can often get along without them by a judicious use of "and so on" or "in this manner." It is usually clear what this means, but the argument is, nevertheless, somewhat fuzzy. The method by mathematical induction is much more clear cut. Also, note again that since we had to divide by b_s, it is necessary to take the coefficients to be in a field.

Theorems 9.1 and 9.1p are very similar as well as being very basic. There seemed no practical way to avoid giving separate proofs for them. But now that we have them, it will be more efficient if we can devise some method of proceeding from this point without having to treat integers and polynomials separately. We really need properties of the set of integers for both, but for polynomials the integers involved are the degrees of the polynomials. We can combine these two properties, give a name to a system that has them, and from this point on for awhile, deal with polynomials and integers together. Furthermore, we shall see that integers and polynomials are not the only systems that come under this category.

Definition.   *An integral domain D is said to be a* Euclidean domain *if for every $d \neq 0$ in D, there is defined a nonnegative integer $n(d)$ such that*

1. *If x and y are nonzero elements of D, $n(x) \leq n(xy)$.*
2. *If x and y are nonzero elements of D, there exist q and r in D such that $x = qy + r$, where either $r = 0$ or $n(r) < n(y)$.*

Call $n(s)$ the n-value† of s. Note that $n(0)$ is not defined.

Theorem 9.1 implies that $\mathbf{Z}$ is a Euclidean domain, where $n(x) = |x|$, the absolute value of x. Theorem 9.1p implies that $F[x]$ is a Euclidean domain, where $n(f)$ is the degree of the polynomial $f(x)$. (Notice that $F[x]$ is an example of a Euclidean domain that is not ordered unless F is ordered.) Now the units in $\mathbf{Z}$ have n-value 1, and those in $F[x]$ all have n-value 0; that is, in each case all the units have the same n-value and all nonunits have greater n-value. This suggests the following theorem.

Theorem 9.2.   In a Euclidean domain, all units have the same n-value as $n(1)$, and all nonunits have greater n-value than $n(1)$.

---

† We could use the term "norm" instead of n-value. But the usual use of norm has to do with vectors or algebraic numbers. What we here call the n-value is closely allied to the idea of a valuation. Valuations are beyond the scope of this book, but the interested reader can find an introduction to this concept in Chapter X of van der Waerden's *Modern Algebra*, vol. I.

Proof: Suppose that a is a unit and that $aq = 1$. Then, from the properties of a Euclidean domain,

$$n(a) \le n(aq) = n(1) \le n(1 \cdot a) = n(a),$$

which shows that $n(a) = n(1)$.

If b is not a unit, then $1 = qb + r$ with $r \ne 0$. Hence $n(r) < n(b)$. But $n(1) \le n(1 \cdot r) = n(r)$ implies that $n(1) < n(b)$, which completes the proof.

Theorem 9.3.   Let a and b be two elements of a Euclidean domain E, not both zero, and S the set $\{ax + by\}$ for all x and y in E. If d is an element of S for which $n(d)$ is a minimum, then d is a factor of all elements of S. (This is an extension of Theorem 10.2 of Chapter I.)

Proof: (Compare the proof of Theorem 10.2 of Chapter I.) First notice that there is a d for which $n(d)$ is a minimum, since the n-values of the elements of E form a well-ordered set, being nonnegative integers. Now let c be any element of S. Using the definition of a Euclidean domain, we see that elements q and r of E exist such that $c = qd + r$, where $r = 0$ or $n(r) < n(d)$. But r is in S, since $c = ax_0 + by_0$ and $d = ax_1 + by_1$ imply that

$$r = a(x_0 - qx_1) + b(y_0 - qy_1).$$

But $n(d)$ is a minimum, and hence $r = 0$ and d is a factor of c.

The two following corollaries are special cases of Theorem 9.3. It is these special cases which we use most.

Corollary 1.   If a and b are two integers, not both zero, there is an integer d in the set $S = \{ax + by\}$, for x and y integers, whose absolute value is a minimum. Then d is a factor of all elements of S.

Corollary 2.   If $a(x)$ and $b(x)$ are two polynomials with coefficients in a field F where not both of the polynomials are zero, then there is a polynomial of least degree $d(x)$ in the set $S = \{a(x)h(x) + b(x)k(x)\}$ for $h(x)$ and $k(x)$ in $F[x]$. Then $d(x)$ divides all polynomials of S.

Another consequence of the above is that we can replace the $\le$ in property 1 of a Euclidean domain by $<$ if y is not a unit. This is affirmed by the following theorem.

Theorem 9.4.   In a Euclidean domain E, if c is not a unit, then for each b in E, $n(b) < n(cb)$.

Proof: Let $S = \{bx\}$, where x is in E, the Euclidean domain. Then $n(b) \le n(s)$ for all elements of S. Suppose that c is not a unit but $n(cb) = n(b)$.

Then $n(cb)$ is also a minimum for S. Hence by the previous theorem all elements of S are multiples of cb. Thus for some y in S, $cby = b$, which implies that $cy = 1$ and hence that c is a unit. This contradiction proves the theorem.

Now we give an example of a Euclidean domain that is neither the set of integers nor a set of polynomials.

Theorem 9.5.   The set of Gaussian integers $S = \{a + bi\}$, where a and b are integers and $i^2 = -1$, form a Euclidean domain.

Proof: First, S is an integral domain. To show this, since S is a subset of the complex numbers, we need merely show that S is closed under addition, subtraction, multiplication, and division except by 0. This you were asked to show in Exercise 3 after Section 5.

To show that E is Euclidean, we let $n(a + bi) = a^2 + b^2$, the square of the absolute value of $a + bi$. Recall that if $s + ti$ is any complex number, where s and t are real (not necessarily integers), the absolute value of $s + ti$ is $\sqrt{s^2 + t^2}$, written $|s + ti|$. Alternatively, a convenient notation is

$$\|s + ti\| = |s + ti|^2 = s^2 + t^2,$$

which is called the *norm* of the complex number. That is, the norm of a complex number is the product of the number and its complex conjugate.

Property 1 holds, since

$$n[(a + bi)(c + di)] = n[ac - bd + i(bc + ad)] = (ac - bd)^2 + (bc + ad)^2$$
$$= (a^2 + b^2)(c^2 + d^2) = n(a + bi)n(c + di),$$

and every $a + bi$ in S has an absolute value of at least 1, except $0 + 0i = 0$, for which the n-value is not defined.

To see Property 2, let x and y be two elements of S. We want to show that there exist elements q and r of S such that

(9.5) $x = qy + r$,   with $r = 0$ or $n(r) < n(y)$.

Translating this into the terminology of norm and absolute value, we see that the condition on r is either that $r = 0$ or that $\|r\| < \|y\|$, that is, $|r| < |y|$. Then (9.5) is equivalent to

$$\frac{x}{y} = q + \frac{r}{y}, \qquad 0 \le \left|\frac{r}{y}\right| < 1,$$

that is,

$$\left|\frac{x}{y} - q\right| < 1.$$

Now if $x = a + bi$ and $y = c + di$, for integers, a, b, c, d with c and d not both zero, we have $x/y = s + ti$ for s and t rational numbers. Hence we need to show that if s and t are in $\mathbf{Q}$, there exist integers u and w in $\mathbf{Z}$ such that $|s + ti - (u + wi)| < 1$. That is, if s and t are rational numbers, we want to show the existence of integers u and w such that

$$(9.6) \qquad\qquad (s - u)^2 + (t - w)^2 < 1.$$

Now it is fairly obvious intuitively and not hard to prove (left as Exercise 8) that for any rational number there is an integer which differs from it by not more than $\frac{1}{2}$. Hence we choose integers u and w so that

$$|s - u| \leq \tfrac{1}{2} \qquad \text{and} \qquad |t - w| \leq \tfrac{1}{2}.$$

This with $(\frac{1}{2})^2 + (\frac{1}{2})^2 = \frac{1}{2} < 1$ proves the theorem.

Note what (9.6) means geometrically. It affirms that if P is any point (s, t) whose coordinates are rational numbers, there is a point $Q = (u, w)$, where u and w are integers, such that P lies strictly within a circle with center at Q and radius 1. Actually, from the proof it may be seen that Q may be chosen so that P will be on or inside the circle with radius $\frac{1}{2}$ and center Q.

Not all sets $\{x + y\sqrt{m}\}$ for x and y integers and m a nonsquare integer are integral domains; in fact, few are. As a step toward showing this, you are asked to show in Exercise 13 at the close of this section that the method of proof of Theorem 9.5 breaks down for the domain $D = \{x + y\sqrt{-5}\}$. Although this is an indication, it is not conclusive, since conceivably for some other definition of n-value one could show the property of being Euclidean to be valid. At the end of Section 11 we shall show conclusively that $\{x + y\sqrt{-5}\}$ is not Euclidean by showing that if it were, one could deduce something that is false. This is discussed further in Sections 11 and 18.

Exercises

1. For what fields $\mathbf{Z}_p$ is $x^3 + 2x^2 + 2x + 4$ divisible by $x^2 + x + 1$?

2. Is there a field $\mathbf{Z}_p$ in which $x^3 + 4x^2 + 4x + 2$ is divisible by $x^2 + x + 1$?

3. In what fields $\mathbf{Z}_p$ is $x^2 + 1$ a factor of $x^3 + x^2 + 22x + 15$?

4. Find necessary and sufficient conditions on integers a, b, and c such that $x^3 + ax^2 + bx + c$ be divisible by $x^2 + 1$ in $\mathbf{Z}_p$.

5. Let a and b be two elements of a Euclidean domain whose n-values are the same. Does this imply that each of a and b is a factor of the other?

6. Prove parts 1 and 2 of Theorem 8.1.

7. Use Theorem 9.1 or otherwise to prove that if a and b are integers with b positive, there exist unique integers q and r such that $a = bq + r$ and $-b < 2r \leq b$.

8. Use Exercise 7 or other means to prove that if a is a rational number, then there is an integer m such that

$$|a - m| \leq \tfrac{1}{2}.$$

9. Let a and b be positive integers with $a > b$. Show that there is an integer c such that $a/b = c + 1/x$, where x is a rational number not less than 1.

10. Find the units of the Gaussian integers.

11. Theorem 9.1p is stated for polynomials whose coefficients are in a field. Suppose that we restrict the polynomials to have integer coefficients. Show that the theorem is false. How can it be modified to become true?

12. Let S be the set $\{a + b\sqrt{-2}\}$, where a and b are integers. Define the n-value of $a + b\sqrt{-2}$ to be $a^2 + 2b^2$, and show that S is a Euclidean domain. Find its units. Is it an ordered domain?

13. Let $S = \{a + b\sqrt{-5}\}$, where a and b are integers, and define the n-value of $a + b\sqrt{-5}$ to be $x^2 + 5y^2$, for x and y rational numbers. Then show that there are rational numbers s and t such that $n(s + t\sqrt{-5} - [u + w\sqrt{-5}]) < 1$ is false for all integers u and w.

*14. Let $w = \tfrac{1}{2}(-1 + \sqrt{-3})$ and consider $S = x + wy$, where x and y are integers. You were asked to show in Exercise 5 of Section 5 that S is an integral domain. Prove that it is Euclidean.

10. Greatest Common Divisor

Looking at Theorem 9.3, we see that if a and b are elements of a Euclidean domain D that are not both zero, then there is an element d that divides both a and b and is also of the form $ax + by$ for x and y in D. Then this element d has two properties:

1. The element d is a factor of both a and b.
2. If g is a factor of both a and b, it is a factor of d.

The latter follows from $d = ax_0 + by_0$ for some x_0 and y_0 in E. Any element that has these two properties we call a *greatest common divisor* (g.c.d.) of a and b. This is a little different from the definition given in Section 10 of Chapter I, for there is nothing said about d's being positive or about its being unique. In fact, the Euclidean domain may not be ordered; in this case there are no positive elements, but still a greatest common divisor can exist.

Suppose that g and g' are two greatest common divisors of a and b. Then by the two properties it follows that each must divide the other. Thus, by Theorem 8.1, their ratio is a unit; that is, $g = ug'$ for some unit u.

We can achieve uniqueness of the g.c.d. for integers and polynomials by imposing another condition. Since 1 and −1 are the units in **Z**, we can make the g.c.d. unique by requiring that it be positive. Since the nonzero elements of F are the units in $F[x]$, we can make the g.c.d. unique by requiring that the polynomial have leading coefficient 1. In either of these cases when we speak of *the* greatest common divisor, it is understood that the extra condition is imposed. For the Gaussian integers, $a + bi$, there are four units and achieving uniqueness is probably not worth the trouble.

Theorem 9.3 shows the existence of a g.c.d. but does not give a very good way of finding it. There is a process, called the *Euclidean algorithm*, which provides a practical way of computing a greatest common divisor of any pair of elements in a Euclidean domain. Before stating and proving the general theorem, we should illustrate the process for a pair of integers. The method consists of using (9.1) again and again. So, here is a calculation that yields the greatest common divisor of 589 and 133:

$$589 = 4 \cdot 133 + 57,$$
$$133 = 2 \cdot 57 \ + 19,$$
$$57 = 3 \cdot 19 \ + \ 0.$$

To prove that 19 is the greatest common divisor of 589 and 133, we may check in turn the two properties. First, to show that 19 is a factor of both 589 and 133, note that the third equation above shows that it is a factor of 57, the second equation that it is a factor of 133, and the first that is a factor of 589. Second, suppose that g is any common factor of 589 and 133; then the first equation shows that it is a factor of 57 and the second equation that it is a factor of 19. Thus 19 has property 2 of the g.c.d. In considering the proof of the following theorem, you may want to see what it means in terms of the illustration that we have just given.

Theorem 10.1.   If we have a Euclidean domain E and two nonzero elements of E, a and b, the following process (called the *Euclidean algorithm*) gives the greatest common divisor:

$$a = q_1 b + r_1 \qquad r_1 \neq 0 \text{ and } n(r_1) < n(b),$$
$$b = q_2 r_1 + r_2 \qquad r_2 \neq 0 \text{ and } n(r_2) < n(r_1),$$
$$r_1 = q_3 r_2 + r_3 \qquad r_3 \neq 0 \text{ and } n(r_3) < n(r_2),$$
$$\cdots\cdots\cdots\cdots\cdots\cdots\cdots\cdots\cdots\cdots$$
$$r_{k-2} = q_k r_{k-1} + r_k \qquad r_k \neq 0 \text{ and } n(r_k) < n(r_{k-1}),$$
$$r_{k-1} = q_{k+1} r_k + 0.$$

Then r_k is the greatest common divisor of a and b. ($r_0 = b$.)

Proof: First, the process must stop because by the definition of a Euclidean domain, each time r_i is either zero or the inequality holds. Then $n(b)$,

$n(r_1), n(r_2), \ldots$ is a decreasing sequence of positive integers. By the well-ordering postulate for integers, this must have a least element, in this case $n(r_k)$. Therefore, in the next step the remainder must be zero.

Now we want to show that r_k is a greatest common divisor. We must check in turn the two properties of a g.c.d. The last equation shows that r_k divides r_{k-1}. In the next-to-last equation r_k is a factor of both terms on the right and hence, by part 2 of Theorem 8.1, is a factor of r_{k-2}. Continuing in this fashion upward, we see that in the second equation r_k divides b and r_1; the first equation shows that r_k divides a as well. So, r_k satisfies the first condition.

For the other condition, suppose that g is a factor of both a and b. Then $r_1 = a - q_1 b$ implies that g is a factor of r_1. The second equation shows that g is a factor of r_2 and so on down to the next-to-last equation, which shows that g is a factor of r_k. Thus both requirements are satisfied, and we have proved what we wanted.

Notice that by use of the concept of a Euclidean domain we have given one proof which suffices to establish the algorithm not only for integers but for polynomials as well. Indeed, in view of Theorem 9.5, the algorithm also gives the g.c.d. for Gaussian integers.

We found in the case of polynomials that a prime polynomial in one field is not necessarily prime in a larger field. It is curious that enlarging the integral domain does not change the greatest common divisor. Thus we have the following result, which we shall find useful later.

Theorem 10.2. If a and b are in a Euclidean domain E and have g as a greatest common divisor, then g is also a g.c.d. of a and b in any Euclidean domain E' that contains E.

Proof: Let g' be a greatest common divisor of a and b in $E' \supset E$. Then $ax + by = g$ for some x and y in E and $ax' + by' = g'$ for some x' and y' in E'. Now g divides a and b in E and hence in E'; since g' is a greatest common divisor in E', it follows that g divides g'. On the other hand, g' is a factor of both a and b, since it is a greatest common divisor; hence from $ax + by = g$ it follows that g' divides g. Thus each of g and g' divides the other and they differ by a unit factor. This completes the proof.

If two elements of a Euclidean domain have 1 as a greatest common divisor, they are called *relatively prime* or *co-prime*. Of course, the identity element of an integral domain is always a unit, and if the greatest common divisor of two elements is a unit, then 1 is also a greatest common divisor. Notice that a unit of a Euclidean domain is not only a factor of all elements of the domain but is relatively prime to them as well.

Corollary 1. If $f(x)$ and $g(x)$ are in $F[x]$ for some field F and are relatively prime in $F[x]$, then they are relatively prime in $F'[x]$ for every field F' that contains F. Conversely, if they are relatively prime in $F'[x]$, they are also in $F[x]$.

Corollary 2. Let $f(x)$ be a polynomial that is irreducible in $F[x]$ and let $h(x)$ be any polynomial in $F[x]$. Then if F' is any field that contains F (including F itself), one of two things can happen (but not both):

1. The polynomials $f(x)$ and $h(x)$ are relatively prime both in $F[x]$ and $F'[x]$.
2. The polynomial $f(x)$ is a factor of $h(x)$ in $F[x]$.

Some comments on corollaries 1 and 2 are in order. Notice that in the former, one of the fields must be contained in the other; otherwise the corollary can be false. For instance, if $F = \mathbf{Q}$ and $F' = \mathbf{Z}_7$, then $f(x) = x^2 - 2$ and $h(x) = x^2 - 5x + 6$ are in both $\mathbf{Q}[x]$ and $\mathbf{Z}_7[x]$, the sets of polynomials in the respective fields. Then $f(x)$ and $h(x)$ are relatively prime in $\mathbf{Q}[x]$, since $x^2 - 2$ is prime in $\mathbf{Q}[x]$, but are not relatively prime in $\mathbf{Z}_7[x]$, since there $f(x) = (x - 3)(x + 3)$ and $h(x) = (x - 2)(x - 3)$. There is, in fact, no field that contains both $\mathbf{Q}$ and $\mathbf{Z}_7$, since the characteristics are different. However, if there should be a field F'' that contains both F and F' and if the polynomial is in both $F[x]$ and $F'[x]$, then Corollary 1 does hold because then being relatively prime in $F[x]$ implies it for $F''[x]$, which, in turn, implies it for $F'[x]$.

For Corollary 2 we should notice that if $f(x)$ and $h(x)$ are not relatively prime in $F[x]$, there is a common factor $g(x)$ in $F[x]$ of positive degree. Then $g(x)$ is a factor of $f(x)$ and since the latter is irreducible, $g(x)u = f(x)$ for some unit u, that is, for some nonzero element u of F. But $g(x)$ also divides $h(x)$, and hence $f(x)$ divides $h(x)$ as well.

It is worthwhile to state Theorem 9.3 using the idea of greatest common divisor. We follow this by the corresponding corollaries.

Theorem 10.3. Let a and b be two elements of a Euclidean domain E, not both zero. If g is a greatest common divisor of a and b, there are elements x and y of E such that $ax + by = g$.

Corollary 1. If a and b are two integers, not both zero, and if g is a greatest common divisor of a and b, then there are integers x and y such that $ax + by = g$. (See Corollary 1 of Theorem 10.2 in Chapter I.)

Corollary 2. If $f(x)$ and $h(x)$ are polynomials in $F[x]$ and if $g(x)$ is a g.c.d. of f and h, then there are polynomials $u(x)$ and $v(x)$ such that

$$f(x)u(x) + h(x)v(x) = g(x).$$

There is another result that we need for the fundamental theorem of the following section.

Theorem 10.4. If a, b, and c are elements of a Euclidean domain, if a and b are relatively prime and if a is a factor of bc, then a is a factor of c.

Proof: Since a and b are relative prime, there are elements x and y of D, the Euclidean domain, such that $ax + by = 1$. Then multiplication by c gives $acx + bcy = c$. But, by hypothesis, a is a factor of bc and hence, by Theorem 8.1, a being a factor of ac and bc implies that a is a factor of c. This completes the proof.

Corollary. If $f(x)$ and $h(x)$ are relatively prime polynomials in $F[x]$ and if $f(x)$ is a factor of $h(x)k(x)$, where $k(x)$ is in $F[x]$, then $f(x)$ is a factor of $k(x)$.

You know from experience that a related concept which is useful is that of the least common multiple (l.c.m.) of two elements. One can define this by paraphrasing our definition of the g.c.d.

Definition. *If a and b are elements of a Euclidean domain, not both zero, an element m is called a* least common multiple *(l.c.m.) of a and b if*

1. *The element m is a multiple of a and b.*
2. *If h is a multiple of both a and b, it is a multiple of m.*

Although there is a kind of algorithm for the computation of the l.c.m., it is rather easier first to compute the g.c.d. and then use the following theorem.

Theorem 10.5. If m is the l.c.m. of two elements a and b of a Euclidean domain and g their g.c.d., then

$$ab = mg.$$

Proof: The proof is left as Exercise 22 after Section 11.

11. Unique Decomposition into Prime Factors

We are familiar with the fact that every integer can be expressed as a product of prime numbers uniquely except for the order of the factors. For instance, $24 = 2 \cdot 2 \cdot 2 \cdot 3$. It can also be written $2 \cdot 3 \cdot 2 \cdot 2$, but however it is written as a product of prime numbers, there will be three 2's and one 3. When we use the idea of a Euclidean domain, we can prove this result all in one operation for integers, polynomials, and any other systems that meet the requirements of a Euclidean domain.

Theorem 11.1.   Let E be any Euclidean domain and c an element of E, not a unit. Then c can be written as

$$c = up_1p_2p_3 \cdots p_k,$$

where the p_i are prime elements, u is a unit, and the product is unique except for the order of the factors and unit multiples. That is, if c is also written $q_1q_2q_3 \cdots q_s$, then $s = k$; by reordering the subscripts, we can make $q_i = u_ip_i$, where each u_i is a unit.

Proof: Since this is a very fundamental theorem, we want to be sure to have a clean-cut proof. Hence we give one by induction on the integer $n(c)$, the n-value of c. First, if $n(c)$ is a minimum for elements of E, from Theorem 9.2, c is a unit and there is nothing to prove. Now we assume for the induction that the theorem holds for *all* elements of E whose n-value is less than $n(c)$.

First, we want to show that c can be expressed as a product of primes if it is not a unit. If c is a prime element, the theorem holds trivially. If c is not a prime, it can be expressed as $c = bd$, where neither b nor d is a unit. By Theorem 9.4, $n(b) < n(c)$ and $n(d) < n(c)$. So, by the induction hypothesis b and d can be expressed as products of prime elements, and hence c is a product of prime elements. Notice that in this part of the proof we had to assume the theorem for *all* smaller n-values—not just for the previous one. (See Exercise 16 after Section 7.)

Second, to show that the decomposition into prime factors is unique except for the order and unit factors, suppose that

(11.1) $$c = up_1p_2p_3 \cdots p_k = wq_1q_2q_3 \cdots q_s,$$

where u and w are units and the p's and q's are prime elements. The first equation of (11.1) shows that p_1 divides c, and hence, from the second equation, p_1 divides the product of the q's. If p_1 does not divide q_1, then it must be relatively prime to q_1, since the only possible common factors are units. Then, by Theorem 10.4, p_1 must divide $q_2q_3q_4 \cdots q_s$. We can repeat this process, decreasing the number of q's in the product each time until we come to a q that p_1 does divide. Then renumber the q's so that the multiple of p_1 is q_1. In this case, since q_1 is a prime, $q_1 = vp_1$ for some unit v. Then $n(c/q_1) < n(c)$, and hence, by the induction hypothesis, c/q_1 is expressible uniquely, except for order and unit multiples, as a product of primes and a unit u'. Thus

$$\frac{c}{q_1} = wq_2q_3 \cdots q_s = u'p_2p_3 \cdots p_k,$$

where the q's are the same as the p's in some order except for unit factors. This shows that $k = s$ and that, after a reordering of the subscripts, $p_iu_i = q_i$, where u_i are units. Thus the theorem is proved.

In **Z**, the set of integers, the prime elements are usually restricted to be positive integers, and in $F[x]$ we can avoid the ambiguity of unit multiples by requiring prime polynomials to have leading coefficient 1. Then the respective corollaries are as follows.

Corollary 1. (Fundamental theorem of arithmetic.) If $E = \mathbf{Z}$ and primes are restricted to be positive integers, then any integer m, except 0, 1, or -1, can be expressed in the form

$$m = \pm p_1 p_2 p_3 \cdots p_k,$$

uniquely except for the order of the primes p_i.

Corollary 2. If $E = F[x]$ and prime polynomials are restricted to have leading coefficient 1, then every polynomial $f(x)$ of positive degree can be expressed in the form

$$f(x) = c p_1(x) p_2(x) \cdots p_k(x)$$

uniquely except for the order of the prime polynomials $p_i(x)$ and c is in F.

In practice, for integers it is often simpler to group together equal primes and write

$$m = p_1^{a_1} p_2^{a_2} \cdots p_r^{a_r},$$

where the primes are distinct (perhaps in increasing order) and the exponents are positive integers. There is a similar expression for polynomials.

You should not forget that polynomials which are prime over one field are not necessarily so over a larger one. Hence the unique decomposition is only true within a particular field. Notice, too, that we have shown that in *any* Euclidean domain the above theorem holds. For instance, it holds for Gaussian integers, $a + bi$. The key theorem is 9.3. Any integral domain for which this theorem holds has unique decomposition into prime factors. It has been shown [T. Motzkin, *Bulletin American Mathematical Society*, vol. 35 (1949), pp. 1142–1146] that there are integral domains for which unique decomposition into prime factors holds but which are not Euclidean. For elaboration on this point see Section 18.

We noticed in Exercise 13 after Section 9 that the integral domain $W = \{x + y\sqrt{-5}\}$, where x and y are integers, appears not to be Euclidean. At least the n-value we tried did not yield the inequality desired. Now we can show that W is not Euclidean; that is, there is *no* n-value that satisfies the requirements of a Euclidean domain. To show this, suppose that W were an integral domain for some n-value. The Theorem 11.1 would affirm that decomposition into prime factors is unique except for the order of the factors and units. Therefore, in order to show that W is not Euclidean, it suffices to find some element w of W such that $w = ab = cd$, where a, b, c, and d are prime elements of W and a is not a unit multiple of either c or d. We state this as a theorem and prove most of it.

Theorem 11.2. In the integral domain $W = \{x + y\sqrt{-5}\}$ with x and y integers, $6 = 2 \cdot 3 = (1 + \sqrt{-5})(1 - \sqrt{-5})$, where 2, 3, $1 + \sqrt{-5}$, and $1 - \sqrt{-5}$ are all prime numbers of W and 2 is not a unit multiple of either $1 + \sqrt{-5}$ or $1 - \sqrt{-5}$.

Proof: Suppose that $a + b\sqrt{-5} = (x + y\sqrt{-5})(z + t\sqrt{-5})$. Then $a - b\sqrt{-5} = (x - y\sqrt{-5})(z - t\sqrt{-5})$ and multiplication shows that

$$(11.2) \qquad a^2 + 5b^2 = (x^2 + 5y^2)(z^2 + 5t^2).$$

If $a = 2$ and $b = 0$, (11.2) becomes $4 = (x^2 + 5y^2)(z^2 + 5t^2)$. Since $2 = x^2 + 5y^2$ is not solvable in integers, one of the indicated factors of 4 must be 4 and the other 1. If we take $4 = x^2 + 5y^2$, it is clear that the only possible solutions are $x = \pm 2$ and $y = 0$. Hence 2 has no factors except 1, -1, 2, and -2; that is, 2 is a prime element of W. This shows also that 2 is not a unit multiple of either $1 + \sqrt{-5}$ or $1 - \sqrt{-5}$. It is almost obvious that it is not a factor either. Also, 2 is not a unit. This almost completes the proof. (See Exercise 23.)

The failure of unique decomposition for many integral domains is a serious defect. We shall show how this can be dealt with in Chapter V.

Exercises

1. Find the greatest common divisor of each pair of integers.
   (a) 189 and 301.   (b) 1261 and 1649.

2. Find the greatest common divisor of each pair of polynomials.
   (a) $x^4 + x^2 + 1$ and $x^3 + 3x^2 + 3x + 2$.
   (b) $x^4 + 4$ and $2x^3 + x^2 - 2x - 6$.

3. Find a g.c.d. of
   (a) $2 + 11i$ and $1 + 3i$.   (b) $-4 + 7i$ and $1 + 7i$.

4. Express each of the numbers in Exercise 1 as a product of prime numbers.

5. Express each of the polynomials of Exercise 2 as a product of prime polynomials (1) in **Q**, (2) in **R**, (3) in **C**, and (4) in $\mathbf{Z}_3$.

6. Let a and b be two integers different from zero which are relatively prime. Prove that $ax + by = c$ is solvable in integers if and only if c is a multiple of the g.c.d. of a and b.

7. Express $x^4 - x^2 - 2$ as a product of prime polynomials in **Q**, **R**, **C**, and $\mathbf{Z}_5$.

8. Let a and b be two integers different from zero that are relatively prime. Let x_0 and y_0 be one pair of integer solutions of $ax + by = c$, for some

integer c. Prove that all solutions of $ax + by = c$ are given by $x = x_0 + kb$, $y = y_0 - ka$, for integers k.

9. If a and b are two nonzero relatively prime integers, show that there are integers x_0 and y_0 such that $ax_0 + by_0 = 1$ and $0 \le x_0 < b$.

10. Let f and g be two polynomials with rational coefficients such that $f(c) = g(c) = 0$ for some real number c. Prove that f and g have a common factor h of positive degree with rational coefficients.

11. Let $f(x)$, $h(x)$, and $k(x)$ be three polynomials in $F[x]$. Prove that there are polynomials $r(x)$ and $s(x)$ in $F[x]$ such that

$$f(x)r(x) + h(x)s(x) = k(x)$$

if and only if $k(x)$ is a multiple of the g.c.d. of $f(x)$ and $h(x)$.

12. Let $f(x)$ and $h(x)$ be two relatively prime polynomials in $F[x]$, and let $r_0(x)$ and $s_0(x)$ be a solution of $f(x)r(x) + h(x)s(x) = 1$. Prove that all solutions are given by $r(x) = r_0(x) + t(x)h(x)$, $s(x) = s_0(x) - t(x)g(x)$, where $t(x)$ ranges over the elements of $F[x]$.

13. If $f(x)$ and $h(x)$ are two relatively prime polynomials of $F[x]$, show that there are polynomials $r(x)$ and $s(x)$, where $r(x)$ has degree less than $f(x)$ and $s(x)$ has degree less than $h(x)$, such that

$$\frac{r(x)}{f(x)} + \frac{s(x)}{h(x)} = \frac{1}{f(x)h(x)}.$$

(This is a result in partial fractions that is useful in calculus.)

14. Let $f(x)$ and $g(x)$ be two polynomials in $D[x]$ for some integral domain D and let the degree of $f(x)$ be equal to or greater than that of $g(x)$. Show that there are polynomials $h(x)$ and $k(x)$, where $k(x)$ is 0 or is of lower degree than $g(x)$, such that

$$\frac{f(x)}{g(x)} = h(x) + \frac{k(x)}{g(x)}.$$

15. Give a proof of Theorem 10.1 using mathematical induction.

16. The usual definition of the greatest common divisor of two numbers a and b is: the largest integer that divides both a and b. Prove that any number that has this property is a greatest common divisor by the definition of Section 10.

17. The usual definition of the least common multiple of two numbers a and b is: the smallest positive integer that is a multiple of both a and b. Prove that any number that has this property is a least common multiple by the definition of Section 10.

18. Prove that two elements a and b of a Euclidean domain E are relatively prime if and only if $ax + by = 1$ for some x and y in E.

19. Prove that in a Euclidean domain if a is relatively prime to both b and c, then it is relatively prime to bc. (See Lemma 2 of Section 5 in Chapter II.)

20. Write integers a and b in the form

$$a = p_1^{a_1} p_2^{a_2} p_3^{a_3} \cdots p_k^{a_k} \quad \text{and} \quad b = p_1^{b_1} p_2^{b_2} p_3^{b_3} \cdots p_k^{b_k},$$

where the a's and b's are nonegative integers and the p's are distinct primes. Why can this always be done? Prove that the greatest common divisor of a and b is

$$g = p_1^{c_1} p_2^{c_2} p_3^{c_3} \cdots p_k^{c_k},$$

where each c_i is the smaller of a_i and b_i (or equal to a_i if $a_i = b_i$).

21. Use the factorizations of a and b in Exercise 20 to find a factored expression for their least common multiple.

22. Use Exercises 20 and 21 or by other means prove Theorem 10.5. (There is a proof independent of the previous exercises.)

23. Complete the proof of Theorem 11.2.

24. Prove the Corollary of Theorem 10.4.

12.   *Factorization in F[x] for F, a General Field*

Recall that every $F[x]$ is a Euclidean domain regardless of whether the field is one of the familiar ones like the rational numbers, the real numbers, or the complex numbers, but also for fields of characteristic p, including finite fields. In this section we begin by finding some properties of polynomials over any field, leaving more special fields until Theorem 12.4 and Section 13.

First, we should give a little attention to polynomials as functions. In Section 4 we developed the idea that a polynomial was a more-or-less formal expression of the form

$$f(x) = a_n x^n + a_{n-1} x^{n-1} + \cdots + a_1 x + a_0,$$

where the coefficients a_i are in some field F and certain common rules of addition and multiplication of polynomials are agreed upon. These rules were set down so that the polynomial $f(x)$ could serve to define a function of x. That is, if we replace x by some element of F, the expression becomes an element of F. In other words, for each c of F, $f(c)$, which we get by replacing x by c, is in F and is uniquely determined by $f(x)$. Furthermore, we chose the rules of addition and multiplication so that $f(c) + g(c)$ is $f(x) + g(x)$ for

$x = c$, and similarly for the product. Of course, it is important that c be in the same field as the coefficients, or at least that there be a field which contains both the coefficients and c, because otherwise product and sum might not be defined. If $f(c) = 0$, we say that c is a *zero* of $f(x)$, that is, a *root* of $f(x) = 0$.

The first theorem follows directly from a special case of Theorem 9.1p.

Theorem 12.1. (Remainder theorem.) If $f(x)$ is a polynomial in $F[x]$ for any field F and $g(x) = x - c$, then

$$(12.1) \qquad f(x) = q(x)(x - c) + r,$$

where r and c are elements of F and $r = f(c)$.

Proof: We know from Theorem 9.1p that r is either 0 or is of lower degree than $x - c$. Hence r is either 0 or some other element of F. Now (12.1) expresses the equality of two polynomials. This means that if we multiply $(x - c)$ and $q(x)$ and add r to the product, we have $f(x)$. This then is also true if we replace x by c throughout. Thus we have $f(c) = q(c)(c - c) + r$, which implies that $r = f(c)$.

Corollary. (Factor theorem.) Above, $x - c$ is a factor of $f(x)$ if and only if $f(c) = 0$.

The factor theorem can be very useful in numerical cases. For instance, if $f(x) = x^3 + x^2 + x - 3$, it is easily verified that $f(1) = 0$. This implies that $x - 1$ is a factor of $f(x)$. The following is an immediate consequence of Theorem 12.1.

Theorem 12.2. If $f(x)$ is a polynomial of positive degree n with coefficients in a field F, then there are at most n distinct elements c of F such that $f(c) = 0$; that is, $f(x)$ has at most n distinct zeros.

Proof: We prove this by induction on the degree of $f(x)$. If $n = 1$, then $f(x) = ax + b$ with $a \neq 0$, and $f(x)$ has the single zero: $-b/a$. Assume the theorem for all degrees less than n. Suppose that $f(x)$ has a zero c in F. Then, by the factor theorem, $f(x) = (x - c)q(x)$ for some polynomial $q(x)$ over F whose degree is $n - 1$. If $f(x)$ has a zero c' different from c, then

$$0 = f(c') = (c' - c)q(c').$$

Thus $c' - c \neq 0$ implies that $q(c') = 0$. Hence all the zeros of $f(x)$ except perhaps c are zeros of $q(x)$. By the induction hypothesis, $q(x)$ has at most $n - 1$ different zeros. Hence $f(x)$ has at most n different zeros and the proof is complete.

Notice that we have actually proved a little more than is stated in the theorem, namely, the following theorem.

Theorem 12.2a. If $f(x)$ is a polynomial of positive degree n with coefficients in a field F and if it has a zero in F, it can be written

$$f(x) = (x - c_1)(x - c_2) \cdots (x - c_r)q_r(x),$$

where q_r is a polynomial of degree $n - r$ in $F[x]$ and the c_i are in F.

An immediate consequence is the following corollary.

Corollary. A quadratic or cubic polynomial $f(x)$ over F is prime (irreducible) unless there is an element c of F such that $f(c) = 0$.

There are two consequences of this theorem that we should notice. First suppose that $f(x)$ and $g(x)$ are two polynomials in $F[x]$. Let f be of positive degree n and g be of degree not more than n. Suppose also that $f(c) = g(c)$ for more than n different numbers c in some field containing F. Then $f(x) - g(x) = h(x)$ is a polynomial of degree n or less which has more than n distinct zeros. Theorem 12.2 then shows that $h(x)$ cannot have positive degree and hence $h(x) = d$, which is independent of x. Then $d = f(c) - g(c)$ implies that $d = 0 = h(x)$, and hence $f(x)$ and $g(x)$ are equal polynomials. Thus we have proved the following theorem.

Theorem 12.3. If $f(x)$ and $g(x)$ are two polynomials in $F[x]$ and $f(x)$ is of positive degree n, if the degree of $g(x)$ is not greater than n, and if $f(c) = g(c)$ for more than n different c, then $f(x) = g(x)$.

On the other hand, in a finite field two polynomials $f(x)$ and $g(x)$ can have the property that $f(c) = g(c)$ for all c in the field, without being the same. For instance, in $F = Z_p$, if $f(x) = x^p$ and $g(x) = x$, then $f(c) = g(c)$ for all c in Z_p by Fermat's theorem, and yet f and g are not the same polynomial. This does not violate the previous theorem, since the degree of $f(x)$ is the same as the number of elements in the field and $x^p = x$ is not true for more than $n = p$ elements of the field.

A second consequence of Theorem 12.2 is rather startling. Let us restate Theorem 5.7 of Chapter II: If G is a finite Abelian group with identity element e such that $x^k = e$ has at most k solutions for every positive integer k, then G is a cyclic group. Now if F is any finite field, F^*, the nonzero elements of F, form a multiplicative group. Theorem 12.2 shows that the hypothesis of the above-quoted theorem is satisfied. Hence we have proved the following theorem.

Theorem 12.4. The set of nonzero elements of any finite field forms a multiplicative cyclic group.

This means that for any finite field F, there is an element g such that every nonzero element of F is a power of g. Since the number of elements in F^* is

$p^n - 1$, we know from Corollary 3 of Theorem 5.3 in Chapter II that the number of generators of F^* is $\phi(p^n - 1)$. We shall discuss finite fields further in Section 15.

13.   Polynomials over Special Fields and Domains

Unfortunately, there is no known proof of the first theorem of this section that does not assume more background than we can muster at this point. So we state it without proof. (A proof is given on p. 101 of the third edition of Birkhoff and Mac Lane's *A Survey of Modern Algebra*.)

Theorem 13.1. (Fundamental theorem of algebra.) If $f(x)$ is any polynomial of positive degree in $\mathbf{C}[x]$, it has a linear factor $x - c$, where c is in $\mathbf{C}$; that is, $f(x)$ has a zero in the field of complex numbers.

When we use the method of proof of Theorem 12.2, we have the following corollary.

Corollary. If $f(x)$ is a polynomial of degree n with complex coefficients and leading coefficient 1, it can be written as a product of n linear factors $(x - c_i)$, where the c_i are complex numbers.

Now suppose that $f(x)$ is a polynomial with real coefficients, with $a + bi$ an imaginary zero; that is, a and b are real and b not zero. Consider

(13.1) $g(x) = (x - a - bi)(x - a + bi) = (x - a)^2 + b^2.$

This polynomial has real coefficients. Now $f(x)$ and $g(x)$ are not relatively prime in $\mathbf{C}$, since they have a factor $x - a - bi$ in common and hence, by Corollary 1 of Theorem 10.2, are not relatively prime in $\mathbf{R}$. But $g(x)$ is a prime polynomial in $\mathbf{R}$ since its zeros are imaginary. This, by Corollary 2 of Theorem 10.2, implies that $g(x)$ is a factor of $f(x)$. Thus we have shown the following theorem.

Theorem 13.2. If $f(x)$ with real coefficients has an imaginary zero $a + bi$, then it has a factor $g(x)$ defined in (13.1).

From this we have, almost as a corollary,

Theorem 13.3. If $f(x)$ is a polynomial with real coefficients, its prime factors are linear or quadratic polynomials.

Proof: We prove this by induction on n, the degree of $f(x)$. If $n = 1$, the theorem is trivial. Therefore, assume it for all degrees up to some integer n.

If $f(x)$ has a real zero c, we can write $f(x) = (x - c)f_1(x)$, and by the induction hypothesis we see that all the factors of $f_1(x)$ are linear or quadratic; hence the same is true for $f(x)$. Otherwise, $f(x)$ has an imaginary zero $a + bi$, and Theorem 13.2 affirms that $f(x)$ has a quadratic factor $g(x)$. Thus $f(x) = q(x)g(x)$, and we can use the same argument as in the previous case to complete the proof.

Corollary 1.   The only irreducible polynomials in the field of real numbers are linear and quadratic polynomials.

Corollary 2.   If $f(x)$ is a real polynomial (that is, in $\mathbf{R}[x]$) of odd degree, it has at least one real zero.

This ties together pretty well the theory about prime polynomials in the real and complex fields. It is much more difficult in the field of rational numbers and other fields. However, there is a theorem which shows that if a polynomial with integral coefficients is irreducible in $\mathbf{Z}[x]$, it is irreducible in $\mathbf{Q}[x]$. This is rather curious, since a prime polynomial in one field does not necessarily remain prime when one enlarges the field. We shall state and prove this theorem for Euclidean domains, but the chief application is to $\mathbf{Z}[x]$, and you may prefer to look at the proof for the set of integers alone. Certainly, there is very little difference.

Theorem 13.4.   (Gauss's lemma.)   Let E be a Euclidean domain, F the field in which it is embedded (see Section 3), and let $f(x)$ be a polynomial in $E[x]$. Then if $f(x)$ is reducible in $F[x]$, it is reducible in $E[x]$.

Proof: From the hypothesis of the theorem, $f(x) = g(x)h(x)$, where $f(x)$ is in $E[x]$ and g and h are in $F[x]$. We want to show that $f(x) = g_0(x)h_0(x)$, where g_0 and h_0 are in $E[x]$. The natural way to start would be to let r be the least common multiple of the denominators of the coefficients of g and s be that for the coefficients of h. Then $rsf(x) = [rg(x)][sh(x)]$, and one might hope that he could prove that $rs - 1$. But this is a vain hope, for if $g(x) = 2x/5 + 2$ and $h(x) = \frac{1}{2}(5x + 5)$, then $r = 5$ and $s = 2$, which denies $rs = 1$. But, nevertheless, $g(x)h(x) = (x + 5)(x + 1) = f(x)$, and the theorem holds. What happens is that $g(x) = (2/5)(x + 5)$, $h(x) = (5/2)(x + 1)$, and $(2/5)(5/2) = 1$.

So we choose relatively prime elements of E, r and r', so that $(r/r')g(x) = g_0(x)$ is in $E[x]$ and is "primitive," that is, has coefficients whose g.c.d. is 1. Similarly, there are relatively prime elements of E, s and s', such that $(s/s')h(x) = h_0(x)$ is in $E[x]$ and is primitive. Now we have

(13.2) $$g_0(x)h_0(x) = \left(\frac{rs}{r's'}\right)f(x) = \left(\frac{t}{t'}\right)f(x),$$

where t and t' are relatively prime. We can without loss of generality assume that $f(x)$ is primitive, since if $f(x) = qf_0(x)$ is in $E[x]$ and q in E, we could replace $g(x)$ by $qg_1(x)$ and start the argument again. Now $f(x)$ primitive implies that t' is a unit, for if t' had a prime factor, it would have to be a factor of every coefficient of $f(x)$. So (13.2) becomes

$$(13.3) \qquad g_0(x)h_0(x) = wf(x), \qquad w \text{ in } E.$$

If we can show that w is a unit of E, we can divide by w and replace g_0 by g_0/w to prove the theorem. We state as a lemma what is left to prove and then prove it.

Lemma. If $g_0(x)$, $h_0(x)$, and $f(x)$ are primitive polynomials over E and (13.3) holds, then w is a unit. In other words, the product of two primitive polynomials in $E[x]$ is a primitive polynomial.

Proof: Let the three polynomials be represented as follows:

$$f(x) = a_n x^n + a_{n-1} x^{n-1} + \cdots + a_1 x + a_0,$$
$$g_0(x) = b_r x^r + b_{r-1} x^{r-1} + \cdots + b_1 x + b_0,$$
$$h_0(x) = c_s x^s + c_{s-1} x^{s-1} + \cdots + c_1 x + c_0.$$

In order to prove the theorem, we assume that p is a prime element dividing w and arrive at a contradiction. Before embarking on the general case, let us see what it looks like for $r = 3$ and $s = 2$. Then we have

$$
\begin{aligned}
wa_0 &= b_0 c_0, \\
wa_1 &= b_1 c_0 + b_0 c_1, \\
wa_2 &= b_2 c_0 + b_1 c_1 + b_0 c_2, \\
wa_3 &= b_3 c_0 + b_2 c_1 + b_1 c_2, \\
wa_4 &= b_3 c_1 + b_2 c_2, \\
wa_5 &= b_3 c_2.
\end{aligned}
$$
(13.4)

The first equation shows that p must divide one of b_0 and c_0. Suppose that p divides b_0 but not c_0. Then the second equation shows that p must divide b_1 and the third equation shows that p divides b_2. Then the fourth equation shows that p divides b_3. Thus we have shown that if p does not divide c_0, then it divides all the b_i; this is false. Then we could begin the argument again, assuming that p divides both b_0 and c_0. Proceeding this way could become rather tiresome. We need to have some argument that avoids considering each separate case. To see how the general argument goes, suppose that p divides b_0 and b_1 but not b_2, while p divides c_0 but not c_1. Then the term in (13.4) that contains $b_2 c_1$ is wa_3. Looking at the right side of this equation, we see that p divides $b_3 c_0$, since it divides c_0, and divides $b_1 c_2$ since it divides b_1. Then since p divides wa_3, it must divide $b_2 c_1$ as well, which is the contradiction we need.

With this as a guide we can consider the general case. Since $g_0(x)$ is primitive, let k be the integer $\leq r$ such that p does not divide b_k but divides b_i for all $i < k$. Similarly, let $u \leq s$ be chosen so that p does not divide c_u but that p divides every c_j for which $j < u$. Then wa_{k+u} is equal to

$$(13.5) \quad b_{k+u}c_0 + b_{k+u-1}c_1 + \cdots + b_{k+1}c_{u-1} + b_k c_u$$
$$+ b_{k-1}c_{u+1} + \cdots + b_0 c_{k+u},$$

where it is understood that if $i > r$, then $b_i = 0$, and if $j > s$, then $c_j = 0$. Notice that in (13.5) $k + u$ is the sum of the subscripts of each term. Now all terms to the left of $b_k c_u$ in (13.5) are divisible by p because the c's are. All terms to the right of $b_k c_u$ are divisible by p because the b's are. Since p is a factor of w, it must be a factor of the sum (13.5) and hence of $b_k c_u$. This is a contradiction. Thus w has no prime factors and must be a unit. This not only proves the lemma but shows that, except for multiplication by a unit of E, $f(x) = g_0(x)h_0(x)$.

Corollary 1. (This follows from the *proof* above.) If $g(x)$ in $F[x]$ is a factor of a primitive polynomial $f(x)$ in $E[x]$, and if $g_0(x) = (r/r')g(x)$ is a primitive polynomial in $E[x]$, then $g_0(x)$ is a factor of $f(x)$ in $E[x]$.

Corollary 2. If $f(x)$ in $Z[x]$ is reducible in $Q[x]$, then it is reducible in $Z[x]$.

The following immediate consequence of Corollary 1 is useful in locating rational zeros of polynomials with integer coefficients.

Theorem 13.5. If $f(x)$ is a polynomial in $Z[x]$ and if r/s is a rational zero of $f(x)$ with r and s relatively prime integers, then r is a factor of $f(0)$ and s is a factor of the leading coefficient of $f(x)$.

Proof: Now r/s being a zero implies that $g(x) = x - r/s$ is a factor of $f(x)$. Hence by Corollary 1, $sx - r$ is a factor and we have

$$f(x) = (sx - r)(c_{n-1}x^{n-1} + c_{n-2}x^{n-2} + \cdots + c_1 x + c_0),$$

where the c_i are integers. Hence the leading coefficient of $f(x)$ is sc_{n-1}, a multiple of s and $f(0) = -rc_0$, a multiple of r.

Corollary. If $f(x)$ is a polynomial in $Z[x]$ with leading coefficient 1, then all its rational zeros are integers.

There is a curious criterion for irreducibility (primality) in $Z[x]$ that not only uses Gauss's lemma but has a proof which is quite similar.

Theorem 13.6   (Eisenstein's criterion for irreducibility.)   Let $f(x)$ be a polynomial in $Z[x]$. Then

$$f(x) = a_n x^n + a_{n-1} x^{n-1} + \cdots + a_1 x + a_0, \qquad n > 0,$$

is irreducible (prime) if, for some prime number p, $a_i \equiv 0 \pmod{p}$ for $i = 0, 1, \ldots, (n-1)$, p does not divide a_n, and p^2 does not divide a_0.

Proof: We assume that $f(x) = g(x)h(x)$, where g and h of positive degree are in $Q[x]$ and seek a contradiction. From Gauss's lemma we may assume that $g(x)$ and $h(x)$ have coefficients that are integers. As above, write

$$g(x) = b_r x^r + b_{r-1} x^{r-1} + \cdots + b_1 x + b_0,$$
$$h(x) = c_s x^s + c_{s-1} x^{s-1} + \cdots + c_1 x + c_0, \qquad b_r c_s \neq 0.$$

Since p^2 is not a factor of $a_0 = b_0 c_0$, p cannot divide both b_0 and c_0. Interchanging g and h if necessary, we can assume that $p \mid c_0$ but $p \nmid b_0$. Since p does not divide $a_n = b_r c_s$, we have $p \nmid c_s$. Then for some w satisfying the condition $0 < w \leq s$, we have

(13.6) $c_0 \equiv c_1 \equiv c_2 \equiv \cdots \equiv c_{w-1} \equiv 0 \pmod{p}, \qquad c_w \not\equiv 0 \pmod{p}.$

Now

(13.7) $a_w = c_0 b_w + c_1 b_{w-1} + \cdots + c_{w-1} b_1 + c_w b_0, \qquad p \nmid b_0,$

where it is understood that any b_i is zero if its subscript is greater than r. We know by the hypothesis of the theorem that a_w is divisible by p, since $w \leq s < r + s = n$. From (13.6) all but the last term on the right of (13.7) is divisible by p. Hence $c_w b_0$ is also divisible by p. This is impossible, since p divides neither c_w [by (13.6)] nor b_0 by (13.7). This is the contradiction that proves the theorem.

Probably the most important consequence (at least for our purposes) of Eisenstein's criterion is the following, which we shall need later. We give an indication of the proof, leaving the details as Exercise 6.

Theorem 13.7.   If p is a prime number, the polynomial $f(x) = (x^p - 1)/(x - 1)$ is a prime polynomial over $Z[x]$.

Indication of the proof: Certainly, $f(x)$ is irreducible if and only if $f(y + 1)$ is. Now $f(0 + 1) = p$ and the polynomial $f(y + 1)$ satisfies the conditions of Eisenstein's criterion. Hence $f(y + 1)$ is irreducible, that is, a prime polynomial.

Exercises

1. Find all the rational roots of each of the following equations.
   (a) $x^4 + x^2 - 5 = 0$.
   (b) $3x^3 + x^2 + x - 2 = 0$.
   (c) $2x^{17} + x - 2 = 0$.
   (d) $3x^4 + 5x^3 + 5x^2 + 5x + 2 = 0$.
   (e) $2x^4 + 3x^3 + 3x^2 + 9x + 6 = 0$.
   (f) $x^4 + x^3 + x^2 + x + 1 = 0$.

2. Here is a modification of an application of Theorem 13.5. Suppose that $f(x) = 2x^3 + 7x^2 + 4x + 12$. To try all the possible fractions with numerator a factor of 12 and denominator 1 or 2 would be rather tedious. But let $x = y - 1$ and see that the leading coefficient of $f(y - 1)$ is 2 and the constant term $f(-1) = 13$. Then the possible rational values of x are -1 ± 13 and $-1 \pm \frac{13}{2}$. Since -1 ± 13 must also be a factor of 12, the only possibility is $+12$. Complete the determination of the rational zeros.

3. Apply the method of Exercise 2 to determine all the rational zeros of $f(x) = 6x^3 + 41x^2 - 46x + 12$.

4. Use the corollary of Theorem 13.5 or otherwise prove that $x^2 - c = 0$ for c an integer has rational solutions if and only if c is the square of an integer. Hence show that $\sqrt{c}$ is irrational unless c is a square.

5. Prove that for any positive integer n, $x^n = c$ has no rational roots unless c is the nth power of an integer.

6. Complete the proof of Theorem 13.7.

7. Show that if $f(x)$ is reducible, then so is $f(x^2)$. Is the converse of this statement true? Prove it or give a counterexample.

8. Prove that if $f(x)$ is written as in Theorem 13.6, where the coefficients are integers, then $f(x)$ is irreducible if some prime p divides all the coefficients except the last and p^2 does not divide a_n.

9. Give an example to show that Eisenstein's criterion fails if the condition $p^2 \nmid a_n$ is omitted.

10. For each finite field F, find a generator of the multiplicative group F^*.
    (a) $F = \mathbf{Z}_7$.
    (b) $F = \mathbf{Z}_{11}$.
    (c) $F = \mathbf{Z}_{13}$.

11. In each of the parts of Exercise 10 give the number of generators.

12. Let $f(x)$ be a polynomial in $F[x]$ and $g(x)$ a quadratic polynomial such that $f(x_i) = g(x_i) = 0$ for $i = 1, 2$. Suppose that $f(x) = q(x)g(x) + cx + d$ for c and d in F and $q(x)$ in $F[x]$. Prove that if $x_1 \neq x_2$, then c and d are both zero.

13. If in Exercise 12, $x_1 = x_2$, what conclusions can be drawn about c and d?

14. State Theorems 13.5 and 13.6 for polynomials with coefficients in a Euclidean domain. What modifications if any need to be made in the proofs?

14.　Algebraic Extensions of Fields

Now we have the equipment to continue the process of construction of fields begun in Section 2. Recall that to construct the field Z_p we started with Z and a prime number p and found the set of integers modulo p. That is, the field Z_p has the following p elements:

$$0, 1, 2, \ldots, (p-1),$$

where addition is modulo p; one forms the ordinary sum and then replaces it by its remainder when divided by p; multiplication modulo p is similarly defined. Analogously, we can start with $F[x]$, the set of polynomials over a field F, some prime polynomial $p(x)$ over F, and replace each polynomial by its remainder when it is divided by $p(x)$. Let us first illustrate the method for a case of special significance.

EXAMPLE 1.　In $R[x]$, the set of polynomials with real coefficients, select a prime polynomial $p(x) = x^2 + 1$. Then form the set of polynomials modulo $p(x)$. That is, to find $f(x)$ modulo $p(x)$, divide $f(x)$ by $p(x)$ and compute the remainder. This remainder will be a polynomial of degree less than 2, hence of the form $ax + b$, where a and b are real numbers. Now we want to show that the set $S = \{ax + b\}$ modulo $(x^2 + 1)$ forms a field. The set S is closed under addition and multiplication modulo $(x^2 + 1)$. The associative and distributive properties hold, since they hold in $R[x]$, the identities are 0 and 1, and the additive inverse of $ax + b$ is $-ax - b$. In nothing so far have we made use of the fact that $x^2 + 1$ is a prime polynomial. It is in showing the existence of a multiplicative inverse that primality is necessary. To this end, let $f(x)$ be any polynomial in $R[x]$ and note from Corollary 2 of Theorem 10.3 that there are polynomials $h(x)$ and $k(x)$ in $R[x]$ such that

$$f(x)k(x) + p(x)h(x) = g(x),$$

where $g(x)$ is a greatest common divisor of $f(x)$ and $p(x) = x^2 + 1$. Now if $p(x)$ is a factor of $f(x)$, then $f(x) \equiv 0 \pmod{x^2 + 1}$. Otherwise $x^2 + 1$ and $f(x)$ are relatively prime, since $x^2 + 1$ has no linear factors in $R[x]$. Hence $g(x)$ can be taken to be 1. This shows that $f(x)k(x) \equiv 1 \pmod{x^2 + 1}$ and hence $k'(x)$, the remainder when $k(x)$ is divided by $x^2 + 1$, is the multiplicative inverse of $f(x) \pmod{x^2 + 1}$. Hence we have shown the following theorem.

Theorem 14.1. The set S of polynomials $\{ax + b\}$, where addition and multiplication is modulo $x^2 + 1$ and a and b are real, is a field.

Now call H the field of polynomials $\{ax + b\}$ of Theorem 14.1 and consider the polynomials

$$h(z) = c_n z^n + c_{n-1} z^{n-1} + \cdots + c_1 z + c_0,$$

where each c_i is in H, that is, has the form $ax + b$, where a and b are real. One polynomial in $H[z]$ is $z^2 + 1$. We now show that in $H[z]$, $z^2 + 1$ has factors. In fact,

$$(z + x)(z - x) = z^2 - x^2 = z^2 + 1,$$

since $x^2 = 1 \cdot (x^2 + 1) - 1$ and hence $x^2 \equiv -1$ modulo $(x^2 + 1)$. Thus $z^2 + 1$ is prime in $R[z]$ but not in $H[z]$.

By this time you probably are aware that we are dealing by a roundabout way with complex numbers. (Cauchy considered them from this point of view.) As is not surprising, we can show that if σ is the correspondence,

$$(ax + b)\sigma = ai + b$$

with a and b real and $i^2 = -1$, then σ is an isomorphism between H in Theorem 14.1 over the field of reals and $\mathbf{C}$, the field of complex numbers. To show this, first note that the correspondence is one-to-one and then verify the following for addition and multiplication:

$$[(ax + b) + (cx + d)]\sigma = (a + c)i + b + d = ai + b + ci + d$$
$$= (ax + b)\sigma + (cx + d)\sigma,$$
$$[(ax + b)(cx + d)]\sigma = [(bc + ad)x + bd - ac]\sigma$$
$$= (bc + ad)i + (bd - ac) = (ai + b)(ci + d)$$
$$= [(ax + b)\sigma][(cx + d)\sigma].$$

Thus we have shown that the field of Theorem 14.1 is isomorphic to the field of complex numbers.

Using the letter i suggests a point of view that has some advantages over division by $x^2 + 1$. Let us contrast the two methods. To compute the product $(ax + b)(cx + d)$ in H, we first multiply to get $acx^2 + bcx + adx + bd$, divide by $x^2 + 1$, and have the remainder, $-ac + bcx + adx + bd$. To compute $(ai + b)(ci + d)$, we multiply as before to get $aci^2 + bci + adi + bd$. But here, instead of dividing by $i^2 + 1$ we can replace i^2 by -1, which is much simpler. Of course, we could have done this with x just as well by noting that since in H, $x^2 + 1 = 0$, then $x^2 = -1$. What we do, in effect, is replace x by i, which is a solution in H of the equation $x^2 + 1 = 0$.

Let us review what we have done. We have shown by use of congruence modulo $(x^2 + 1)$ that there *is* a field and that it can be considered to be the set $\{(ai + b)\}$, where $i^2 = -1$. We describe this by saying that H is found by "adjoining i to **R**" and write $H = \mathbf{R}(i)$. In H the polynomial $z^2 + 1$ is re-ducible. One could avoid much of this by postulating that there is a number whose square is -1 or by saying "let there be such a number"; but showing in effect that there is such a "number" is rather more satisfactory. Indeed it would be very unsatisfactory if, whenever we wished to extend a field, a new postulate would be necessary.

Notice that if instead of considering $x^2 + 1$, we had taken $x^2 + 2$, for example, the process would work equally well and yield a field $\mathbf{R}(j)$ where $j^2 = -2$. If the field were **Q** instead of **R**, $x^2 - 2$ would also be satisfactory. But $x^2 - 2$ would not do in **R**, since there it is factorable into $(x + \sqrt{2}) \cdot (x - \sqrt{2})$; hence one could not divide by $x + \sqrt{2}$.

Since these ideas are very fundamental, let us consider another example before going on to the general case.

EXAMPLE 2.   Start now with the field $\mathbf{Z}_2$ consisting of two elements, 0 and 1, where $1 + 1 = 0$. Then $p(x) = x^3 + x + 1$ is a prime polynomial over $\mathbf{Z}_2$, since $p(1) = 1 = p(0)$ shows, by the factor theorem, that $p(x)$ has no linear factors. Now let H be the set of polynomials with coefficients in $\mathbf{Z}_2$ modulo $p(x)$. Since the remainders will be at most of the second degree,

$$H = \{ax^2 + bx + c\} \text{ modulo } p(x),$$

where a, b, and c are in $\mathbf{Z}_2$; that is, each is 1 or 0. Also, H is a field by the same argument used in Example 1. To show how products are formed, consider

$$(x^2 + x + 1)(x^2 + 1) = \begin{cases} x^4 + x^3 + (1+1)x^2 + x + 1 & \text{in } \mathbf{Q}, \\ x^4 + x^3 + x + 1 & \text{in } \mathbf{Z}_2. \end{cases}$$

To compute this in H, we divide by $x^3 + x + 1$ and take the remainder, or replace x by c, where $c^3 + c + 1 = 0$; that is, $c^3 = c + 1$. Thus

$$c^4 + c^3 + c + 1 = c^4 = c(c + 1) = c^2 + c$$

in Z_2. Then H can be considered to be obtained from $\mathbf{Z}_2$ by adjoining c, a zero of $x^3 + x + 1$. Of course, one can carry through the same process without replacing x by c, but in the beginning, the change of notation has the advantage of highlighting the change in point of view. As in Example 1, $p(z)$ in $H[z]$ has a factor $z - c$, for c in H.

With the above examples behind us we now state and prove the general theorem.

Theorem 14.2.   Let F be any field and $p(x)$ a prime polynomial over that field. Then the set of polynomials $f(x)$ modulo $p(x)$ forms a field $F' = F_{p(x)}$. In $F'[z]$ the polynomial $p(z)$ has a linear factor.

Proof: Now F' is the set of polynomials $f(x)$ modulo $p(x)$ over F. That is, if $p(x)$ has degree n, then each $f(x)$ has degree less than n and addition and multiplication is modulo $p(x)$. To show closure under addition is easy; to show it under multiplication, let $r(x)$ and $s(x)$ be two polynomials of F'. First, form their product in $F[x]$; then the remainder after division by $p(x)$ is the product of r and s in F'. Addition and multiplication are associative in F', since they are in $F[x]$ and the distributive property holds. It remains to show the existence of a multiplicative inverse. Let $f(x)$ be any polynomial, except 0, of F'. It is also in $F[x]$. As above, if $f(x)$ is not divisible by $p(x)$ in $F[x]$, it is relatively prime to $p(x)$. Then we know that there are polynomials $k(x)$ and $h(x)$ in $F[x]$ such that

$$f(x)k(x) + p(x)h(x) = 1.$$

This means that $f(x)k(x) \equiv 1 \pmod{p(x)}$. If $k'(x)$ is the remainder when $k(x)$ is divided by $p(x)$, then $k'(x)$ is the multiplicative inverse of $f(x)$ in F'. Finally, $p(z)$ in $F'[z]$ has a factor $z - x$ from the factor theorem, since $p(x) = 0$ in F'. This completes the proof.

To illustrate the finding of the inverse, we show the computation leading to the inverse of $x^2 + x + 1$ in Example 2:

$$(x^2 + x + 1)x^2 + (x^3 + x + 1)(x + 1) = 1.$$

Hence $(x^2 + x + 1)^{-1} = x^2$ in F'.

Now replace x by c, to signify the second point of view—that c is a "number" that is a zero of $p(z)$ in F'. Then consider the set of elements

(14.1) $$a_{n-1}c^{n-1} + a_{n-2}c^{n-2} + \cdots + a_1 c + a_0,$$

where the a_i are in F, and in multiplying we remember that $p(c) = 0$. Now

$$f(x)\sigma = f(c)$$

is a one-to-one correspondence between the elements of F' and the set (14.1). To show that it is an isomorphism, we must show that the image of the sum of any two elements of F' is the sum of the images, and similarly for the product. The former is almost obvious. To show the latter, let $g(x)$ be any polynomial in $F[x]$. Then, for some polynomials $q(x)$ and $r(x)$ in $F[x]$:

(14.2) $$g(x) = q(x)p(x) + r(x),$$

where $r(x) = 0$ or is a polynomial of degree less than n. If we divide $g(x)$ by $p(x)$, the remainder is $r(x)$. So, to find the product of two elements (polynomials) of F', we multiply in $F[x]$ to get $g(x)$ and divide by $p(x)$, which yields $r(x)$, the product of the polynomials in F'. To find the corresponding product of expressions (14.1), since multiplication is the same in c as in x, (14.2) becomes

$$g(c) = q(c)p(c) + r(c) = r(c),$$

since $p(c) = 0$. Therefore, we have the polynomial $r(c)$ instead of $r(x)$, and the correspondence σ is an isomorphism. The isomorphism shows that the polynomials (14.1) in c form a field. We have proved the following theorem:

Theorem 14.3. Let $p(x)$ be a prime polynomial in $F[x]$. The set of elements (14.1) with $p(c) = 0$ is a field isomorphic to the field F' of Theorem 14.2.

Definition. *The field consisting of elements* (14.1) *with* $p(c) = 0$ *is called a simple algebraic extension of F and written* $F(c)$. *The degree of* $p(x)$ *is called the* degree *of the extension and also the* degree *of c. If* F' *is a simple algebraic extension of F of degree n, we write* $[F':F] = n$.

Now, $p(x) = 0$ might have another root c' in F' or some other extension of F. Then $F(c')$ would also be isomorphic to $F_{p(x)}$ and hence to $F(c)$. Then we have the following theorem.

Theorem 14.4. If c and c' are two roots of $p(x) = 0$, where $p(x)$ is prime in F, then $F(c)$ and $F(c')$ are isomorphic.

Is there any other polynomial $p'(x)$ that is prime over F and has c as a zero? The following theorem asserts that "no" is the answer.

Theorem 14.5. Given a field F and a prime polynomial $p(x)$ over F with zero c and leading coefficient 1, there is no other polynomial that has these properties.

Proof: Suppose that $p'(x)$ is prime in F, has leading coefficient 1, and has c as a zero with c in F', an extension of F. Then in F', $p(x)$ and $p'(x)$ have a common factor $x - c$. Thus in F' they are not relatively prime. Hence by Corollary 1 of Theorem 10.2 they are not relatively prime in F either. This means that their primality implies that each must divide the other. Hence, since their leading coefficients are 1, they must be the same polynomial. This completes the proof. This theorem justifies our speaking of *the* degree of an algebraic extension.

Before proceeding, let us illustrate the above for $F = \mathbf{Q}$ and $p(x) = x^3 - 2$. This polynomial $p(x)$ is prime over $\mathbf{Q}$, since it has no rational zero. In $\mathbf{C}$ the roots of $p(x)$ are $c = \sqrt[3]{2}$, wc, and w^2c, where w is an imaginary cube root of 1, that is, a root of $z^2 + z + 1 = 0$. By Theorem 14.4, the extensions $\mathbf{Q}(c)$ and $\mathbf{Q}(cw)$ are isomorphic. Notice, however, that they are not the same field, since all elements of $\mathbf{Q}(c)$ are real and $\mathbf{Q}(cw)$ contains the imaginary number cw, for instance.

Now let us use this same example to lead into the next concept. Start with $F' = \mathbf{Q}(c)$, where c is the real root of $x^3 - 2 = 0$. The polynomial $p(x) =$

$x^2 + x + 1$ is prime in $\mathbf{Q}(c)$, since it is prime in $\mathbf{R}$, which contains $\mathbf{Q}(c)$. Thus we can adjoin to F' a zero w of $p(x)$ to get the field $F'' = F'(w)$. In this field F'', the polynomial $z^3 - 2$, is a product of three linear factors $(z - c) \cdot (z - cw)(z - cw^2)$. That is, F'' is a field that contains all the zeros of $z^3 - 2$. We call this the splitting field or root field of the polynomial (see below). Now we state and prove the result of which the above is an example.

Theorem 14.6.   Let F be a field and $g(x)$ any polynomial in $F[x]$. There is a field F_0 obtained from F by a finite sequence of simple algebraic extensions such that the polynomial $g(x)$ is a product of linear factors in F_0.

Proof: If $g(x)$ is composite, we can express it as a product of prime factors and then deal with each one separately. We assume that $g(x)$ is prime in F. Then we find an extension $F(c) = F'$ so that in $F'[z]$, $g(z)$ has a linear factor $z - c$. That is, $g(z)$ is composite in F'. Then, by Theorem 12.2a we may write

$$g(z) = (z - c)(z - c_2) \cdots (z - c_r)q(z),$$

where $q(z)$ has no linear factors, and so on.

Let us tidy up the proof by using mathematical induction on n the degree of $g(x)$. For $n = 1$ there is no problem. Assume the theorem for all degrees less than n. Then, as above, there is an extension F' such that in $F'[z]$, $g(z)$ has a linear factor. Thus

$$g(z) = (z - c)h(z),$$

where $h(z)$ is of degree $n - 1$ over F'. Hence there is, by the induction hypothesis, a finite sequence of simple algebraic extensions of F' yielding a field F_0 in which $h(x)$ is a product of linear factors. Hence in F_0, $g(z)$ is a product of linear factors. This completes the proof.

Definition.   *A polynomial is said to be* completely decomposible *in a field F if in that field it can be expressed as a product of linear factors.*

Definition.   *The* root field *or* splitting field *of a polynomial $g(z)$ in $F[z]$ is the field obtained by adjoining to F all the zeros of $g(z)$.*

REMARK: The first word "the" in the above definition needs justification. After all, we had a special way of finding a field that splits the polynomial into linear factors. Who knows, there might well be other ways of doing this. It can be proved (see the books by Herstein and Birkhoff and Mac Lane listed in the Bibliography) that any two root fields of a polynomial over a field are isomorphic. Now, except for a relatively minor use in Theorem 15.3, the root fields that we will consider are subfields of the complex field, where a root field is unique from the construction. Hence we omit the proof of this result.

Above we have found it useful to consider a sequence of simple algebraic extensions one after the other. We need a new term.

Definition.   *If a field H can be obtained by a finite sequence of simple algebraic extensions of a field F, then H is called an* algebraic extension *of F.*

A natural question to ask is: Suppose that a field H is obtained from a field F by a finite sequence of simple algebraic extensions; is there an element c of H such that $H = F(c)$; that is, is every algebraic extension a simple algebraic extension? In Chapter VI this question is answered in the affirmative for many fields.

We shall find the following theorem useful.

Theorem 14.7.   Let $f(x)$ be an irreducible polynomial in $F[x]$ for a field F and $h(x)$ a polynomial over F. If in some extension F' of F, there is an element c such that $f(c) = h(c) = 0$, then $f(x)$ is a factor of $h(x)$.

Proof: Since $f(c) = h(c) = 0$ in F', the factor theorem (the Corollary of Theorem 12.1) affirms that $(x - c)$ is a factor of both $f(x)$ and $h(x)$ in F'. Hence $f(x)$ and $h(x)$ are not relatively prime in F' and thus by Corollary 1 of Theorem 10.2 not relatively prime in F. Then Corollary 2 of Theorem 10.2 shows us that $f(x)$ is a factor of $h(x)$, completing the proof.

We should broaden the idea of an extension of a field F. Suppose that t is an element that is not in F and let H be the smallest field that contains t and F. Since H contains t and is closed under addition and multiplication, every polynomial in t must be in H. As in Section 2, there are two possibilities. First, there may be some polynomial in t over F which is zero. In that case, let $g(t)$ be such a polynomial of least degree. It must be irreducible over F and H is then an algebraic extension of F.

Second, there may be no polynomial in t over F that is zero except the zero polynomial. In this case H must contain all polynomials in t and, since closure under division is necessary, H must contain all rational functions of t. Then the set of all rational functions of t is called a *transcendental extension* of F and is written $F(t)$. The word "transcendental" signifies nonalgebraic. The field $\mathbf{Q}(\pi)$ is a transcendental extension of $\mathbf{Q}$ since it can be proved, although we do not do it here, that there is no polynomial in $\mathbf{Q}[x]$ that has π as a zero. (For a proof, see p. 173 of the book by Hardy and Wright listed in the Bibliography.)

Let us recapitulate the principal results of this important section. We have shown without use of the fundamental theorem of algebra that, starting with any field F and any polynomial $g(x)$ of degree n in F, there is a field F_0 in which $g(x)$ is a product of n linear factors. Knowing this, we can describe F_0 as obtainable by a sequence of adjunctions of roots of equations. However, the fundamental theorem of algebra tells us something more, namely, that for the complex field **C**, *every* polynomial in **C** is completely decomposable. In other words, we do not have to extend **C** to get complete decomposition. This field **C** is an example of an algebraically complete field. That is, a field is *algebraically complete* if every polynomial with coefficients in that field is completely decomposable in that field. In other words, in an algebraically complete field every polynomial can be expressed as a product of linear factors. Notice that this property of being algebraically complete is a rather special one. For instance, if $F = \mathbf{Q}(\sqrt{2})$, then $z^2 - 2$ is factorable in $F[z]$. But F is not algebraically complete since, for example, $z^2 - 3$ has coefficients in F but has no linear factors in $F[z]$. In Chapter VI we shall be carrying many of these ideas much further.

15.   Finite Fields

Let F be a finite field, that is, one with a finite number of elements. We showed in Section 2 that it contains $\mathbf{Z}_p$ as a subfield, that is, its characteristic is p for some prime p, and that it must contain p^n elements for some positive integer n. To construct such a field using the method of Section 14, we could seek a polynomial $p(x)$ in $\mathbf{Z}_p[x]$ of degree n that is prime over $\mathbf{Z}_p$. Then the polynomials in $\mathbf{Z}_p[x]$ modulo $p(x)$ form a field of order p^n. But we do not know at this point that for every n there is such a prime polynomial. Therefore, it is easier to approach the problem in another way, leaving the question of the existence of prime polynomials until later.

If there were a finite field F of degree n over $\mathbf{Z}_p$, the number of elements would be $q = p^n$ and the number of nonzero elements $q - 1$; that is, the multiplicative group F^* would have order $q - 1$. Thus the order of every element of F^* would be a divisor of $q - 1$ and hence would satisfy the equation

$$(15.1) \qquad\qquad x^{q-1} = 1, \qquad \text{where } q = p^n,$$

over the field F. That is, if there is to be a field with $q = p^n$ elements, then all the elements of the field, including zero, would be roots of the equation

$$(15.2) \qquad\qquad x^q = x, \qquad \text{where } q = p^n,$$

for n a positive integer. Now the equation (15.2) has roots in the root field F'. To show the existence of a field with q elements, we must show that the roots of (15.2) form a field.

Theorem 15.1.   The roots of (15.2) form a field with p^n elements.

Proof: Since the set of roots of (15.2) are all in the root field F', by Theorem 1.1 we need to show closure under addition, subtraction, multiplication, and division, except by zero. Since closure under addition is the most difficult to prove, we leave that until the last.

1. If b and c are roots of (15.2), then bc is a root, for $(bc)^q = b^q c^q = bc$.

2. If b is a root of (15.2), then $(-b)^q = (-1)^q b^q = (-1)^q b = (-1)^{q-1}(-b)$. Now, if p is odd, then q is also and $(-1)^{q-1} = 1$. If $p = 2$, then $-1 = 1$ and $(-1)^{q-1} = 1$. In both cases $-b$ is a root of (15.2).

3. If b is a root of (15.2) different from zero, then $b^{q-1} = 1$ and

$$(b^{-1})^{q-1}(b^{q-1}) = 1$$

imply that $(b^{-1})^{q-1} = 1$.

4. To show that the zeros of (15.2) form a field, it remains to show that if b and c are roots of (15.2), then $b + c$ is. Here we make heavy use of the fact that p is the characteristic of the field. We want to prove that

$$(15.3) \qquad (b + c)^q = b^q + c^q,$$

since $b^q + c^q = b + c$. This we show by induction on n in $q = p^n$. First, for $n = 1$ we have

$$(15.4) \quad (b + c)^p = b^p + pb^{p-1}c + \tfrac{1}{2}p(p - 1)b^{p-2}c^2 + \cdots$$
$$+ \tfrac{1}{2}p(p - 1)b^2c^{p-2} + pbc^{p-1} + c^p,$$

by the binomial theorem. Now all but the first and last coefficients are divisible by p, since the coefficient of $b^{p-r}c^r$, for $0 < r < p$, is

$$\frac{p!}{(p - r)!\,r!}$$

and no denominator has p as a factor. Thus (15.3) holds for $q = p$. Now assume (15.3) for $q' = p^{n-1}$. Then

$$(b + c)^q = [(b + c)^{q'}]^p = [b^{q'} + c^{q'}]^p = b^{q'p} + c^{q'p} = b^q + c^q.$$

In the second equality we used the induction hypothesis, and for the third we used (15.4) with $b^{q'}$ and $c^{q'}$ in place of b and c. We have shown closure under addition, subtraction, division, and multiplication. Hence the set of roots of (15.2) form a field.

To show that the field has exactly q elements, note first that $x^q - x$ is a product of exactly q linear factors with coefficients in F'. Thus it remains to show that no two roots of (15.2), that is, that no two linear factors, are the same, in other words, that $x^q - x$ has no factor of the form $(x - c)^2$. There are two ways to prove this. One is to use the idea of a derivative from the calculus. Doing this without preparation would present difficulties, since the derivative is based on the existence of limits and we do not want to restrict the field F. We shall show in Section 16 how these difficulties can be dealt with. To avoid all this here, we shall use a little trick. We suppose that

$$(15.5) \qquad\qquad x^q - x = (x - c)^2 g(x)$$

for some polynomial $g(x)$ and arrive at a contradiction. The trick (not a very clever one) is to replace $x - c$ by y and see that (15.5) becomes

$$(15.6) \qquad\qquad (y + c)^q - (y + c) = y^2 g(y + c).$$

This means that y^2 must be a factor of the left side of (15.6). But on the left-hand side of (15.6) the only terms that are not divisible by y^2 are

$$(15.7) \qquad\qquad qyc^{q-1} + c^q - y - c.$$

But q is a power of p, and hence $qyc^{q-1} = 0$ in Z_p. Furthermore, since c is a root of (15.2), $c^q = c$. Hence (15.7) reduces to $-y$, which is certainly not a multiple of y^2. This is the contradiction we sought, which shows that the roots of (15.2) are distinct. This completes the proof of the theorem.

Now, it is possible for curiosity's sake to prove that for every Z_p and every positive integer n, there is a polynomial of degree n that is prime over Z_p. We know from Theorem 12.4 that the nonzero elements of F', the root field of (15.2), form a multiplicative cyclic group. Let g be a generator of this group. We know that g is a zero of some prime polynomial $p(x)$ with coefficients in Z_p. Suppose that the degree of $p(x)$ is k. Then every element of the field is expressible in the form

$$(15.8) \qquad\qquad a_{k-1}g^{k-1} + a_{k-2}g^{k-2} + \cdots + a_1g + a_0,$$

where the a_i are in Z_p. Furthermore, all the powers of g are expressible in the form (15.8). No two of (15.8) are equal, since g satisfies no polynomial equation of degree less than k. But there are p^k elements of the form (15.8). Hence $k = n$ and the prime polynomial $p(x)$ has degree n. Thus we have shown that g is a zero of a prime polynomial of degree n over Z_p, which is what we set out to do. We have proved the following theorem.

Theorem 15.2. For every prime p and positive integer n, there is a prime polynomial over Z_p of degree n.

If we assume that the root field of a polynomial is unique up to an iso-morphism (see the Remark after the definition of root field, Section 14), then since the elements of a finite field form the root field of the polynomial $x^q - x$, where $q = p^n$, we can state with confidence the following theorem.

Theorem 15.3. For each prime p and positive integer n, there is, up to an isomorphism, a unique field with p^n elements.

This theorem is a little surprising because the irreducible equation that leads to the definition of a finite field is not necessarily unique. For instance, over $\mathbf{Z}_2$ the polynomials $x^3 + x + 1$ and $x^3 + x^2 + 1$ are both irreducible and both lead to a field with $2^3 = 8$ elements.

Let us illustrate the results of this section by considering in $\mathbf{Z}_5[x]$ the equation $x^4 = 2$. First, we show that $x^4 - 2$ is prime. (Note that we cannot use Eisenstein's criterion here.) Now $x^4 - 2$ has no linear factor since, by Fermat's theorem, $x^4 = 1$ in $\mathbf{Z}_5$ for all nonzero elements. Suppose that $x^4 - 2 = (x^2 + ax + b)(x^2 + cx + d)$. Then $bd = -2$ and $a + c = 0$ from the coefficients of x^0 and x^3. Thus the coefficient of x in the product is $bc + ad = a(d - b)$. If $a \neq 0$, then $b = d$ and $bd = b^2$, which cannot be -2 in $\mathbf{Z}_5$. If $a = 0$, the coefficient of x^2 is $b + d = 0$ and $bd = -b^2$, which also cannot be -2 in $\mathbf{Z}_5$. Hence $x^4 - 2$ is prime in $\mathbf{Z}_5$. Let c be a zero of $x^4 - 2$ and $F = \mathbf{Z}_5(c)$. Then

$$(jc)^4 = j^4 c^4 = 2 \qquad \text{for } j = 1, 2, 3, 4.$$

Hence in $F[z]$, $z^4 - 2 = (z - c)(z - 2c)(z - 3c)(z - 4c)$ and F is the root field of $x^4 - 2$.

Exercises

1. Write the multiplication table of the field of polynomials
   (a) In $\mathbf{Z}_3$ modulo $x^2 + x + 2$.   (b) In $\mathbf{Z}_2$ modulo $x^3 + x + 1$.
   (c) In $\mathbf{Z}_2$ modulo $x^3 + x^2 + 1$.

2. Find the inverse of $x + 1$ in each field F.
   (a) $F = \mathbf{Q}[x]$ modulo $x^2 + 1$.   (b) $F = \mathbf{Q}[x]$ modulo $x^2 - 2$.
   (c) $F = \mathbf{Z}_3[x]$ modulo $x^3 - x + 1$.
   (d) $F = \mathbf{Z}_5[x]$ modulo $x^3 - x + 2$.

3. Find the inverse of $c + 1$ in $\mathbf{Q}(c)$.
   (a) For $c^2 + 1 = 0$.   (b) For $c^2 = 2$.

4. Compare the results of parts (a) and (b) of Exercises 2 and 3.

5. Let $F' = \mathbf{Q}(\sqrt{2})$ and $F'' = F'(\sqrt{3})$. Find a quadratic polynomial in F' whose zero is $\sqrt{2} + \sqrt{3}$ and then a polynomial of degree 4 over $\mathbf{Q}$ which has $\sqrt{2} + \sqrt{3}$ as a zero.

6. Show that $\sqrt{2}$ and $\sqrt{3}$ are in the field $\mathbf{Q}(\sqrt{2} + \sqrt{3}) = F'$ and hence that if $F = \mathbf{Q}(\sqrt{2})$, then $F' = F(\sqrt{3})$.

7. For which of the following fields is $x^2 - 2$ a prime polynomial: $\mathbf{Z}_3$, $\mathbf{Z}_5$, $\mathbf{Z}_{11}$?

8. Show in detail why the field of polynomials in x over $\mathbf{Q}$ modulo $x^2 - 2$ is isomorphic to the field $\mathbf{Q}(\sqrt{2})$.

9. Show that $x^3 - 2$ is prime in $\mathbf{Z}_7$. Let c be a zero of $x^3 - 2$ in the field F with 7^3 elements. Show that $\mathbf{Z}_7(c)$ is the splitting field of $x^3 - 2$.

10. Show that $x^3 - 3$ is prime in $\mathbf{Z}_7$. Let d be a zero of $x^3 - 3$ in the field F' with 7^3 elements. Find an isomorphism between F' and the field F of Exercise 9. Are F and F' the same field?

11. Let $f(x) = 3x^4 + 2x^3 + x^2 + 2x + 3 = x^2[3(x^2 + 1/x^2) + 2(x + 1/x) + 1]$. Let $y = x + 1/x$ and show that $f(x) = x^2g(y)$ for some quadratic polynomial in y. Find a splitting field for $f(x)$.

12. Find a splitting field of $f(x) = x^4 + x^2 + 1$.

13. Let $p(x)$ be a quadratic prime polynomial over a field F. Show that the splitting field of $p(x)$ is an algebraic extension of degree 2.

14. Find the splitting field of $f(x) = (x^5 - 1)/(x - 1)$ over $\mathbf{Q}$. Show that it is an algebraic extension of degree 4.

15. By Theorem 13.7, $f(x) = (x^p - 1)/(x - 1)$, for p a prime number, is a prime polynomial. Show that the splitting field of $f(x)$ is of degree $p - 1$.

16. Show that $a + b\sqrt{2} + c\sqrt{3} + d\sqrt{6}$, for a, b, c, and d in $\mathbf{Q}$ is a zero of a quadratic polynomial with coefficients in $\mathbf{Q}(\sqrt{2})$.

17. Show that $c = \sqrt[3]{2}$ is a zero of a quadratic polynomial with coefficients in $\mathbf{Q}(cw)$, where w is a zero of $y^2 + y + 1$.

*18. Let F be a field containing p^n elements for p a prime number and n an integer greater than 1. Prove that for each nonzero x in F, x^u is in $\mathbf{Z}_p$, if u is a multiple of $(p^n - 1)/(p - 1)$.

16. *Derivatives and Separability*

In proving Theorem 15.1, we avoided the use of derivatives. But in Section 5 of Chapter VI and later we shall refer to a result that is proved most easily by use of derivatives. Since it is also interesting to see how one can define a derivative over a general field that is not even ordered, in this section we show how it can be done. Actually, what we do is to define a derivative without limits so that the result is the same as if limits were involved. We start with polynomials and then show that the process can be extended to include rational functions, that is, quotients of polynomials. The method fails if the functions are not rational ones, for instance, for sin x.

Let $g(x)$ be a polynomial with coefficients in any field F. From the factor theorem (the corollary of Theorem 12.1)

(16.1) $$G(x, a) = \frac{g(x) - g(a)}{x - a}$$

is a polynomial in x and a since the numerator is zero when $x = a$. Then a derivative is defined as follows.

Definition.   *If $g(x)$ is a polynomial over a field F and $G(x, a)$ is defined as in (16.1), then the derivative of $g(x)$, written $g'(x)$, is $G(x, x)$.*

For example, if $g(x) = x^3 + 2x + 1$. Then

$$G(x, a) = \frac{x^3 + 2x + 1 - (a^3 + 2a + 1)}{x - a} = \frac{x^3 - a^3 + 2(x - a)}{x - a}$$

$$= x^2 + ax + a^2 + 2$$

and $G(x, x) = 3x^2 + 2$, which is, not surprisingly, what one would get by using the usual formulas for differentiation.

The formulas for differentiation follow just as in beginning calculus. We list them here for convenience in terms of polynomials $h(x)$ and $g(x)$:

1. $[g(x) + h(x)]' = g'(x) + h'(x)$.
2. $[g(x)h(x)]' = g(x)h'(x) + g'(x)h(x)$.
3. $g'(cx) + cg'(x)$.
4. If $g(x) = b_n x^n + b_{n-1}x^{n-1} + \cdots + b_1 x + b_0$, then

$$g'(x) = nb_n x^{n-1} + (n - 1)b_{n-1}x^{n-2} + \cdots + b_1.$$

The formula for the derivative of a composite function also holds, but that requires more careful notation; since we shall not use this formula, we do not include it. (If you are interested, you might see how it goes.) Indeed, since the proofs of the four properties listed go practically as in the calculus, there is no need to repeat them here. But just to show what it looks like in this setting, we given the proof of formula 2.

Let $k(x) = g(x)h(x)$. Then

$$K(x, a) = \frac{k(x) - k(a)}{x - a} = \frac{g(x)h(x) - g(x)h(a) + g(x)h(a) - g(a)h(a)}{x - a}$$

$$= g(x)H(x, a) + h(a)G(x, a).$$

Hence $K(x, x) = g(x)h'(x) + g'(x)h(x)$.

Although we shall not have occasion to use it, we can also define the derivative of a rational function. Suppose that $r(x) = g(x)/h(x)$, where g and h are polynomials. Then if $h(a) \neq 0$, we have

$$(16.2) \qquad \frac{r(x) - r(a)}{x - a} = \frac{g(x)/h(x) - g(a)/h(a)}{x - a} = \frac{g(x)h(a) - h(x)g(a)}{h(x)h(a)(x - a)}$$

$$= \frac{L(x, a)}{h(x)h(a)},$$

where, from the factor theorem, $L(x, a)$ is a polynomial in x and a. Using the method that we used to prove formula 2, we can derive the formula for the derivative of a quotient:

5. If $r(x) = g(x)/h(x)$, then

$$r'(x) = \frac{h(x)g'(x) - g(x)h'(x)}{h^2(x)}.$$

Our chief purpose in defining derivatives is to enable us to prove the following two theorems.

Theorem 16.1.   Let $f(x)$ be a polynomial with coefficients in a field F and let c be an element of F', a field containing F, such that $f(c) = 0$. Then in F', $(x - c)^2$ is a factor of $f(x)$ if and only if $f(c) = f'(c) = 0$.

Proof: Suppose that $f(x) = (x - c)^2 h(x)$. Then

$$f'(x) = 2(x - c)h(x) + (x - c)^2 h'(x),$$

which implies that $f'(c) = 0$. Conversely, suppose that $f(c) = f'(c) = 0$. From the factor theorem, $f(x) = (x - c)g(x)$. Then

$$f'(x) = g(x) + (x - c)g'(x) \quad \text{and} \quad f'(c) = g(c) = 0.$$

The factor theorem then shows that $x - c$ is a factor of $g(x)$ and completes the proof of the theorem.

Theorem 16.2.   Let $f(x)$ be an irreducible polynomial over a field F and let c be a zero of $f(x)$ in a field F', an algebraic extension of F. Then $(x - c)^2$ is a factor of $f(x)$ if and only if $f'(x)$ is the zero polynomial.

Proof: Suppose that $(x - c)^2$ is a factor of $f(x)$. Then, by Theorem 16.1, $x - c$ is a factor of $f'(x)$. Now $f'(x)$ has coefficients in F. Since $f(x)$ is irreducible over F, there is no polynomial over F of smaller degree which has c as a zero. Since $f'(x)$ either has smaller degree than $f(x)$ or is the zero polynomial, the latter is the only possibility. On the other hand, if $f'(x)$ is the zero polynomial, $x - c$ is trivially a factor of $f'(x)$ and, by Theorem 16.1, $(x - c)^2$ is a factor of $f(x)$. This completes the proof.

Corollary.   If F is a field of characteristic zero, an irreducible polynomial $f(x)$ over F has no repeated factors in any extension field F'; that is, $(x - c)^2$ is not a factor of $f(x)$.

The corollary follows, since in a field of characteristic zero, the derivative of a polynomial is zero if and only if it is in the field of coefficients.

Although we shall use only Theorem 16.2, there is some point in putting this result in perspective. To do this, we carry the topic a little further, first by giving an example of an irreducible polynomial with multiple zeros and then exploring a little the conditions under which the phenomenon occurs.

EXAMPLE. Let F be the field consisting of all rational functions of t with coefficients in $\mathbf{Z}_p$ for some odd prime number p. Then consider the polynomial

$$f(x) = x^p - t.$$

This polynomial has no zero in F, since there is no rational function of t whose pth power is t. But if we adjoin to F a zero w of $f(x)$, we see that

$$(x - w)^p = x^p - w^p = x^p - t.$$

This shows first that $f(x) = (x - w)^p$ in the extension of F and also that $f(x)$ is irreducible in F, since w is not in F. This is in accord with Theorem 16.2, since $f'(x)$ is the zero function in F.

Definition. *An element w of an extension F' of a field F is called* separable *if it is a zero of some polynomial over F that has no repeated factors in F'. If all the elements of F' are separable over F, F' is called a* separable extension. *A polynomial over F is called* separable *if all its zeros in extensions of F are separable. If an element or an extension is not separable, they are called* inseparable.

The term may seem a little curious. One can think of it intuitively to mean that two equal zeros cannot be "separated." We can restate the preceding corollary by use of this terminology.

Corollary. All extensions of a field of characteristic zero are separable.

Let us see what inseparable polynomials look like. Suppose that $f(x)$ is over some field F_p of characteristic p and that $f'(x)$ is the zero polynomial. Then the only coefficients of $f(x)$ that are not zero are coefficients of powers of x that are multiples of p. That is, $f(x) = g(x^p)$ for some polynomial g over F_p. Thus we can write

$$f(x) = b_r(x^p)^r + b_{r-1}(x^p)^{r-1} + \cdots + b_1 x^p + b_0,$$
$$g(x) = b_r x^r + b_{r-1} x^{r-1} + \cdots + b_1 x + b_0.$$

Then, using the binomial theorem as in the proof of Theorem 15.1, we see that $f(x) = [g(x)]^p$. So we have proved the following theorem.

Theorem 16.3. A polynomial $f(x)$ over F_p, a field of characteristic p, is inseparable if and only if it is the pth power of a polynomial over the same field.

Corollary 1. If c is a zero of an inseparable polynomial $f(x)$ over F_p, then

(16.3) $$f(x) = (x - c)^{pm}[h(x)]^p$$

for some positive integer m and $h(x)$ a polynomial for which $h(c) \neq 0$.

The corollary follows, since if $f(x) = [g(x)]^p$, then c must be a zero of $g(x)$ and we can write $g(x) = (x - c)^m h(x)$, where $h(c) \neq 0$.

Corollary 2. If F_p is a field of characteristic p and $f(x)$ is an irreducible polynomial over this field, then $f(x)$ is separable unless its degree is a multiple of p.

We have seen, then, that we must work rather hard to find an inseparable extension of a field. As noted, we shall be concerned from this point on only with separable fields. There is more attention paid to inseparable fields in the books by van der Waerden, Herstein, and Birkhoff and Mac Lane listed in the Bibliography.

17. Rings

There is another algebraic structure, called a *ring*, which is obtained by relaxing some of the conditions of a field. We have not introduced it up to this point because our concern was with various kinds of integral domains and fields. Since the concept of a ring comes to mind when we deal with matrices, let us review some of the properties of the set S of matrices of order 2 considered in Section 4 of Chapter I. We defined there the binary operations of addition and multiplication. We found that S is an Abelian group under addition. With regard to multiplication S is associative and has an identity element, the identity matrix. But multiplication is not commutative nor does a multiplicative inverse always exist for nonzero matrices. With this in mind we define a ring as follows.

Definition.   *A* ring *is a set S of one or more elements and two binary operations, in notation addition and multiplication, such that*

1. *The elements of S form an additive Abelian group.*
2. *Products of elements of S are associative.*
3. *The distributive property holds.*

The properties of a ring are just the properties described above for matrices of order 2 (or, in fact, of any fixed order) except that the set of matrices has a multiplicative identity. So the set of matrices of order 2 with elements in any *field* is a ring with an identity element. In fact, the same statement holds if "field" is replaced by "integral domain" or, indeed, by "ring."

A ring in which multiplication is commutative is called a *commutative ring*. The best example we have of such a ring is a set of integers modulo m, where m need not be a prime number. This particular ring is not only commutative but also has a multiplicative identity. In fact, the set of integers modulo m has all the properties of an integral domain except that it has divisors of zero.

A ring that has all five properties of a field listed in Section 1, except for commutativity of multiplication, is called a *skew field* or *division ring*. (Actually, as was indicated at the beginning of this chapter, some authors prefer not to require commutativity in a field.) The simplest example of a division ring that is not a field is the set of "quaternions," which we defined in terms of matrices in Exercise 4(e) of Section 4 in Chapter I, as follows: $Q = \{a_1I + a_2J + a_3K + a_4L\}$, where the a's are real numbers, I is the identity matrix of order 2, and

$$J = \begin{bmatrix} i & 0 \\ 0 & -i \end{bmatrix}, \qquad K = \begin{bmatrix} 0 & 1 \\ -1 & 0 \end{bmatrix}, \qquad L = JK, \qquad i^2 = -1.$$

Let us see how this works. Multiplication of matrices gives the following:

(17.1) $$J^2 = K^2 = L^2 = -I \qquad \text{and} \qquad JK = -KJ.$$

We can disregard, if we wish, the definition of J, K, and L in terms of matrices, replace I by 1, and verify, from (17.1), the following:

(17.2) $$(a_1 + a_2J + a_3K + a_4L)(a_1 - a_2J - a_3K - a_4L) = a_1^2 + a_2^2 + a_3^2 + a_4^2.$$

Since the sum of squares on the right of (17.2) is a positive real number unless all four a's are zero, it follows that every quaternion not zero has a multiplicative inverse. Hence the set of quaternions so defined is a division ring.

Recall that an integral domain with a finite number of elements is a field. A remarkable theorem of Wedderburn states that a division ring with a finite number of elements is a field. Two proofs and a reference to the original proof may be found in Chapter 7 of Herstein's *Topics in Algebra*.

We can define an isomorphism between two rings just as we did for fields. Homomorphisms of rings are similarly defined. Since we shall have more to say about rings in Chapter V, we shall here give just one example of a ring homomorphism and a second "example," which looks like a ring homomorphism but is not.

EXAMPLE 1. Let M be the set of matrices of order 2 whose elements are integers and T the set of matrices of order 2 whose elements are integers between 0 and $m - 1$ inclusive, for some integer $m > 1$. In T, addition and multiplication is modulo m. For each matrix A in M, define a correspondence σ by

$$A\sigma = R_A,$$

where R_A is in T and is congruent to A modulo m; that is, corresponding elements of the two matrices are congruent. Then σ is a homomorphism of M onto T, for

$$(A + B)\sigma = R_{A+B} = R_A + R_B = A\sigma + B\sigma,$$
$$(AB)\sigma = R_{AB} = R_A R_B = (A\sigma)(B\sigma).$$

It should be noted that $R_A R_B$ is not just the product of R_A and R_B but the matrix of T, which is congruent to their product. Just as for groups, the kernel of a homomorphism is defined to be the set of elements of M whose image is the zero matrix in T. Here the kernel of σ is the set of matrices whose elements are all multiples of m.

EXAMPLE 2.   This is an example that looks like a ring homomorphism but is not. In Exercise 9 you will be asked to show why it is not. Let M be the set of matrices of order 2 with elements in a field F. Let N be the subset of M such that every matrix in N has 0 in the upper right corner; N is a sub-ring of M (?). Define a correspondence σ so that the image of any matrix A of M is the matrix obtained by replacing its upper right element by 0. Then σ is a homomorphism of M onto N (?).

If one looks at a proof of the binomial theorem (Section 7), one sees that the expansion of $(b + c)^n$ for n a positive integer is valid for b and c in any ring provided only that $bc = cb$. The whole ring need not be commutative. Hence the expansion would hold for two commutative matrices. In fact, in a way, one need not confine n to be an integer under certain conditions. Let us look at a curious result. Suppose that N is some matrix of order 2 (or, in fact, of any order) whose square is the zero matrix. Then look at the *formal* expansion of $(I + N)^{1/2}$ by the binomial theorem. That is, we see what we would get if the binomial theorem were true for the exponent $\frac{1}{2}$, assuming that the field is not of characteristic 2. Then the expansion would be

$$B = I^{1/2} + \tfrac{1}{2}I^{-1/2}N,$$

which stops there, since all higher powers of N are zero. But $I^{1/2}$ is I, since $I^2 - I$ and B may be written $I + \frac{1}{2}N$. This "ought to" be the square root of $I + N$. That this is indeed the case is shown by computing

$$B^2 = (I + \tfrac{1}{2}N)^2 = I + N + \tfrac{1}{4}N^2 = I + N.$$

What we have done is to use the binomial expansion to make an "educated guess" and verified that the guess is true.

Exercises

1. Suppose that N is a matrix such that $N^3 = 0$. By use of a formal binomial expansion, find a cube root of $I + N$.

2. Suppose that N is a matrix such that $N^3 = 0$. By use of a formal binomial expansion, find a cube root of $I + cN$ where c is a nonzero element of the field in which the coefficients of the matrix lie.

3. When we examined the square root of $I + N$ with $N^2 = 0$, the assumption was made that in the field of elements of the matrices $2 \neq 0$; that is, the characteristic is different from 2. Is there a square root if the field is of characteristic 2?

4. Suppose that N is a matrix for which $N^2 = 0$. Let n be a positive integer and find a matrix B such that $B^2 = nI + N$.

5. Let R_m denote the ring of integers modulo m for some integer m. Find all the automorphisms of R_m.

6. Is the ring of integers homomorphic to the ring of integers modulo m? Why or why not?

7. Show that in $\mathbf{Z}_p$

$$\frac{x^p - 1}{x - 1} = (x - 1)^{p-1}.$$

8. Show that in $\mathbf{Z}_p$,

$$\frac{x^p - u^p}{x - u} = (x - u)^{p-1}.$$

9. Show why the correspondence in Example 2 is not a homomorphism.

10. By multiplication and use of (17.1), show that (17.2) holds.

11. Use the results of Section 16 to prove that over a field of characteristic zero the polynomial $x^q - x$ has no repeated factors for q an integer greater than 1.

12. Let $f(x)$ be a polynomial of degree n over a field F and c an element in F or an extension of F. Prove that

$$f(x) = f(c) + f'(c)(x - c) + \tfrac{1}{2}f''(c)(x - c)^2 + \cdots$$

$$+ \frac{f^{(r)}(c)}{r!}(x - c)^r + \cdots + \frac{f^{(n)}(c)}{n!}(x - c)^n.$$

[$f^{(r)}$ denotes the rth derivative.] This is called *Taylor's series*.

13. Use Exercise 12 or other means to prove that if, over a field F,

$$f(x) = (x - c)^t g(x), \qquad \text{with } g(c) \neq 0,$$

then

$$f(c) = f'(c) = \cdots = f^{(t)}(c) = 0 \qquad \text{and} \qquad f^{(t+1)}(c) \neq 0.$$

18.   Hasse's Generalization of a Euclidean Domain

A question that naturally arises in connection with Euclidean domains as defined in Section 9 is: Are there integral domains that are not Euclidean domains and yet in which unique decomposition into prime factors does hold? As we stated in Section 11, T. Moskin has shown that the answer to the question is "yes." In this section we discuss, for those interested, some of the ramifications of this question. No use is made of this section later in the book.

Dedekind (Dirichlet-Dedekind, *Vorlesungen der Zahlentheorie*, reprint by Chelsea Publishing Company, New York, 1968, p. 451) showed that the integral domain $L = \{x + y\alpha\}$, where $\alpha = \frac{1}{2}(1 + \sqrt{-19})$ and x and y are integers, has the property of unique decomposition into prime factors and yet L is not a Euclidean domain for $n(x + y\alpha) = (x + y\alpha)(x + y\bar{\alpha})$, where $\bar{\alpha}$ is the complex conjugate of α; that is,

$$n(x + y\alpha) = x^2 - xy + 5y^2.$$

Moskin's contribution was to show that there is no n-value whatsoever (he called it a norm) for which L is Euclidean.

Helmut Hasse [*Journal für Mathematik*, vol. 159 (1926), pp. 3–12] showed that by modifying somewhat the conditions defining an n-value, one can find sufficient conditions for unique decomposition. For many integral domains the conditions are also necessary, but to deal with this would take us too far afield. Now we lead up to Hasse's statement of a sufficiency condition and prove it.

Hasse starts with an integral domain V and in place of the n-value uses another function $\chi(a)$ for a in V. In the first place, $\chi(a)$ is not assumed to be an integer, as was the case for the n-value, but it is to be real-valued and non-negative. This is part of condition 1 below. However, the function χ is restricted to have no infinite sequence

$$\chi(a_1) > \chi(a_2) > \chi(a_3) > \cdots.$$

One can think of this as a condition of being well-ordered; this is condition 3 below. Therefore, these relax somewhat the properties of an n-value.

But χ is restricted to be multiplicative; that is, $\chi(ab) = \chi(a)\chi(b)$. This holds not only for norms but also for absolute values that may not be integers. We can make the function multiplicative for polynomials by defining $\chi(f) = 2^k$, for instance, where k is the degree of the polynomial f, and $\chi(0) = 0$. Then $\chi(fg) = \chi(f)\chi(g)$. This is condition 2 below.

We have modified a little for simplicity's sake, conditions 3 and 4 of Hasse. The third condition is stated somewhat differently, although the two statements are almost equivalent. For the fourth condition we have used a sharpened form of his, but he gives a proof that the sharpened condition can be deduced from a seemingly weaker one. In the statement of the theorem, there are two italicized terms that we shall define immediately afterward.

Theorem 18.1. (Hasse.) Let V be an integral domain. Then V has the property of *unique decomposition* if there exists a real-valued function $\chi(y)$ for y in V with the following properties for all a, b, y, and z in V.

1. $\chi(a) \geq 0$ with $\chi(a) = 0$ if and only if $a = 0$.
2. $\chi(ab) = \chi(a)\chi(b)$.
3. Every decreasing sequence $\chi(a_1) > \chi(a_2) > \chi(a_3) > \cdots$ is finite in extent, that is, has a least element.
4. If $\chi(a) \geq \chi(b) > 0$, then there exist z and y in V with y *prime to* b such that $\chi(ya - zb) < \chi(b)$.

First, we repeat for convenience a few terms that we defined in Section 8.

Definitions. *A* unit *of V is an element of V whose reciprocal (multiplicative inverse) is in V. We call b a* factor *or divisor of c if there is a d in V such that $bd = c$; we use the notation $b \mid c$. An element p of V is called a* prime *if its only divisors are units and unit multiples of itself. The last definition can be made a little more concise by calling two elements b and c of V associates if each divides the other; that is, $b = uc$, where u is a unit. In these terms we can call an element of V a* prime *if its only divisors are units and associates.*

Now we define the two terms underlined in the statement of the theorem.

Definition. *If a, b, and c are in V, call b prime to a if $a \mid bc$ implies $a \mid c$. (This condition should be used with care. Notice that b prime to a does not imply that a is prime to b.)*

Definition. *An element not a unit of an integral domain is said to have the property of* unique decomposition *if it can be written as a product of prime elements and if*

$$n = p_1 p_2 \cdots p_k = q_1 q_2 \cdots q_t$$

for primes p_i and q_j implies that $k = t$ and, after reordering the q's if necessary, p_i is an associate of q_i for $1 \leq i \leq k$. An integral domain has the property of unique decomposition *if all its elements that are not units have this property.*

We should emphasize the point that if we compare χ with the n-value, we see that the former is less restricted than the latter in that its values need not be integers, but more restricted in that it is assumed to be multiplicative. Notice that if V is a Euclidean domain whose n-value is multiplicative, then it satisfies the conditions of Theorem 18.1 for χ the n-value and $y = 1$ in condition 4. Now we give the proof.

Proof: We wish to show that if conditions 1, 2, 3, and 4 hold, then every nonunit element of V has the property of unique decomposition. First, we prove the following (compare the corresponding discussion for n-values).

5. $\chi(a) \geq 1$ for $a \neq 0$. For suppose that $\chi(a) < 1$; then $\chi(a) > \chi(a^2) > \chi(a^3) > \cdots$, which denies property 3.

6. An element u of V is a unit if and only if $\chi(u) = 1$. Let e be the identity element of V. Then properties 1 and 2 imply that $\chi(e) = 1$. If u is a unit, then $uv = e$ for some element v of V and $\chi(u)\chi(v) = 1$. Since by property 5, $\chi(u) \geq 1 \leq \chi(v)$, it follows that $\chi(u) = \chi(v) = 1$. Conversely, if $\chi(b) = 1$, then, by property 4, there exist y and z in V with y prime to b such that $\chi(ye - zb) < \chi(b) = 1$. By 5 this implies that $ye - zb = 0$. Thus $b \mid ye$ and, since y is prime to b, $b \mid e$. Hence b is a unit. We have proved property 6.

Next we prove that every element a, not a unit, in V is expressible as a product of primes. Suppose that a is not a prime; then $a = bc$, where neither b nor c is a unit. Now $\chi(b) < \chi(a)$ by 2, 5, and 6. If b is not a prime, express it as a product of elements of V. So continue or use an induction argument, noting that the process must stop from condition 3.

To show unique decomposition, suppose that this property fails for a nonunit element c in V. Since by property 3 there cannot be an infinite sequence of such elements c with decreasing values of $\chi(c)$, it follows that there must be such a c for which $\chi(c)$ is a minimum. So let m be a nonunit element of V that does not have the property of unique decomposition and such that if d is a nonunit element of V for which $\chi(d) < \chi(m)$, then d has the property of unique decomposition. Now m cannot be prime, since for it unique decomposition does not hold. Suppose that

$$m = p_1 p_2 \cdots p_r = q_1 q_2 \cdots q_s,$$

where the p_i and q_j are primes and, since m is not a prime, r and s are greater than 1. If p_1 were an associate of q_1, then m/p_1 would have the property of unique decomposition, since $\chi(m/p_1) < \chi(m)$ and m/p_1 is not a unit. This would imply that m has the property of unique decomposition. So we suppose that no p_i is an associate of any q_j. Take $\chi(p_1) \geq \chi(q_1)$ and choose z and y with y prime to q_1, so that $\chi(yp_1 - zq_1) < \chi(q_1)$. Define

$$
\begin{aligned}
m' &= (yp_1 - zq_1)p_2 p_3 \cdots p_r = yp_1 p_2 \cdots p_r - zq_1 p_2 p_3 \cdots p_r \\
&= yq_1 q_2 \cdots q_s - zq_1 p_2 p_3 \cdots p_r \\
&= q_1(yq_2 q_3 \cdots q_s - zp_2 p_3 \cdots p_r).
\end{aligned}
$$

Since q_1 is not an associate of any of $p_2, p_3, \ldots, p_r$ and since $\chi(m') < \chi(m)$ implies that m' has the property of unique decomposition, it follows that q_1 is a factor of $yp_1 - zq_1$ and hence $q_1 \mid yp_1$. Since y is prime to q_1, we see that $q_1 \mid p_1$, which is a contradiction. This completes the proof.

For the integral domain $L = \{x + y\alpha\}$ discussed in the second paragraph of this section, there is a function χ having the properties listed in Theorem 18.1. In fact, χ can be taken to be the n-value referred to in that paragraph. However, there are integral domains in which unique decomposition holds and yet in which a function does not exist having the properties listed in Theorem 18.1. For further details see Hasse's article.

IV

Vector Spaces and Matrices

1. Introduction

This could be thought of as a "service chapter," in that the selection of topics is largely dictated by what we need in the final chapters of this book. (More specifically, see the Preface.) This is not to say that these topics are an unimportant part of algebra but that they are usually dealt with under the title "linear algebra." Since many who read this book will already have had some experience in this branch of mathematics, there is little point in trying to repeat the whole subject here.

On the other hand, it would be inconvenient to have to refer you to results in other books, as they are needed in the following chapters. Furthermore, there is some advantage in looking at vector spaces, matrices, and determinants from the point of view of what has gone before in this book.

2. Vector Spaces

In Section 3 of Chapter I and at various other spots we have concerned ourselves with ordered pairs: (a, b), where a and b are in some field. We have called such a pair a vector, since it could be thought of as the coordinates of the terminal point of a physical vector originating at the origin. Another reason for calling it a vector is that one can define the sum of two number pairs in accord with vector addition. Although we shall shortly look at vectors from a more abstract point of view, our first definition is the following specialized one.

Definition. *A sequence of n elements of a field F, $\alpha = (a_1, a_2, \ldots, a_n)$, that is, an ordered n-tuple of elements of F, is called a* vector. *The a_i are called the* components *of the vector. We say that the vector is* over *a field F if its components are in F.* (In this chapter we usually denote vectors by Greek letters.)

Two vectors are *equal* if and only if corresponding components are equal. Just as for number pairs, we define the sum of two vectors as follows.

Definition. *If $\alpha = (a_1, a_2, \ldots, a_n)$ and $\beta = (b_1, b_2, \ldots, b_n)$ are two vectors over a field F, their sum is the vector*

$$(a_1 + b_1, a_2 + b_2, \ldots, a_n + b_n).$$

We showed in Section 3 of Chapter I that the set of ordered number pairs over a field form an additive Abelian group. We leave as Exercise 17 the proof of the following theorem.

Theorem 2.1. The set of all vectors $\xi = (x_1, x_2, \ldots, x_n)$, where the x_i are elements of a field F, form a group under addition. The additive identity is the vector $(0, 0, \ldots, 0)$ with n components and the additive inverse of ξ is $(-x_1, -x_2, \ldots, -x_n)$.

Definition. *We call the additive identity vector the* zero *or* null *vector and designate it by θ.*

Another operation on vectors is multiplying by an element of the field, often called multiplication by a scalar ("scalar" being another word for an element of the field). It is defined as follows:

Definition. *If $\alpha = (a_1, a_2, \ldots, a_n)$ and if c and all the a_i are in a field F, then*

$$c\alpha = (ca_1, ca_2, \ldots, ca_n).$$

Now the set of all n-tuples over a field F is closed under addition and multiplication by a scalar. (Such a set is often denoted by V^n.) But V^n can also have subsets that are closed under the two operations. For instance, the set $\{x\alpha\}$ as x ranges over the elements of a field is also closed under addition and multiplication by a scalar. Any such subset is called a vector space, in accord with the following definition.

Definition. *A set of ordered n-tuples over a field F that is closed under addition and multiplication by a scalar is called a* vector space.

For instance, let $\alpha = (1, 1, 1)$ and $\beta = (1, 1, 0)$. Then if a vector space contains both α and β, it must contain all vectors

$$x\alpha + y\beta = x(1, 1, 1) + y(1, 1, 0) = (x + y, x + y, x).$$

This set is a vector space S. Also $\{x\alpha\}$ and $\{y\beta\}$ are vector spaces that are proper subspaces of S. Of course, S is a proper subspace of V^3.

Based on the above experience, we can give an abstract definition that has the advantage of freeing us from thinking of n-tuples. It may be that at this point you do not crave such freedom. In that case you may remain in bondage for a time, but you should not allow this to continue too long, for there are advantages in the more abstract point of view.

Definition. *A* vector space V *over any field F is a nonempty set of entities S, called* vectors, *if for every α, β, and γ in S two operations are defined: addition, $\alpha + \beta$, and multiplication by a scalar, $c\alpha$, for c in F, with the following properties:*

1. *The set S is an Abelian group under addition; the additive identity is called the* zero *or* null vector *and denoted by θ.*
2. *If c is in F and α in S, then cα is in S and cα = αc* (closure and commutativity of multiplication by a scalar).
3. *For c and c' in F, (cc')α = c(c'α)* (associativity for scalar multiplication).
4. *For c and c' in F, c(α + β) = cα + cβ and (c + c')α = cα + c'α* (distributivity for multiplication by a scalar).
5. $1 \cdot \alpha = \alpha$.

Sometimes in the definition of a vector space it is not assumed that $c\alpha = \alpha c$; this would be true, for example, if we were considering a vector space over a noncommutative ring instead of a field. But *we* shall assume commutativity.

There are two properties of a vector space that follow immediately from the five listed above. We prove the first and leave the proof of the second as Exercise 16.

6. $0 \cdot \alpha = \theta$ for all α in S.
7. $(-1)\alpha = (-\alpha)$.

To prove property 6, write the following sequence of equations:

$$\alpha + 0 \cdot \alpha = 1 \cdot \alpha + 0 \cdot \alpha = (1 + 0)\alpha = 1 \cdot \alpha = \alpha.$$

This shows that if $0 \cdot \alpha$ is added to α, the result is α. From the uniqueness of the additive identity element of a group we know that $0 \cdot \alpha = \theta$.

Since we set up the abstract definition of a vector space on the basis of our experience with n-tuples of elements of a field, it is scarcely surprising that a set of n-tuples closed under addition and multiplication by a scalar is a vector space. But we should state this formally.

Theorem 2.2. Let F be a field and S a nonempty set of n-tuples whose components are in F. If S is closed under addition and multiplication by scalars of F, then S is a vector space.

Here are three examples of vector spaces.

EXAMPLE 1.   Let $\alpha_1, \alpha_2, \ldots, \alpha_k$ be a set of n-tuples of elements of a field F. Recall from Section 3 of Chapter I that

$$x_1\alpha_1 + x_2\alpha_2 + \cdots + x_k\alpha_k$$

is called a *linear combination* of the α_i with the x_i as *coefficients*. The set S of all such linear combinations with the x_i in F is closed under addition and multiplication by a scalar in F and hence, by Theorem 2.2, is a vector space.

The next two examples involve linear combinations but are not concerned with n-tuples.

EXAMPLE 2.   Consider all linear combinations of

$$1, x, x^2, \ldots, x^r, \ldots$$

with coefficients in a field F. This set is $F[x]$, the set of all polynomials in x over F. To see that it is also a vector space, notice first that since $F[x]$ is an integral domain (see Section 4 of Chapter III), it follows that it is an Abelian group under addition. This is the first of the five properties defining a vector space, where the null vector of this space is the zero polynomial. Property 2 holds, since if $f(x)$ is in $F[x]$ and c in F, then $cf(x)$ is also in $F[x]$ and $cf(x) = f(x)c$. Similarly, properties 3, 4, and 5 hold as well. So we know that $F[x]$ is a vector space.

EXAMPLE 3.   Consider a simple algebraic extension $F(c)$ of a field F. (See Section 14 of Chapter III.) Let n be the degree of $F(c)$ over F. Then the elements of $F(c)$ are the linear combinations of

$$1, c, c^2, \ldots, c^{n-1}$$

with coefficients in F. Just as for polynomials the set of all such linear combinations is a vector space. That is, $F(c)$ is a vector space over F.

Since the remaining chapters of this book are vitally concerned with extensions of fields, this connection with vector spaces is basic. In fact, except for this relation we would have little use for vector spaces in the chapters that follow.

3.   Spanning a Vector Space

In dealing with groups, we found that generators were useful. The corresponding concept for vector spaces goes by a different name. For instance, let ϵ_i denote the n-tuple all of whose elements are zero except for a 1 in the ith place. Then every vector in V^n can be written as follows:

$$(3.1) \qquad x_1\epsilon_1 + x_2\epsilon_2 + \cdots + x_n\epsilon_n = (x_1, x_2, \ldots, x_n), \qquad x_i \text{ in } F.$$

Thus every vector of V^n is a linear combination of the vectors ϵ_i with coefficients in F. We say that the set of vectors $\epsilon_1, \epsilon_2, \ldots, \epsilon_n$ *span* (instead of generate) V^n. More generally, we have the following definition.

Definition.   *A set of vectors S is said to* span *a vector space V over F if V consists of all linear combinations of the vectors of S with coefficients in F.*

In Example 1, the α's span the vector space. In Example 2, the powers of x span $F[x]$ over F. In Example 3, $1, c, c^2, \ldots, c^{n-1}$ span $F(c)$ over F.

Of course, for any vector space, there can be a number of sets that span it. For instance, V^3 is spanned by each of the following sets:

$$S_1: (1, 0, 0), (0, 1, 0), (0, 0, 1), (1, 1, 1).$$
$$S_2: (1, 1, 1), (1, -1, 1), (-1, 1, 1).$$
$$S_3: (1, 0, 1), (0, 1, 1), (1, 1, 0).$$

You are asked to show in the exercises that these indeed span V^3.

Exercises

1. In each case when the indicated sum exists, express it as a single vector; when it does not exist, explain why.
   (a) $(1, 2, 3) + (3, 4, 5)$.   (b) $3(1, 0, 5) + 2(1, 0)$.
   (c) $3(1, 0, 5) + 2(\frac{1}{2}, 3, 4)$.   (d) $5(1, 0, 7) - 3(4, \frac{1}{2}, 7)$.
   (e) $(1, 0, -6, 7) + (1, 2, 3)$.
   (f) $2(1, 0, 4, 8) + (2, 3, -2, 7) + 0(5, 7, 6)$.

2. In the following sets let a and b range over all real numbers. Which of the sets are vector spaces? Give reasons for your answers.
   (a) $\{(a, a + 1)\}$.   (b) $\{(a, b)\}$, where $a + b = 0$.
   (c) $\{(a, b)\}$, where $a^3 + b^3 = 0$.

3. Find the vector α satisfying each of the following equations.
   (a) $\alpha + (3, 4, 5) = (7, 1, -7)$.   (b) $3\alpha = (1, 2, 4)$.
   (c) $\alpha + 2(3, 1, 0) = (2, 0, 5)$.   (d) $5\alpha + (7, -2, 5) = 7\alpha$.

4. Find the pairs of vectors α and β that satisfy the following pairs of equations, when such vectors exist. When they do not exist, explain why not.
   (a) $3\alpha + 4\beta = (1, 0, 5)$ and $\alpha + 2\beta = (2, 3, 0)$.
   (b) $2\alpha + \beta = (1, 2)$ and $4\alpha + 2\beta = (2, 3)$.
   (c) $3\alpha + \beta = (4, 4)$ and $6\alpha + 2\beta = (8, 8)$.

5. Show that all the solutions (x, y) of the following pair of equations form a vector space:

$$3x - 2y = 0,$$
$$6x - 4y = 0.$$

6. If in Exercise 5 the zeros on the right were replaced by ones, would the solutions form a vector space?

7. Show that the set of solutions (x, y, z) of the following pair of equations forms a vector space:

$$3x + 4y + z = 0,$$
$$x - 2y + 4z = 0.$$

8. If in Exercise 7 the two zeros on the right were replaced by ones, would the set of solutions form a vector space?

9. Let S be the vector space spanned by the following set of three vectors: $(1, 2, 4), (1, -1, 3), (2, 1, 7)$. Does any proper subset of these three vectors span S? Why or why not?

10. Answer the same question in Exercise 9 when the third vector of the set is replaced by $(1, 0, 0)$.

11. Let $\alpha_1 = (1, 1, 1)$ and $\alpha_2 = (1, 0, -1)$. For what vectors α will the three vectors α_1, α_2, and α span the vector space V^3?

12. Prove that each of the sets S_1, S_2, and S_3 immediately before this set of exercises span V^3.

13. Is there any proper subset of S_1 just before this set of exercises that spans V^3?

14. Let V be the set of vectors (x, y, x), where x and y are in a field F. Find a set of two vectors that span V.

15. Suppose that a set S of vectors spans a vector space V. Show that every set S' containing S also spans V.

16. Prove property 7 of a vector space.

17. Prove Theorem 2.1.

18. Suppose that in Theorem 2.1 we restrict the components of the n-tuples to be in an integral domain D instead of a field and require the scalar multiples also to be in D. Is the theorem still valid? Why or why not?

4. Linear Dependence and Dimension

In finding a set of elements that span a vector space, there would be obvious advantages in a minimal set, in the sense that no proper subset could span the space. For instance, as we saw in Exercise 13 after Section 3, the set

$$S_1: \epsilon_1 = (1, 0, 0), \, \epsilon_2 = (0, 1, 0), \, \epsilon_3 = (0, 0, 1), \, \gamma = (1, 1, 1)$$

is not a minimal set, since the first three span V^3. The fourth is superfluous. As a matter of fact, any three of the four suffice to span the space V^3; that is, each of the four vectors is in the space spanned by the other three.

Another example is the set

$$\alpha = (1, 2, 3), \qquad \beta = (3, 1, 5), \qquad \gamma = (4, 3, 8).$$

This set is not minimal, since $\alpha + \beta = \gamma$. Thus

$$x\alpha + y\beta + z\gamma = (x + z)\alpha + (y + z)\beta,$$

which shows that any linear combination of the three is a linear combination of the first two. Again, as it happens, each of the vectors is in the space spanned by the other two. But it is not always as simple as this. For instance, if $\delta = (2, 4, 6)$, then γ is not a linear combination of α and δ, although α and γ span the same space as α, δ, and γ. So, in testing for minimality we might have to try several combinations.

To deal more systematically with the problem of finding a minimal set, suppose that the vectors $\alpha_1, \alpha_2, \ldots, \alpha_k$ span a vector space V. We want to know if a proper subset spans V. We can look at the expression

(4.1) $$x_1\alpha_1 + x_2\alpha_2 + \cdots + x_k\alpha_k.$$

Suppose that a set of x_i could be found, not all zero, such that the vector (4.1) were the null vector; then if $k > 1$, one of the vectors α_i would have to be a linear combination of the others. For instance, if $x_1 \neq 0$ with (4.1) the null vector, then

$$\alpha_1 = \frac{-(x_2\alpha_2 + x_3\alpha_3 + \cdots + x_k\alpha_k)}{x_1}$$

and α_1 is a linear combination of the other α_i. We define the following concept.

Definition. *A set of vectors* $\alpha_1, \alpha_2, \ldots, \alpha_k$ *is called* linearly dependent *over a field F if there is a set of elements* x_i *of F, not all zero, such that*

$$x_1\alpha_1 + x_2\alpha_2 + \cdots + x_k\alpha_k = \theta.$$

If there is no such set of x_i, *the vectors* $\alpha_1, \alpha_2, \ldots, \alpha_k$ *are called* linearly independent *over F*.

In passing we should notice that in considering linear dependence we must have regard for the field. For instance, the vectors $(1, 1)$ and $(\sqrt{3}, \sqrt{3})$ are linearly dependent over the field of real numbers but linearly independent over the field of rational numbers. Another comment is due with regard to the word "set." The *set* $\{(1, 1), (0, 1), (1, 1)\}$ is the same as the set $\{(0, 1), (1, 1)\}$. When we list the elements of any set, there is no point in repetitions, nor are we concerned with the order in which the elements appear. We thus have the understanding that when we refer to a set of vectors, points, or whatever, we do not have repetitions, and unless we use the word "order" we are not concerned with the order in which elements are listed.

Now we have shown that if a set of k vectors spanning a space V is linearly dependent, we can find a proper subset that spans V unless $k = 1$. In that case, if the set is to be linearly dependent, α_1 must be the zero vector and the space V the *null space*, that is, the vector space consisting of the zero vector alone.

We collect these results in a theorem.

Theorem 4.1. Let S be a set of k vectors over a field F that span a space V, where $k > 1$. A proper subset of S spans V if and only if the set S is linearly dependent over F.

Corollary. If a set S of vectors is linearly independent over a field F, then the space spanned by S is spanned by no proper subset of S.

In the corollary we do not need to specify that $k > 1$, for a set consisting of a single vector α is linearly dependent if and only if $\alpha = \theta$.

Let us see what all this means in geometrical terms. If $\alpha = (1, 2)$, the space over **R** spanned by α is the set $(x, 2x)$, that is, all the points on the line $y = 2x$. If $\beta = (2, 4)$, then α and β are linearly dependent over **R**; that is, each is on the line spanned by the other. The space is one-dimensional. Similarly, the space spanned by $(1, 2, 3)$ is a line in three dimensions. On the other hand, as we saw in Section 4 of Chapter I, if $\gamma = (0, 1)$, the space spanned by α and γ is two-dimensional; that is, all vectors (x, y) are linear combinations of α and γ.

As another example, consider the solutions of $x + 2y + 3z = 0$. These are the points of a plane through the origin. We can write them in vector form as

$$(-2y - 3z, y, z) = y(-2, 1, 0) + z(-3, 0, 1).$$

Hence the points of the plane $x + 2y + 3z = 0$ are the points of a vector space spanned by $(-2, 1, 0)$ and $(-3, 0, 1)$.

It would appear that the dimension of a space should be equal to the number of vectors in a set of linearly independent vectors that span it. To define it in this way presents one difficulty. We know that there certainly would be a number of sets of linearly independent vectors that span any given space. It would be unfortunate if the number of vectors in two such spanning sets were different. With faith that this difficulty can be resolved, we make the following definition.

Definition.   *A vector space V over a field F is of dimension k if there is a set of k linearly independent vectors over F that spans it. A set of linearly independent vectors is called a* basis *of the space that they span.*

Now we hasten to state and prove the theorem that justifies the definition of dimension.

Theorem 4.2.   If $\alpha_1, \alpha_2, \ldots, \alpha_r$ and $\beta_1, \beta_2, \ldots, \beta_s$ are two linearly independent sets of vectors that span a vector space over a field F, then $r = s$.

Proof: Interchanging the α's and the β's if necessary, we may assume that $r \leq s$. We want to prove equality. Since β_1 is in V, it is a linear combination of the α's and hence

$$a_1\alpha_1 + a_2\alpha_2 + \cdots + a_r\alpha_r - \beta_1 = \theta$$

for some set of a_i in F. If $r = 1$, then $\beta_1 = a_1\alpha_1$ with $a_1 \neq 0$; hence β_1 spans V and $r = s = 1$. If $r > 1$, then not all the a_i are zero, since $\beta_1 \neq \theta$. Permute subscripts if necessary to assume that $a_1 \neq 0$ and see that $\alpha_1 = b_1\beta_1 + b_2\alpha_2 + \cdots + b_r\alpha_r$ for b_i in F. Hence

$$\sum_{i=1}^{r} x_i\alpha_i = x_1(b_1\beta_1 + b_2\alpha_2 + \cdots + b_r\alpha_r) + \sum_{i=2}^{r} x_i\alpha_i.$$

This shows that V is spanned by

(4.2) $\beta_1, \alpha_2, \alpha_3, \ldots, \alpha_r$.

Since the set (4.2) spans V, β_2, as well as α_1, is a linear combination of the vectors of (4.2) and hence $c_1\beta_1 + c_2\alpha_2 + \cdots + c_r\alpha_r - \beta_2 = \theta$ for c_i in F. Since β_1 and β_2 are linearly independent, not all the c's beginning with c_2 are zero. We may assume that $c_2 \neq 0$. If $r = 2$, then α_1 and α_2 are linear combinations of β_1 and β_2 and hence $r = s = 2$. If $r > 2$, then V is spanned by

(4.3) $\beta_1, \beta_2, \alpha_3, \ldots, \alpha_r$.

If we continue in this fashion, we find that $\beta_1, \beta_2, \ldots, \beta_r$ span V and hence $r = s$. The proof is complete.

We noted at the beginning of Section 3 that the set of n-tuples ϵ_i, $1 \leq i \leq n$, span V^n over any field F. Furthermore, (3.1) implies that the ϵ_i are linearly independent over any field F, provided of course that its multiplicative identity is denoted by 1. So the vectors ϵ_i are a basis of V^n and the dimension, naturally, is n. We call the ϵ_i a *canonical basis* of V^n. It is important to notice that by this construction, the components of the n-tuples are in the same field as the field of coefficients of the linear combinations. Suppose, on the contrary, that we consider over the field of rational numbers the space spanned by the set of vectors

$$(1, 1), (\sqrt{3}, \sqrt{3}), (\sqrt{5}, \sqrt{5}).$$

This space has dimension 3. This is one difficulty one avoids by using the abstract definition of a vector space instead of the particular example of n-tuples.

In light of the above, let us look at Examples 2 and 3 of the vector spaces given in Section 2. The set $F[x]$ of polynomials in x over a field F is a vector space with the following basis:

$$1, x, x^2, \ldots, x^k, \ldots.$$

Here the basis has an infinite number of elements. If $F(c)$ is an algebraic extension of degree n over F, its elements form a vector space over F with the following basis of n elements:

$$1, c, c^2, \ldots, c^{n-1}.$$

There is another result that not only throws more light on linear independence but is useful in the next section.

Theorem 4.3.   Let $\alpha_1, \alpha_2, \ldots, \alpha_r$ be a set of vectors spanning a space V. Then the set is linearly independent if and only if every element of V is a *unique* linear combination of the α's.

Proof: If $\beta = x_1\alpha_1 + \cdots + x_r\alpha_r = y_1\alpha_1 + \cdots + y_r\alpha_r$, then

$$\theta = (x_1 - y_1)\alpha_1 + \cdots + (x_r - y_r)\alpha_r.$$

If there are two distinct representations of β as a linear combination of the α's, then the α's are linearly dependent. That is, if the α's are linearly independent, the representations are the same.

If every element of V is a unique linear combination of the α's, then $\theta = x_1\alpha_1 + x_2\alpha_2 + \cdots + x_r\alpha_r$ implies that each x_i is zero. Thus the set of α's is linearly independent. This completes the proof.

Now consider the question: If V is a vector space of dimension n, is *every* set of n linearly independent vectors of V a basis of V? One certainly feels that the answer must be "yes." In showing this, it is easier to prove a theorem first and then deduce the answer as a corollary.

Theorem 4.4.   (Completion theorem.)   If $\alpha_1, \alpha_2, \ldots, \alpha_n$ is a basis of a vector space V over a field F and $\beta_1, \ldots, \beta_s$, a set of linearly independent vectors of V, then either $s = n$ and the β's are a basis of V, or $s < n$ and the set of β's together with $n - s$ of the α's constitute a basis of V.

Proof: Let W be the vector space spanned by the β's and see that since each β_i is in V, $W \subseteq V$. If each α_i is in W, then $V \subseteq W$ and hence $V = W$, which shows that the β's span V and, by Theorem 4.2, $n = s$. If some α_i is not in W, then $W \subset V$, and by change of notation if necessary, we may take α_1 not in W. Then if

$$b_1\beta_1 + b_2\beta_2 + \cdots + b_s\beta_s + a\alpha_1 = 0,$$

$a = 0$, since α_1 is not in W and then every $b_i = 0$, since the β's are linearly independent. Hence

$$\beta_1, \beta_2, \ldots, \beta_s, \alpha_1$$

are linearly independent and span a space W_1 of dimension $s + 1$.

If $\alpha_2, \ldots, \alpha_n$ are all in W_1, then $W_1 = V$, $s + 1 = n$ and the theorem is proved. If this is not the case, we may take α_2 not in W_1 and see as above that the set

$$\beta_1, \ldots, \beta_s, \alpha_1, \alpha_2$$

is linearly independent and spans a vector space W_2 that is contained in V and has dimension $s + 2$.

If we continue this process, it must stop short of α_n, since

$$\beta_1, \ldots, \beta_s, \alpha_1, \ldots, \alpha_n$$

is linearly dependent. Hence the theorem is established.

Corollary.   If $\beta_1, \beta_2, \ldots, \beta_n$ is a set of n linearly independent vectors of a vector space of dimension n over F, then the set of β's is a basis of the vector space.

5.   *Isomorphisms of Vector Spaces*

In several places in this book we have met the idea of an isomorphism. This concept is important also for vector spaces. As you would expect, it is a one-to-one correspondence between two vector spaces that preserves the effect of the two operations.

Definition.   *Two vector spaces V and V' over a field F are called* isomorphic
*if there is a one-to-one correspondence between the vectors of V and V' such
that if α and β in V correspond to α' and β' in V', respectively, then*

1. *$\alpha + \beta$ corresponds to $\alpha' + \beta'$.*
2. *$c\alpha$ corresponds to $c\alpha'$ for all c in F.*

One could relax this definition by not requiring that the field be the same
for V and V' or that only one of V and V' be a vector space, but in these
cases one would have to make other restrictions to make it work. The
definition given is sufficient for our purposes.

Notice that under an isomorphism, the zero vectors of V and V' must
correspond, since $\alpha + \theta = \alpha$ if and only if $\alpha' + \theta' = \alpha'$.

If a vector space has a finite basis, an isomorphism is determined by the
correspondence for the elements of the basis. (This is also true when the
space has an infinite basis, but we do not consider this contingency here.) To
see how this goes, let $\alpha_1, \alpha_2, \ldots, \alpha_r$ be a basis of V. Then let $\alpha'_1, \alpha'_2, \ldots, \alpha'_r$ be
the vectors of V' corresponding to the α's. From the isomorphism

$$c_1\alpha'_1 + c_2\alpha'_2 + \cdots + c_r\alpha'_r = \theta' \quad \text{if and only if} \quad c_1\alpha_1 + c_2\alpha_2 + \cdots + c_r\alpha_r = \theta.$$

Thus $\alpha'_1, \alpha'_2, \ldots, \alpha'_r$ is a linearly independent set. To prove that the α'_i span V',
let β' be any vector of V'. Because of the one-to-one correspondence, there is
a vector in V, call it β, to which β' corresponds. Now β, being in V, can be
expressed as a linear combination of the α_i. Denote by β'' the same linear
combination of the α'_i. Then β'' corresponds under the isomorphism to β. But
since the correspondence is one-to-one, β' and β'' must be the same vector.
Thus β' is a linear combination of the α'_i. So we have proved half of the
following theorem.

Theorem 5.1.   Let V and V' be two vector spaces over a field F and let V have
dimension r. Then V and V' are isomorphic if and only if V' also has dimension r.

Proof: We have proved above that if the spaces are isomorphic, their
dimensions are the same. It remains to show that if the dimensions are the
same, they are isomorphic. To this end, let $\alpha_1, \alpha_2, \ldots, \alpha_r$ be a basis of V and
$\alpha'_1, \alpha'_2, \ldots, \alpha'_s$ a basis of V'. Since the dimensions are the same, $r = s$. Then
set up the correspondence

$$\sum_{i=1}^{r} x_i\alpha_i \leftrightarrow \sum_{i=1}^{r} x_i\alpha'_i.$$

Since the α_i form a basis, Theorem 4.3 shows that every element of V is a
unique linear combination of the α_i, and similarly for V' and the α'_i. Hence
the correspondence is one-to-one. From the way we set up the correspond-

ence, the conditions of an isomorphism are satisfied. This completes the proof.

Corollary. If V is a vector space of dimension n over a field F, then it is isomorphic to V^n over F.

This corollary shows that for vector spaces with finite bases, that is, with bases of a finite number of elements, nothing is really lost if we consider n-tuples. But there are vector spaces without a finite basis. We saw, for example in Section 4, that $F[x]$ is a vector space over F with an infinite basis.

There are many other interesting and important properties of vector spaces but since the above is all we need for the chapters that follow, we content ourselves with the results obtained up to this point. For further properties you are referred to books on linear algebra.

Exercises

1. Which of the following sets of vectors are linearly dependent and which are independent?
   (a) $(4, 6)$ and $(6, 9)$.   (b) $(7, 3)$ and $(1, 6)$.
   (c) $(\frac{1}{2}, -1)$ and $(-4, 8)$.   (d) $(2, 4)$ and $(6, 3)$.
   (e) $(1, 4, 5)$, $(-1, 3, 6)$, and $(2, 0, 7)$.
   (f) $(2, 4, 6)$, $(1, 0, 7)$, and $(0, 1, -2)$.
   (g) $(1, 0, 0, 0)$, $(0, 1, 0, 0)$, $(0, 0, 1, 0)$, and $(a, b, c, 0)$.

2. Show that the pair of vectors $\alpha = (1, 3)$, $\beta = (5, 7)$ is linearly independent and find numbers r and s such that $r\alpha + s\beta = (-1, 5)$.

3. Show graphically the meaning of Exercise 2.

4. For what numbers x will the pair of vectors $(2, 3)$, $(x, 5)$ be linearly dependent?

5. Prove that if the pair of vectors (a, b), (c, d) are linearly dependent, if (a, b) and (e, f) are linearly dependent, and if a and b are not both zero, then (c, d) and (e, f) are linearly dependent.

6. If in Exercise 5 the word "dependent" is replaced by "independent," is the conclusion true?

7. A 1-gallon can is filled with a 40 per cent solution of alcohol and another with a 70 per cent solution. Prove that one may get a gallon of any strength between 40 and 70 per cent by a proper mixture of the contents of the two cans. Why is it not possible to get a strength of 30 per cent? What connection has this problem with the preceding sections?

8. Generalize the result of Exercise 7 to an *a* per cent solution and a *b* per cent solution.

9. Let α and β be two linearly independent vectors having two components each and consider the set of vectors: $r\alpha + s\beta$, where $r + s = 1$ and r and s are real. Show that all the points corresponding to this set of vectors lie on the line determined by the points corresponding to the vectors α and β.

10. In Exercise 9 under the conditions on r and s, what points correspond to r and s both positive? To r positive and s negative? To r negative and s positive?

11. Let α_1, α_2, α_3, and α_4 be four vectors with four components each. Suppose that the first component of α_1 is different from zero, the first two components of α_2 are 0 and 2, the first three of α_3 are 0, 0, 3, and the first three components of α_4 are all zero. Under what conditions will the set be linearly dependent? Under what conditions will it be linearly independent?

12. Find a basis for the space spanned by each set of vectors.
    (a) $(1, 2, 3)$, $(1, 0, 5)$, $(1, 1, 4)$, $(3, 2, 1)$.
    (b) $(1, 2, 3)$, $(2, 4, 6)$, $(1, 0, 1)$, $(2, 2, 4)$.
    (c) $(3, 0, 5)$, $(1, 2, 7)$, $(2, 1, 6)$, $(1, -1, -1)$.

13. Consider the set of vectors (x, y, z) such that $x + y + z = 0$ and $x + 2y - z = 0$. Show that these form a vector space and find a basis of this space.

14. In each part the lowercase letters denote elements of a field F and the indicated set is over all elements of F. In each case S and T are vector spaces. For each part indicate whether S and T are isomorphic. Where there is an isomorphism give a one-to-one correspondence and show why it satisfies the requirements of an isomorphism.
    (a) $S = \{(a, b)\}$ and $T = \{(r + s, s)\}$.
    (b) $S = \{(a, a, b)\}$ and $T = \{(r, s)\}$.
    (c) $S = \{(a, b, c)\}$ and $T = \{(r\alpha + s\beta)\}$, where α and β are two linearly independent vectors of V^2.
    (d) $S = \{(0, a)\}$ and $T = \{(b, 0)\}$.
    (e) $S = \{(a + b, a - b)\}$ and $T = \{(a, b)\}$.

15. In parts (a), (d), and (e) of Exercise 14, show what the correspondence means geometrically.

16. Let α, β, γ, and δ be a set of vectors and suppose that 3 is the dimension of the space that they span. Let W be the set of 4-tuples (a, b, c, d) for which $a\alpha + b\beta + c\gamma + d\delta = \theta$. Prove that W is a vector space of dimension 1.

17. Do Exercise 16 for the case in which the space spanned by the four vectors is of dimension 2. What is the dimension of W?

18. Let α_1, α_2, α_3 and β_1, β_2, β_3 be two bases of V^3. In how many ways can a one-to-one correspondence be set up between the two bases?

19. Consider the following one-to-one correspondence between two bases of V^3:

$$(1, 0, 0) \leftrightarrow (0, 1, 1); \quad (0, 1, 0) \leftrightarrow (1, 0, 1); \quad (0, 0, 1) \leftrightarrow (1, 1, 0).$$

The vector (a, b, c) is a linear combination of the first basis. Write it as a linear combination of the second basis.

20. Let W and V be vector spaces of dimensions r and s, respectively, with $r < s$. Show that W is isomorphic to a subspace of V.

21. Prove that if α, β, and γ are linearly independent, so are $\alpha + k\beta$, β, and γ for any element k of the field.

22. Let $\alpha_1 = (x_1, y_1, z_1)$ and $\alpha_2 = (x_2, y_2, z_2)$ be two vectors such that

$$ax_i + by_i + cz_i = 0 \qquad \text{for } i = 1, 2.$$

Prove that for all numbers r and s, the components of the vector $r\alpha_1 + s\alpha_2$ satisfy the equation.

23. Suppose that in Exercise 22 the equations are $ax_i + by_i + cz_i = d_i$ for $i = 1, 2$, and the restriction is imposed that $r + s = 1$. Show that the same conclusion can be drawn.

24. Find a geometric interpretation of the result of Exercise 23.

25. Let α_1, α_2, and α_3 be a basis of a vector space F. Find a correspondence between V and V^3 which is an isomorphism.

6. Matrices

Our chief use of Sections 6, 7, 8, and 9 is in Section 6 of Chapter V. A matrix is a rectangular array of elements of a field F. (One can also consider other kinds of elements, but we confine our attention to fields.) If a matrix has s rows and n columns, we call it an s by n matrix. We noted in Section 17 of Chapter III that the set of all matrices of order 2 over a field forms a ring with identity element. The object of this section is to show that for corresponding definitions of equality, sum, and product, it is true that for any positive n, all n by n matrices with elements in a field form a ring with identity element.

While we are considering equality and addition, we shall not limit ourselves to square matrices; but we shall find that if both addition and multiplication are to be possible, it is necessary that matrices be square.

Just as for matrices of order 2, we call two matrices A and B *equal* if they have the same number of rows and columns and if corresponding elements are equal. Notationally, it is convenient to write a matrix as

(6.1)
$$A = \begin{bmatrix} a_{11} & a_{12} & a_{13} & \cdots & a_{1n} \\ a_{21} & a_{22} & a_{23} & \cdots & a_{2n} \\ \multicolumn{5}{c}{\dotfill} \\ a_{s1} & a_{s2} & a_{s3} & \cdots & a_{sn} \end{bmatrix},$$

where the first subscripts indicate the row in which the element occurs and the second subscripts designate the column. We sometimes abbreviate this by writing $A = (a_{ij})$ with indication of the range of values of i and j. By definition, two matrices A and B are equal if $A = (a_{ij})$ and $B = (b_{ij})$, where i and j have the same ranges in both and $a_{ij} = b_{ij}$ for all i and j. If a matrix is square and has n rows, we say that it is of *order n*.

The *sum* of two matrices is defined in a natural way as follows:

Definition. *If A and B are two matrices having s rows and n columns, then the matrix $A + B$ is that obtained from A and B by adding corresponding elements. That is, the element of the ith row and jth column of $A + B$ is $a_{ij} + b_{ij}$, using the notation above.*

If we consider a vector to be a matrix with one row, this definition is consistent with vector sums. Just as we can add an r-tuple and an s-tuple only if $r = s$, so we can add matrices only if they have the same number of rows and columns. The properties of addition can be summarized in the following theorem.

Theorem 6.1. The set of all s by n matrices with coefficients in a field F is an additive group.

Partial proof: Closure, commutativity, and associativity follow from the same properties for elements of a field. The additive identity is "the" zero matrix, that is, the matrix all of whose elements are zero and which has s rows and n columns. (The zero matrix must conform to the other matrices of the set in the number of rows and columns.) The additive inverse of A is the matrix whose elements are the negatives of the elements of A. We designate this inverse by $-A$. This is the essence of the proof. It seems scarcely necessary to go into detail.

We define multiplication of matrices just as we did in Section 4 of Chapter I. For instance, to get the element in the first row and first column of the matrix product AB, we compute the sum of the products of the elements of the first row of A and corresponding elements in the first column of B. If there are to be corresponding elements, it follows that the number of elements in the first row of A must be the same as the number of elements in the first column of B. That is, for the product AB to have meaning, the number of columns of A must be the same as the number of rows of B. This means that if A is an s by n matrix, then for AB to have meaning, B must be an n by r matrix for some r.

Definition.   *If A is an s by n matrix and B an n by r matrix, the matrix AB is computed in the following way: To get the element in the ith row and jth column of AB, compute the sum of the products of the elements of the ith row of A and the corresponding elements in the jth column of B.*

In notation, write $A = (a_{ij})$ and $B = (b_{ij})$. Then the element in ith row and jth column of AB is

$$a_{i1}b_{1j} + a_{i2}b_{2j} + \cdots + a_{in}b_{nj} = \sum_{k=1}^{n} a_{ik}b_{kj} \qquad \text{for } 1 \leq i \leq s, 1 \leq j \leq r.$$

The product AB has s rows and r columns. For instance, if

$$A = \begin{bmatrix} 1 & 2 & 3 \\ 4 & 5 & 6 \end{bmatrix} \quad \text{and} \quad B = \begin{bmatrix} 1 & 1 \\ 2 & 0 \\ 3 & 1 \end{bmatrix},$$

then

$$AB = \begin{bmatrix} 14 & 4 \\ 32 & 10 \end{bmatrix} \quad \text{and} \quad BA = \begin{bmatrix} 5 & 7 & 9 \\ 2 & 4 & 6 \\ 7 & 11 & 15 \end{bmatrix}.$$

Notice that, as for matrices of order 2, AB is not necessarily equal to BA. In fact, if $C = (1\ 2\ 3)$, the product CB would have meaning but BC would not.

Now let A be an s by n matrix. If $A + B$ is to have meaning, B must be an s by n matrix. If either AB or BA is also to have meaning, s and n must be equal. In other words, if both product and sum are possible, both matrices must be square. Therefore, from this point on, except for occasional lapses, we shall *assume that the matrices are square.* We want to prove the following theorem, which summarizes properties of matrices.

Theorem 6.2.   The set of all square matrices $M_n(F)$ of order n over a field F is a ring with identity element.

Proof: The additive properties hold, by Theorem 6.1. Since the set is closed under multiplication, this operation is a binary operation for $M_n(F)$. The multiplicative identity is the matrix I of order n whose elements are all 0 except for 1's along the principal diagonal (the sequence of elements a_{ii}).

To prove associativity for multiplication, there are two possible roads to follow. One can show an isomorphism between matrices and linear transformations, as we did in Chapter I, or one can prove it by brute force. Since we do not need to consider linear transformations here, it is shorter to take the second road. Let $A = (a_{ij})$, $B = (b_{ij})$, and $C = (c_{ij})$ be three matrices of order n. Denote by $(AB)_{ik}$ the element in the ith row and kth column of AB, and similarly for other products. Then

$$(AB)_{ik} = \sum_r a_{ir}b_{rk} \quad \text{and} \quad [(AB)C]_{iu} = \sum_k \left[\sum_r a_{ir}b_{rk} \right] c_{ku}.$$

Also

$$(BC)_{ru} = \sum_k b_{rk}c_{ku} \quad \text{and} \quad [A(BC)]_{iu} = \sum_r a_{ir} \left[\sum_k b_{rk}c_{ku} \right],$$

where each subscript ranges over the integers from 1 to n inclusive. For the product $(AB)C$ one sums first over r and then over k. For the product $A(BC)$ one sums first over k and then over r. Since all sums are finite, the order of summing makes no difference in the result. Hence we have shown that multiplication of matrices is associative.

It remains to show the distributive property; that is,

$$A(B + C) = AB + AC.$$

We leave this as an exercise. This will complete the proof.

With matrices, just as for vectors, there is an operation called *multiplication by a scalar* or *scalar multiplication* defined as follows: If $A = (a_{ij})$, where the a_{ij} are in a field F and if c is any element of F, cA is defined to be (ca_{ij}); that is, cA is the matrix derived from A by multiplying each element by c. In particular, cI is the matrix all of whose elements are zero except that every element on the principal diagonal is c. That is, if $cI = (c_{ij})$, then $c_{ii} = c$ and $c_{ij} = 0$ if $i \neq j$. Now

$$cA = Ac = (cI)A = A(cI).$$

For this reason, cI is sometimes called a *scalar matrix*.

7.   Inverses of Matrices

As we found in Chapter I, some matrices have (multiplicative) inverses and some do not.

Definition.   *Let A be a matrix of order n and I the identity matrix of the same order. If there is a matrix C such that $CA = I = AC$, then C is called the* inverse *of A and written* $C = A^{-1}$. *If a matrix has an inverse, it is called* nonsingular; *otherwise it is called* singular.

NOTE: We are confining our attention to square matrices.

Let us explore the conditions under which a matrix has an inverse. Since each row of A is an n-tuple of elements of F, we can write A as

$$A = \begin{bmatrix} \rho_1 \\ \rho_2 \\ \vdots \\ \rho_n \end{bmatrix},$$

where ρ_k is the kth row of A. If $CA = I$ with $C = (c_{ij})$, we have

$$c_{11}\rho_1 + c_{12}\rho_2 + \cdots + c_{1n}\rho_n = \epsilon_1,$$

where ϵ_1 is the first row of I. More generally,

$$c_{i1}\rho_1 + c_{i2}\rho_2 + \cdots + c_{in}\rho_n = \epsilon_i, \qquad 1 \le i \le n.$$

This means that every ϵ_i and hence every element of V^n over F, is a linear combination of the rows of A. Since the n row vectors of A span V^n and V^n has dimension n, the rows of A are linearly independent. Thus we have almost proved the following theorem.

Theorem 7.1.   There is a matrix C such that $CA = I$ if and only if the rows of A are linearly independent.

Proof: It remains to prove that if the rows of A are linearly independent, then there is a matrix C such that $CA = I$. Now if the rows of A are linearly independent, they form a basis for V^n. Hence, for each i,

(7.1) $x_{i1}\rho_1 + x_{i2}\rho_2 + \cdots + x_{in}\rho_n = \epsilon_i$

is solvable. Then C can be chosen to be the matrix whose rows are the solutions of (7.1). This completes the proof.

Now, for A to have an inverse not only must CA be I, but AC must also be I. Our next theorem states that this is so.

Theorem 7.2.   If there is a matrix C such that $CA = I$ and A is square, then $AC = I$ also.

Proof: The equation $CA = I$ implies that $ACA = A$, that is,

$$(AC - I)A = (0),$$

where (0) denotes the zero matrix. We need to show that if $DA = (0)$, then D is the zero matrix. (Recall that we are assuming that $CA = I$ and that A is square.) Now if $DA = (0)$, then the ith row of D will be a solution of

(7.2) $x_{i1}\rho_1 + x_{i2}\rho_2 + \cdots + x_{in}\rho_n = \theta,$ $1 \leq i \leq n,$

since every row of (0) is θ. Since the rows of A are linearly independent we see that every x_{ij} must be zero. Hence $DA = (0)$ implies that $D = (0)$, and thus $(AC - I)A = (0)$ implies that $AC - I = (0)$ and $AC = I$. This completes the proof.

Theorems 7.1 and 7.2 combine to give the following theorem.

Theorem 7.3.   A square matrix A is nonsingular if and only if its rows are linearly independent.

Similarly, it can be shown that if there is a matrix E such that $AE = I$, then $EA = I$, for A square. Furthermore, one can apply the same reasoning to show that $AC = I$ implies that the column vectors of A are linearly independent. We leave this as an exercise. We also leave as exercises the proofs of the following theorems. (See Section 4 of Chapter I.)

Theorem 7.4.   The product of two matrices is singular if and only if at least one of them is singular.

Theorem 7.5.   The set of nonsingular matrices of order n is a multiplicative group.

Exercises

1. Given the matrices

$$A = \begin{bmatrix} 1 & 5 \\ 2 & 3 \\ 4 & 6 \end{bmatrix}, \qquad B = \begin{bmatrix} 1 & 5 & -2 \\ 2 & 0 & 1 \end{bmatrix},$$

$$C = \begin{bmatrix} 7 & 2 & 6 \\ 2 & 0 & 4 \end{bmatrix}, \qquad D = [1 \quad 2 \quad 3],$$

for which pairs is the sum defined? When it is defined, find the sum.

2. For what pairs of matrices in Exercise 1 is the product defined in one order or the other? Find the products for these pairs.

3. For the following matrix A, find A^2, A^3, and $A + A^2 + A^3$:

$$\begin{bmatrix} 0 & 0 & 1 \\ 1 & 0 & 0 \\ 0 & 1 & 0 \end{bmatrix}.$$

4. Find the square of the matrix

$$\begin{bmatrix} \frac{1}{3} & \frac{2}{3} & -\frac{2}{3} \\ \frac{2}{3} & \frac{1}{3} & \frac{2}{3} \\ -\frac{2}{3} & \frac{2}{3} & \frac{1}{3} \end{bmatrix}.$$

5. Let A be the square matrix (a_{ij}) of order 2. Find PA and AP for each of the following matrices P:

$$\begin{bmatrix} 0 & 1 \\ 1 & 0 \end{bmatrix}, \quad \begin{bmatrix} 3 & 0 \\ 0 & 1 \end{bmatrix}, \quad \begin{bmatrix} 1 & 4 \\ 0 & 1 \end{bmatrix}.$$

6. For each of the matrices P in Exercise 5, find P^{-1}.

7. Find the inverse of the matrix

$$B = \begin{bmatrix} 1 & 3 \\ 2 & 5 \end{bmatrix}.$$

8. Which of the following matrices have inverses? Justify your answers and find the inverses that exist.

(a) $\begin{bmatrix} 1 & 2 & 3 \\ 4 & 5 & 6 \\ 7 & 8 & 9 \end{bmatrix}$.

(b) $\begin{bmatrix} -1 & 1 & 1 \\ 1 & -1 & 1 \\ 1 & 1 & -1 \end{bmatrix}$.

9. Use the result of Exercise 7 to solve each of the following matric equations: $BX = C$ and $YB = C$, where $C = \begin{bmatrix} 1 \\ -1 \end{bmatrix}$. Are X and Y equal?

10. If $A^r = I$ for r an integer greater than 1 and if $A - I$ is nonsingular, show that

$$A^{r-1} + A^{r-2} + \cdots + A = -I.$$

Show that A has an inverse that is a polynomial in A.

11. Let A be a matrix of order n and ξ be the vector $(x_1, x_2, \ldots, x_n)$. Prove that the set of solutions of $\xi A = 0$ is a vector space.

12. Given a square matrix A of order n, and let N be the set of vectors γ such that $A\xi = \gamma$ is *not* solvable. Does N constitute a vector space?

13. Let A be a matrix of order n and suppose that for every column matrix C with n elements but not the zero matrix, it is true that $AX = C$ has no

more than one solution. Then prove that for each such C, $AX = C$ does have a solution and it is unique.

14. If A is a matrix of order n, define its *trace*, written tr(A), to be the sum of the elements on the principal diagonal, that is,

$$\text{tr}(A) = \sum_{i=1}^{n} a_{ii}$$

for $A = (a_{ij})$. Prove that tr$(AB) =$ tr(BA) for every matrix B of order n.

15. Prove the distributive property for matrices. Why does not $A(B + C) = AB + AC$ imply that $(B + C)A = BA + CA$?

16. If A and B are square matrices of order n and

$$A(AB - BA) = (AB - BA)A,$$

prove that $A^r B - BA^r = r(AB - BA)A^{r-1}$, first for $r = 2$ and $r = 3$, and then for all positive integers r. (An argument by induction would seem appropriate.)

17. Prove that the set of matrices

$$\begin{bmatrix} a & b \\ -b & a \end{bmatrix}$$

under addition and multiplication by a scalar is isomorphic to the set of vectors (a, b).

18. If the matrix of Exercise 17 has real elements, show that the set of all matrices of this form is isomorphic to the set of complex numbers $a + bi$ as to both addition and multiplication.

19. Prove Theorem 7.4.

20. Prove Theorem 7.5.

21. If A is a matrix of order n, prove that if there is a matrix C such that $AC = I$, then the columns of A are linearly independent vectors.

22. Prove that the set of matrices of order n over a field F is a vector space with a basis E_{ij}, where E_{ij} is the matrix of order n whose only nonzero element is a 1 in the ijth place.

8. Determinants

The only place in the next two chapters of this book where we shall use determinants is in Section 6 of Chapter V. There determinants are crucial. If you are not to study that section, you may prefer to skip this section completely or merely assume the results. However, for reference purposes we develop in this section the definition of a determinant and prove the properties that we shall need, plus a few more.

In Section 4 of Chapter I we defined the determinant (abbreviated "det") of a matrix of order 2 as follows:

$$\text{If } A = \begin{bmatrix} a_1 & a_2 \\ b_1 & b_2 \end{bmatrix}, \qquad \text{then det}(A) = a_1b_2 - b_1a_2.$$

We showed there that if A is singular, then $\det(A) = 0$, and conversely. We also showed in Exercise 23 of Section 4 of Chapter I that determinants are multiplicative, that is,

$$(8.1) \qquad\qquad \det(AB) = \det(A)\det(B).$$

So (8.1) is one desirable property of a determinant.

Now certainly the determinant of a matrix must depend on its elements. Thus it is natural to make it an algebraic expression in the elements. To emphasize this, we call the matrix $X = (x_{ij})$ and consider the determinant a function of the n^2 elements x_{ij}. Then if A is a matrix (a_{ij}) with elements in a field, we can compute the determinant of A by replacing the x_{ij} by a_{ij} in the expression for the determinant of X.

A third desirable property comes from comparing the following two phenomena: (1) From Theorem 7.4 the product of two matrices is singular if and only if at least one of them is singular; (2) the product of two elements of a field is zero if and only if one of them is zero. Hence the property (8.1) would give a connection between these two phenomena if we define a determinant so that a matrix is singular if and only if its determinant is zero. So, now we have three desirable properties of a determinant:

1. $\text{Det}(X)$ should be a function of the elements of X.
2. $\text{Det}(XY)$ should be $\det(X)\det(Y)$.
3. $\text{Det}(A)$ should be 0 if and only if A is singular.

Let us explore what requirements these properties would impose on a determinant. If a row of a matrix B consists of zeros, then its rows are linearly dependent and, by Theorem 7.3, B is singular. So, if the third desirable property is to hold, the determinant of a matrix with a row of zeros must be zero. If a row is zero, the other elements of the matrix do not matter; therefore, every term of the function $\det(X)$ should contain an element from each row. Similarly, from the remarks after Theorem 7.3, if B contains a column of zeros, its determinant should also be zero. Therefore, we have a right to hope that the determinant of $X = (x_{ij})$ might be a linear combination of products such as the following:

$$(8.2) \qquad\qquad x_{1i_1}x_{2i_2}\cdots x_{ni_n},$$

where $i_1, i_2, \ldots, i_n$ is some permutation of the numbers $1, 2, \ldots, n$.

A matrix is also singular if two of its rows are equal. This property would be reflected in the determinant if we knew that whenever we interchange two rows of a matrix X, we change the sign of its determinant. (This is more precisely expressed in the statement of the next theorem.) It may seem rather startling that this requirement, together with the one that the determinant is a sum of products such as (8.2), almost fixes what a determinant must be.

Theorem 8.1.   Let $X = (x_{ij})$ be a matrix of order n and let

$$(8.3) \qquad D(X) = \sum f(i_1, i_2, \ldots, i_n) x_{1i_1} x_{2i_2} \cdots x_{ni_n},$$

where the sum is over all permutations $i_1, i_2, \ldots, i_n$ of the integers from 1 to n and f is a function of the permutation. If for each matrix X, $D(X)$ has the property that interchanging two rows of X replaces $D(X)$ by $-D(X)$, then

$$(8.4) \qquad D(X) = f(1, 2, \ldots, n) \sum (-1)^p x_{1i_1} x_{2i_2} \cdots x_{ni_n},$$

where the sum is over all permutations: $i_1, i_2, \ldots, i_n$ of $1, 2, \ldots, n$, and p is 0 or 1 according as the permutation is even or odd. (See Section 6 of Chapter II.)

Proof: Let Y be the matrix obtained from X by interchanging the first two rows, that is,

$$x_{1j} = y_{2j}, \quad x_{2j} = y_{1j}, \quad \text{and} \quad x_{kj} = y_{kj},$$

for all j and for $3 \le k \le n$. Then

$$x_{1i_1} x_{2i_2} x_{3i_3} \cdots x_{ni_n} = y_{1i_2} y_{2i_1} y_{3i_3} \cdots y_{ni_n}.$$

The respective coefficients are $f(i_1, i_2, i_3, \ldots, i_n)$ and $f(i_2, i_1, i_3, \ldots, i_n)$. If $\det(Y)$ is to be the negative of $\det(X)$, we see that

$$f(i_1, i_2, i_3, \ldots, i_n) = -f(i_2, i_1, i_3, \ldots, i_n).$$

That is, if we interchange the first two i's in f, we change the sign of f. The same argument can be used to show that if we interchange *any* two i's, we merely replace f by its negative.

We know from Theorem 6.1 of Chapter II that every permutation of the numbers from 1 through n can be accomplished by a sequence of transpositions, that is, interchange of pairs. Furthermore, the even permutations are those which can be obtained by an even number of interchanges of pairs, odd permutations are those which can be obtained by an odd number of interchanges. An even permutation preserves the sign of the determinant and an odd permutation changes it. Thus we have proved the theorem.

REMARK: In Theorem 8.1, $f(1, 2, \ldots, n)$ is the coefficient of $x_{11} x_{22} x_{33} \cdots x_{nn}$ in the expression for $D(X)$. If $X = I$, the identity, the only nonzero term in $D(X)$, is the product of the elements of the diagonal of X. Thus $f(1, 2, \ldots, n)$ is the determinant of X when $X = I$. But if the determinant is to be multiplicative, then $IA = A$ implies that $\det(I) \det(A) = \det(A)$, and hence $\det(I)$ *should be* 1. Therefore, with a look backward at (8.4) we have evolved the following definition of a determinant.

Definition.   *If $X = (x_{ij})$ is a matrix of order n, then*

$$\det(X) = \sum (-1)^p x_{1i_1} x_{2i_2} \cdots x_{ni_n},$$

where the sum is over all permutations $i_1, i_2, \ldots, i_n$ of $1, 2, \ldots, n$ and p is 0 or 1 according as the permutation is even or odd. Each product of the sum is called a term *of the expansion of $\det(X)$.*

Two conclusions immediately follow from the definition. First, if all the elements in some row (or column) of a matrix are zero, then its determinant is zero, since every term of $\det(X)$ contains an element from each row (and each column). Second, $\det(I) = 1$, for the only nonzero elements in I are $a_{ii} = 1$, and for the identity permutation $p = 0$.

Let us see what this definition gives us for $n = 2$ and $n = 3$. For $n = 2$ the two terms are $a_{11}a_{22}$ and $a_{12}a_{21}$. Since $(1, 2)$ is an even permutation of $(1, 2)$ and $(2, 1)$ is an odd permutation, the determinant is

$$a_{11}a_{22} - a_{12}a_{21}.$$

(See Section 4 of Chapter I.)

For $n = 3$, notice that the even permutations are

$$\begin{pmatrix} 1 & 2 & 3 \\ 1 & 2 & 3 \end{pmatrix}, \quad \begin{pmatrix} 1 & 2 & 3 \\ 2 & 3 & 1 \end{pmatrix}, \quad \begin{pmatrix} 1 & 2 & 3 \\ 3 & 1 & 2 \end{pmatrix}.$$

Hence

$$\det(A) = a_{11}a_{22}a_{33} + a_{12}a_{23}a_{31} + a_{13}a_{21}a_{32}$$
$$- a_{11}a_{23}a_{32} - a_{12}a_{21}a_{33} - a_{13}a_{22}a_{31}.$$

We defined the determinant so that it would have certain properties. Now we should see what properties it indeed has. From the way it was defined, interchanging two rows of a matrix changes the sign of its determinant. Next we need to verify the multiplicative property.

Theorem 8.2.   If X and Y are two matrices of order n, then

$$\det(XY) = \det(X) \det(Y).$$

Proof: From the definition of the product of matrices, the element in the ith row and jth column of XY is

$$x_{i1}y_{1j} + x_{i2}y_{2j} + \cdots + x_{in}y_{nj}.$$

We can think of each term of $\det(XY)$ to be of the form

$$g_p(y_{11}, y_{12}, \ldots, y_{nn})x_{1i_1}x_{2i_2} \cdots x_{ni_n},$$

where the function g_p depends on the y_{ij} and on the permutation represented by the second subscripts of the x's, but is independent of the x_{ij}.

Since interchanging two rows of XY, that is, interchanging two rows of X, changes the sign of $\det(XY)$, Theorem 8.1 shows that

$$(8.5) \qquad \det(XY) = g_1(y_{11}, y_{12}, \ldots, y_{nn}) \sum (-1)^p x_{1i_1}x_{2i_2} \cdots x_{ni_n},$$

where g_1 is g_p for the identity permutation. Now g_1, as for all the g's, is independent of X. Replacing X by I in (8.5) does not alter g_1 and reduces the summation part of (8.5) to 1. Thus

$$\det(IY) = \det(Y) = g_1.$$

Equation (8.5) implies that $\det(XY) = \det(X)\det(Y)$, and the proof is complete.

We have shown that the determinant, as defined, has the first two of the three properties listed as desirable. Our next task is to show that it satisfies the third. Now $\det(I) = 1$, as we noted after the definition of a determinant. Then if A is a nonsingular matrix, we have

$$\det(A)\det(A^{-1}) = \det(I) = 1,$$

which shows that $\det(A)$ cannot be zero. If A is nonsingular, its determinant is not zero. It remains to show that if $\det(A) \neq 0$, then A is nonsingular. We state the third property as a theorem.

Theorem 8.3. The determinant of a matrix is zero if and only if the matrix is singular.

Proof: In view of the preceding we only need to prove that if A is singular, then $\det(A) = 0$. As in Section 7, denote by ρ_i the ith row of A. Then, if A is singular, its rows are linearly dependent; that is,

$$c_1\rho_1 + c_2\rho_2 + \cdots + c_n\rho_n = \theta,$$

where not all the c's are zero. To simplify notation, assume that $c_1 \neq 0$. Let $d_i = c_i/c_1$ for $2 \leq i \leq n$ and see that

$$\rho_1 + d_2\rho_2 + \cdots + d_n\rho_n = \theta.$$

Then let D be the matrix whose first row is $(1, d_2, \ldots, d_n)$ and whose ith row is ϵ_i, the ith row of the identity matrix, for $2 \leq i \leq n$. If you write this out, you will see that the first row of the product DA is the n-tuple θ. Since DA has a zero row, as we noted after the definition of a determinant, its determinant is 0. If we can show that D is nonsingular, then $\det(D) \neq 0$ and $0 = \det(DA)$ $= \det(D)\det(A)$ implies that $\det(A) = 0$.

It remains to show that D is nonsingular, that is, that its rows are linearly independent. This results from the following computation:

$$x_1(1, d_2, \ldots, d_n) + x_2\epsilon_2 + \cdots + x_n\epsilon_n = (x_1, x_2 + d_2x_1, \ldots, x_n + d_nx_1).$$

This vector is zero if and only if every x_i is zero. As noted in the previous paragraph, this shows that $\det(A) = 0$ and completes the proof.

We have found all the results about determinants that we shall need in the rest of this book. In Section 9 we derive some theorems that are helpful in the computation of a determinant. In Section 10 we show how matrices relate to some of the topics in previous chapters.

9. Computation of Determinants

Using the definition of a determinant is not usually the easiest way to compute it. Although we cannot afford the space to develop "expansion by minors," which has a number of advantages, we shall derive results that lead to a practical way of finding the determinant of a numerical matrix. Our process consists of using certain operations on the matrix that do not change the determinant and yet reduce it to more manageable form. A convenient form to which one can reduce a matrix is the triangular form defined as follows.

Definition. *A square matrix is called* triangular *if either all the elements below the principal diagonal are zero or all the elements above it are zero.*

We first find the determinant of a triangular matrix and then show how we can reduce any matrix to such a form without altering its determinant.

Theorem 9.1. If $A = (a_{ij})$ is a triangular matrix, then $\det(A) = a_{11}a_{22}a_{33} \cdots a_{nn}$; that is, $\det(A)$ is the product of the elements along its principal diagonal.

Proof: Suppose that in A it is true that $a_{ij} = 0$ whenever $i < j$, that is, for all elements above the principal diagonal. Then the term

$$a_{1i_1}a_{2i_2} \cdots a_{ni_n}$$

is different from zero only if

(9.1) $1 \geq i_1, 2 \geq i_2, 3 \geq i_3, \ldots,$ and $n \geq i_n.$

Adding the left and right sides of the inequalities, we see that

(9.2) $1 + 2 + 3 + \cdots + n \geq i_1 + i_2 + i_3 + \cdots + i_n,$

with the equality holding if and only if all the equalities hold in (9.1). But since $i_1, i_2, i_3, \ldots, i_n$ is a permutation of $1, 2, 3, \ldots, n$, the equality must hold in (9.2). Thus the only nonzero term in $\det(A)$ is $a_{11}a_{22}a_{33} \cdots a_{nn}$, the product of the elements on the diagonal, as we wished to show.

Now consider two operations on a matrix.

Theorem 9.2. Interchanging two rows of a matrix changes the sign of its determinant. Adding a multiple of one row to another does not change the determinant of a matrix.

Proof: We have noted that the definition of the determinant implies the first sentence. To prove the second, let $I_{ij}(k)$ denote the matrix obtained from the identity matrix by adding k times the ith row to the jth row. If you write this out, you will see that $I_{ij}(k)$ has zeros everywhere except for 1's along the principal diagonal and a k for the ith element of the jth row. Thus $I_{ij}(k)$ is a triangular matrix whose diagonal elements are all 1; hence, by Theorem 9.1, its determinant is 1. Now if $B = I_{ij}(k)A$, the matrix B is obtained from A by adding k times the ith row to the jth row. By the multiplicative property of determinants, A and B have the same determinant. This completes the proof.

The two operations in Theorem 9.2 can be used to reduce any matrix to triangular form. Rather than show this formally, we illustrate the process for a particular matrix of order 3.

$$\det \begin{bmatrix} 0 & 1 & 5 \\ 3 & 6 & 3 \\ 2 & 1 & 0 \end{bmatrix} = -\det \begin{bmatrix} 3 & 6 & 3 \\ 0 & 1 & 5 \\ 2 & 1 & 0 \end{bmatrix} = -\det \begin{bmatrix} 3 & 6 & 3 \\ 0 & 1 & 5 \\ 0 & -3 & -2 \end{bmatrix}$$

$$= -\det \begin{bmatrix} 3 & 6 & 3 \\ 0 & 1 & 5 \\ 0 & 0 & 13 \end{bmatrix} = -39.$$

The first step is to interchange the first two rows; the second step is to add $-\tfrac{2}{3}$ of the first row to the third; and the third step is to add 3 times the second row to the third. The final form is triangular, and hence the determinant is the product of the diagonal elements. Actually, quite often one is not so much concerned with the value of the determinant as with whether or not it is zero. For instance, a good way to determine if a given set of vectors is linearly independent is often to form a matrix with the vectors as its rows, convert the matrix into triangular form, and then note that the given set is linearly independent if and only if all the diagonal elements of the triangular form are different from zero.

Exercises

1. Find the determinants of each of the matrices

$$\begin{bmatrix} 2 & 3 \\ -5 & 4 \end{bmatrix}, \quad \begin{bmatrix} 1 & 2 \\ 3 & 6 \end{bmatrix}, \quad \begin{bmatrix} 1 & 0 & 4 \\ 0 & -1 & 3 \\ 1 & 4 & 5 \end{bmatrix}, \quad \begin{bmatrix} 7 & 8 & 9 \\ 4 & 5 & 6 \\ 1 & 2 & 3 \end{bmatrix}.$$

2. Find the determinant of

$$\begin{bmatrix} 1 & 2 & 0 & 0 \\ 2 & 3 & 0 & 0 \\ 0 & 0 & 4 & 5 \\ 0 & 0 & 6 & 1 \end{bmatrix}.$$

3. How many terms are there in the determinant of X if X is a matrix of order n?

4. Find the sign of the term $x_{12}x_{31}x_{43}x_{24}$ in the determinant of $X = (x_{ij})$ of order 4.

5. Find a basis for the vector space spanned by

$$(1, 0, 1), \quad (0, 1, 1), \quad (2, 3, 5), \quad (1, -1, 0).$$

6. Prove that if A is a nonsingular matrix, then $[\det(A)]^{-1} = \det(A^{-1})$.

7. Prove that if C is nonsingular, then A and CAC^{-1} have the same determinants.

8. Prove that $\det(cA) = c^n \det(A)$, where n is the order of the matrix A.

9. Show that the following is the equation of a line through the points (x_1, y_1) and (x_2, y_2):

$$\det \begin{bmatrix} x & y & 1 \\ x_1 & y_1 & 1 \\ x_2 & y_2 & 1 \end{bmatrix} = 0.$$

10. Prove that the area of the parallelogram three of whose vertices are $(0, 0)$, (x_1, y_1), and (x_2, y_2) is the absolute value of

$$\det \begin{bmatrix} x_1 & y_1 \\ x_2 & y_2 \end{bmatrix}.$$

11. Generalize the result of Exercise 10 to three dimensions.

12. Let (x_1, y_1), (x_2, y_2), and (x_3, y_3) be pairs of coordinates of three non-collinear points. Show that if the determinant of the following matrix is

set equal to zero, it is the equation of a circle through the three points indicated:

$$\begin{bmatrix} x^2 + y^2 & x & y & 1 \\ x_1^2 + y_1^2 & x_1 & y_1 & 1 \\ x_2^2 + y_2^2 & x_2 & y_2 & 1 \\ x_3^2 + y_3^2 & x_3 & y_3 & 1 \end{bmatrix}.$$

13. What happens to the determinant described in Exercise 12 if the three points are collinear?

14. Find a matrix such that if its determinant is set equal to zero, the resulting equation is that of a parabola of the form $y = ax^2 + bx + c$ passing through the three points (x_i, y_i) for $i = 1, 2, 3$. What restrictions must be made on the three points?

15. Prove Theorem 9.1 for the case in which all the elements of the triangular matrix below the principal diagonal are zero.

16. Show that if $B = AI_{i,j}(k)$, where $I_{i,j}(k)$ is defined in the proof of Theorem 9.2, then B is obtained from A by adding k times the jth column to the ith column.

17. Prove that the two operations of Theorem 9.2 do not change the vector space spanned by the rows of the matrix.

18. The following is called *Vandermonde's determinant*:

$$\det \begin{bmatrix} 1 & x_1 & x_1^2 & \cdots & x_1^{n-1} \\ 1 & x_2 & x_2^2 & \cdots & x_2^{n-1} \\ \multicolumn{5}{c}{\dots\dots\dots\dots\dots\dots} \\ 1 & x_n & x_n^2 & \cdots & x_n^{n-1} \end{bmatrix}.$$

Prove that this determinant is zero if $x_i = x_j$ for any two i and j and hence that it is divisible by $(x_i - x_j)$ for $i \neq j$. Prove that the determinant is equal to the following product:

$$\prod (x_i - x_j),$$

where the product is over all i and j such that $1 \leq j < i \leq n$.

10. Homomorphisms, Automorphisms, and Cosets

In this section we show how some of the material in Chapter II applies to matrices. We have already noticed that the nonsingular matrices of order n over a field F constitute a multiplicative group. Call this group M. What is the center of M?

Theorem 10.1.   The center of M, the multiplicative group of matrices of order n over a field F, is the set of scalar matrices of order n, that is, cI, where c is in F.

Proof: Let N denote the center of M and D a matrix of N. Certainly, every scalar matrix is in N, since both $(cI)A$ and $A(cI)$ are obtained from the matrix A by multiplying each element by c. We want to show that N has no other elements. As in Section 9, let $I_{ij}(1)$ denote the matrix obtained from the identity matrix by adding the ith row to the jth. Then if D is in N,

$$(10.1) \qquad I_{ij}(1)D = DI_{ij}(1) \qquad \text{for all } i \text{ and } j, \text{ } i \neq j.$$

Now $I_{ij}(1)D$ and D are the same except that the jth row of the former is

$$(10.2) \qquad j\text{th row: } d_{j1} + d_{i1}, d_{j2} + d_{i2}, \ldots, d_{ji} + d_{ii}, \ldots, d_{jn} + d_{in}.$$

On the other hand, $DI_{ij}(1)$ is obtained from D by adding the jth column to the ith. Hence $DI_{ij}(1)$ and D are the same except for the ith column of the former, which is

$$(10.3) \qquad i\text{th column: } d_{1i} + d_{1j}, d_{2i} + d_{2j}, \ldots, d_{ji} + d_{jj}, \ldots, d_{ni} + d_{nj}.$$

Now (10.2) shows that $d_{iu} = 0$ except for $u = i$, and (10.3) shows that $d_{vj} = 0$ except for $v = j$. From (10.2) the element in the jth row and ith column of $I_{ij}(1)D$ is $d_{ji} + d_{ii}$, and from (10.3) the element in the jth row and ith column of $DI_{ij}(1)$ is $d_{ji} + d_{jj}$, which imply $d_{ii} = d_{jj}$. Hence we have shown that (10.1) implies that

$$(10.4) \quad d_{iu} = 0 \quad \text{for } u \neq i, \quad d_{vj} = 0 \quad \text{for } v \neq j, \quad d_{ii} = d_{jj} \quad \text{for all } i \text{ and } j.$$

Hence $D = dI$ for some element d of F, and the proof is complete.

Now, as in Section 8 of Chapter II, the inner automorphisms of M are those defined by

$$X\sigma_C = C^{-1}XC \qquad \text{for } C \text{ any nonsingular matrix in } M.$$

Two matrices C and C' define the same inner automorphism if and only if $C = C'd$ for some nonzero element d of F, as you will be asked to show in Exercise 1. We have the same connection between this and the "alibi versus alias" point of view that was described in Section 8 of Chapter II.

The idea of a coset comes into play in the solutions of sets of linear equations as in Section 4 of Chapter II. The set of equations

$$(10.5) \qquad \begin{aligned} a_{11}x_1 + a_{12}x_2 + \cdots + a_{1n}x_n &= c_1 \\ a_{21}x_1 + a_{22}x_2 + \cdots + a_{2n}x_n &= c_2 \\ &\cdots\cdots\cdots\cdots\cdots\cdots \\ a_{n1}x_1 + a_{n2}x_2 + \cdots + a_{nn}x_n &= c_n \end{aligned}$$

can be written in matrix form as follows: $A\xi = \gamma$, where $A = (a_{ij})$, ξ is the column matrix whose elements are x_i, and γ is the column matrix whose elements are c_i, for $1 \leq i \leq n$. The set (10.5) is solvable if and only if γ is in the vector space spanned by the columns of A, for if γ_i denotes the ith column of A, (10.5) becomes

$$A\xi = [\gamma_1, \gamma_2, \ldots, \gamma_n]\xi = \gamma_1 x_1 + \gamma_2 x_2 + \cdots + \gamma_n x_n.$$

If A is nonsingular, the γ_i are linearly independent and, by Theorem 4.3 there is just one solution. If A is singular, there may be many or none.

It helps in considering the solutions when A is singular if we look at the corresponding set of homogeneous equations

$$A\xi = \theta.$$

As you were asked to show in Exercise 11 of Section 7, this set of solutions constitutes a vector space S. Then, referring back to the set (10.5), we see that if $A\xi_1 = \gamma$ and $A\xi_2 = \gamma$, then $A(\xi_1 - \xi_2) = \theta$. Hence the difference of any two solutions of (10.5) is a vector in S. Thus if ξ_0 is one solution of (10.5), all solutions are given by the coset

$$S + \xi_0.$$

When we write of a coset in this connection, we are referring to a coset of the additive group S. If A is nonsingular, S is the null space.

Exercises

1. Prove that two matrices C and C' define the same inner automorphism of the multiplicative group of nonsingular matrices of order n over a field it and only if $C = dC'$ for some nonzero element d of F.

2. What quotient group is isomorphic to the group of inner automorphisms of the group of Exercise 1?

3. Let $F = \mathbf{Z}_3$ and find the number of matrices in $M_2^*(F)$, the multiplicative group of nonsingular matrices of order 2 over F. Find the number of distinct inner automorphisms of $M_2^*(F)$.

4. Do Exercise 3 for $F = \mathbf{Z}_p$, where p is a prime number.

5. Do Exercise 3 for $M_3^*(F)$.

6. Prove that the correspondence $M\sigma = \det(M)$ is a homomorphism of the multiplicative group of nonsingular matrices of order n over a field F onto F^*, the multiplicative group of nonzero elements of F. What is the kernel of this homomorphism?

V

Ideals

---

1. Introduction

In this chapter we shall propose a problem, develop in an exploratory manner the mathematics necessary to solve it, and, in the end, show some ramifications of what we have found.

In Section 11 of Chapter III we showed that every integer can be expressed as a product of prime factors uniquely except for the order of the factors. In fact, the corresponding result holds for an integral domain if it is a Euclidean domain. But we noted in Theorem 11.2 of Chapter III that in the integral domain $W = \{x + y\sqrt{-5}\}$, one does not have the property of factorization into prime factors uniquely except for order. The problem of this chapter is to show that if we go about it in the right way, we can have unique factorization into some kind of prime quantities for algebraic fields.

To solve our problem, we develop in the first three sections some properties of algebraic extensions of fields, especially of the field $\mathbf{Q}$ of rational numbers. In Section 4 we define over a commutative ring with identity element the concept of an ideal, which, for algebraic fields, is the basic concept for the solution of our problem. In Section 5 we see that we can make some progress toward a solution of the problem for a general commutative ring. In Sections 6 and 7 we restrict ourselves to algebraic fields to come through with a solution to the problem. (Since two of the proofs in Section 6 are difficult, you may prefer to skip them.) In Section 8 we return for the most part to consideration of commutative rings with identity element, although toward the end we use the more general results to derive relationships about ideals in algebraic fields. In Section 9 we discuss how much of the previous work carries over to noncommutative rings. As often happens in mathematics, the tools developed to solve a problem turn out to be more important than the solution of the problem itself because of the answers they give to other problems and the contribution they make to our knowledge of structure.

To point out the problem, let us consider an example different from that in Theorem 11.2 of Chapter III which also has the disagreeable property of not having unique decomposition into prime factors. Let W be the integral domain $\{x + y\sqrt{-5}\}$, where x and y are integers, and see that 9 has two factorizations, as follows:

$$9 = 3 \cdot 3 = (2 + \sqrt{-5})(2 - \sqrt{-5}).$$

We want to show that 3 and $2 + \sqrt{-5}$ are primes in W and that neither is a unit multiple of the other. First, suppose that $2 + \sqrt{-5}$ factored as follows:

$$(1.1) \qquad (a + b\sqrt{-5})(c + d\sqrt{-5}) = 2 + \sqrt{-5}$$

for $a, b, c,$ and d integers. If we multiply the left side of (1.1), we get

$$(1.2) \qquad (ac - 5bd) + (bc + ad)\sqrt{-5} = 2 + \sqrt{-5}.$$

If in (1.2) we replace b and d by $-b$ and $-d$, respectively, on the left, the right side becomes $2 - \sqrt{-5}$. Doing the same in (1.1), therefore, gives us

$$(1.3) \qquad (a - b\sqrt{-5})(c - d\sqrt{-5}) = 2 - \sqrt{-5}.$$

Equating the product of the left sides of (1.1) and (1.3) to the product of the right sides, we obtain

$$(a^2 + 5b^2)(c^2 + 5d^2) = (2 + \sqrt{-5})(2 - \sqrt{-5}) = 9.$$

Thus $a^2 + 5b^2$ and $c^2 + 5d^2$ are both integers whose product is 9. Neither of them can be 3, since $3 = x^2 + 5y^2$ is not solvable in integers x and y. Hence one of $a^2 + 5b^2$ and $c^2 + 5d^2$ is 1 and the other is 9. Suppose that $a^2 + 5b^2 = 1$. Then $a + b\sqrt{-5}$ is a unit because its reciprocal is $a - b\sqrt{-5}$, which is also in W. Thus we have shown that if (1.1) holds, one of the factors on the left must be a unit. This means that $2 + \sqrt{-5}$ must be a prime in W.

Similarly (see Exercise 2 after Section 3), 3 is a prime in W. Also, $(2 + \sqrt{-5})/3$ is not a unit, since it is not in W. Thus we have shown that 9 can be expressed in two different ways as products of primes in W.

The problem of this chapter is to find a way to arrive at some kind of unique decomposition for "algebraic integers." Our first step is to develop in the next section what we mean by an algebraic integer.

2.   Algebraic Integers

In Section 14 of Chapter III we showed that, starting with any field F and an irreducible polynomial $f(x)$ of degree n in that field, we can form an extension of F by considering all the polynomials in θ, where θ is a zero of the irreducible polynomial. This is equivalent to considering all the expressions of the form

$$(2.1) \qquad a_{n-1}\theta^{n-1} + a_{n-2}\theta^{n-2} + \cdots + a_1\theta + a_0,$$

where $f(\theta) = 0$ and the a_i are in F. Any zero of a polynomial with coefficients in F is called an *algebraic number* over F, and the zeros of any irreducible polynomial are called *conjugates* of each other. Unless we indicate something to the contrary, we shall limit ourselves to extensions of Q, the field of rational numbers. However, much of the theory would carry over without much difficulty to any field F whose integral domain is Euclidean.

So now, taking F to be the field of rational numbers, we ask what numbers of the form (2.1) should we consider to be algebraic integers? The most natural definition would be to require that the coefficients a_i be integers. But there is difficulty with this. For example, if $c = \frac{1}{2}(1 + \sqrt{-3})$, then $Q(\sqrt{-3}) = Q(c)$, but c is not expressible in the form $a + b\sqrt{-3}$, with a and b integers. We shall shortly see that c has a good claim to being an algebraic integer. The point is that requiring that the a_i be integers ties us too much to the number θ, which defines the field. As is so often the case with definitions, it is better to define an algebraic integer by how it behaves rather than by how it looks. To this end, recall Theorem 13.5 of Chapter III. For convenience we restate it here: If $f(x)$ is a polynomial in $Z[x]$ and r/s is a rational zero of $f(x)$, where r and s are relatively prime integers, then r is a factor of $f(0)$ and s is a factor of the leading coefficient of $f(x)$.

An immediate consequence of this result is that if $f(x)$ in $Z[x]$ has leading coefficient 1, then *all* its rational zeros are integers. From this point of view the following definition should seem natural.

Definition.   *An algebraic number in $Q(\theta)$ is called an* algebraic integer *if it is a zero of some polynomial in $Z[x]$ with leading coefficient* 1.

Notice that every integer k is a zero of some such polynomial, namely, $x - k$, and hence is an algebraic integer. But $\sqrt{2}$, being a zero of $x^2 - 2$, is also an algebraic integer. So the set of algebraic integers includes all the integers and many other numbers besides. In the definition the word "some" is important. Not all polynomials with an algebraic integer as a zero have leading coefficient 1. For instance, if $f(x)$ is in $Z[x]$ with leading coefficient 1, then $g(x) = (2x - 1)f(x)$ has leading coefficient 2 but has as zeros all the zeros of $f(x)$. But if the polynomial having θ as a zero is *irreducible* over Q, then the following theorem asserts, with a restriction or two, that its leading coefficient must be 1. Recall that a polynomial in x with integer coefficients is called *primitive* if 1 is the g.c.d. of its coefficients.

Theorem 2.1. If θ is an algebraic integer, there is a unique primitive irreducible polynomial $f(x)$ with integral coefficients, whose leading coefficient is positive and for which $f(\theta) = 0$. Then the leading coefficient of $f(x)$ is 1.

Proof: We know from Theorem 14.5 of Chapter III that there is a unique irreducible polynomial $f(x)$ over $\mathbf{Q}$ whose leading coefficient is 1 such that $f(\theta) = 0$. Since θ is an algebraic integer, there is a polynomial $g(x)$ with integer coefficients and leading coefficient 1 such that $g(\theta) = 0$. Now $g(\theta) = 0 = f(\theta)$ implies by Theorem 14.7 of Chapter III that $f(x) \mid g(x)$. Hence

$$(2.2) \qquad\qquad g(x) = f(x)h(x),$$

where $h(x)$ is a polynomial and the coefficients of f and h are rational numbers. Let r and s be the least common multiples of the denominators of the coefficients of $f(x)$ and $h(x)$, respectively. Then

$$rsg(x) = [rf(x)][sh(x)],$$

where $rf(x)$ and $sh(x)$ are primitive polynomials in $\mathbf{Z}[x]$. Now, $g(x)$ is also primitive, since its leading coefficient is 1. Hence by the lemma after Gauss's lemma (Theorem 13.4 of Chapter III), rs is a unit. This shows that $f(x)$ and $h(x)$ have integer coefficients. Also, $f(x)$ and $h(x)$ have leading coefficient 1 or -1. Thus by multiplying $f(x)$ by -1 if necessary, we complete the proof of the theorem.

Corollary. If $f(x)$ is the irreducible polynomial over $\mathbf{Q}$ with leading coefficient 1 such that $f(\theta) = 0$ for θ an algebraic integer, then the coefficients of $f(x)$ are integers.

The corollary follows, since the theorem asserts that there is an irreducible polynomial $f(x)$ with integral coefficients and leading coefficient 1 of which θ is a zero. Theorem 14.5 of Chapter III asserts that there is only one irreducible polynomial $g(x)$ over $\mathbf{Q}$ with leading coefficient 1 of which θ is a zero. Hence $g(x) = f(x)$.

Now if an algebraic integer has the form (2.1), the coefficients, a_i, are not necessarily integers, as the following example shows. Let $\theta = \sqrt{-3}$ and $c = \frac{1}{2}(1 + \sqrt{-3})$. Then c is an algebraic integer because it is a root of the equation $x^2 - x + 1 = 0$, even though it is not a linear combination of 1 and $\sqrt{-3}$ with integer coefficients. However, it can be shown that all algebraic integers in $\mathbf{Q}(\sqrt{-3})$ can be written in the form $x + yc$, where x and y are integers. (In Corollary 2 of Theorem 6.2 we consider a more general case.)

We shall find useful later the following result, which is easily proved.

Theorem 2.2. If $f(x) = a_n x^n + a_{n-1} x^{n-1} + \cdots + a_1 x + a_0$, where the a_i are integers and if $f(\theta) = 0$, then $a_n \theta$ is an algebraic integer.

Proof: The proof is shown by the following calculation:

$$a_n^n f(\theta) = (a_n \theta)^n + a_{n-1} a_n (a_n \theta)^{n-1} + \cdots + a_1 a_n^{n-1}(a_n \theta) + a_0 a_n^n = 0.$$

Before proceeding, we should remark that in this book whenever we use the word "integer" we mean one of the numbers $0, \pm 1, \pm 2, \ldots$. When we refer to algebraic integers, we retain the adjective. You will find that in some texts the word "integer" is understood to include the algebraic integers and that when one wishes to refer to ordinary integers, they are called rational integers. This is perfectly consistent practice, but we shall not use such nomenclature here.

While we are about it, we should note that an *algebraic unit* is naturally defined to be an algebraic integer whose reciprocal (multiplicative inverse) is also an algebraic integer. To identify the algebraic units, suppose that $f(x)$ is written in the form of Theorem 2.2. Then

$$(2.3) \quad x^n f\left(\frac{1}{x}\right) = a_n + a_{n-1} x + \cdots + a_1 x^{n-1} + a_0 x^n = g(x) \quad (a_i \text{ integers}).$$

This shows that if c is a nonzero root of $f(x) = 0$, then $1/c$ is a root of $g(x) = 0$. Then if c is an algebraic integer, a_n may be assumed to be 1 and if $a_0 = \pm 1$, then $1/c$ is also an algebraic integer. Thus we have proved part of the following theorem.

Theorem 2.3. If $f(x)$ is a polynomial in x whose coefficients are integers, whose leading coefficient is 1, and such that $f(0) = \pm 1$, then its zeros are algebraic units. Furthermore, if θ is an algebraic unit, it is a zero of some $f(x)$, as described in the previous sentence.

Proof: We need to prove the second sentence of the theorem. Since θ is an algebraic integer, there is by Theorem 2.1 a unique primitive irreducible polynomial $f(x)$ with integral coefficients and positive leading coefficient such that $f(\theta) = 0$. The leading coefficient of $f(x)$ is then 1. Then if $g(x)$ is defined as in (2.3), it is irreducible by Exercise 20 after Section 3 and, since $\theta \neq 0$, has $1/\theta$ as a zero. Furthermore, the sign may be chosen so that $\pm g(x)$ has a positive leading coefficient. Then Theorem 2.1 asserts that 1 is the leading coefficient of $\pm g(x)$. Thus $a_0 = \pm 1 = f(0)$, and the proof is complete.

When we refer to the previous theorem, we see that $c = \frac{1}{2}(1 + \sqrt{-3})$ is a unit in $Q(\sqrt{-3})$, since $c^2 - c + 1 = 0$. To apply the above, let us characterize the algebraic integers in quadratic fields $F = Q(\sqrt{r})$, where r is an integer but not the square of an integer. We can, without loss, assume that r

is "square-free," that is, has no square factors except 1, since if $r = v^2 r'$, then $a + b\sqrt{r} = a + bv\sqrt{r'}$. To find the condition that $a + b\sqrt{r}$ be an algebraic integer for a and b in $\mathbf{Q}$, note that it is a root of the equation

$$h(x) = (x - a - b\sqrt{r})(x - a + b\sqrt{r}) = x^2 - 2ax + a^2 - rb^2.$$

Since $h(x)$ is irreducible over $\mathbf{Q}$, the corollary of Theorem 2.1 implies that $a + b\sqrt{r}$ will be an algebraic integer if and only if $2a$ and $a^2 - rb^2$ are both integers. First, if a is an integer, then $a^2 - rb^2$ an integer implies that rb^2 is an integer and, since r is square-free, b is an integer, by Exercise 21 after Section 3. Second, if $2a = d$ is an odd integer,

$$d^2 - 4rb^2 = d^2 - r(2b)^2 = 4k$$

for some integer k. Since d is an odd integer, so are r and $2b$. Letting $2b = e$, we have

(2.4) $$d^2 - re^2 = 4k.$$

Now, using the same reasoning as above, we see that (2.4) implies that e is an integer. Also, since d is odd, re^2 and hence e must be odd. Since the square of an odd integer is congruent to 1 modulo 4, (2.4) implies that $1 - r \equiv 0$ (mod 4). Thus we have almost proved the following theorem.

Theorem 2.4. *If r is a square-free integer, then the algebraic integers in $\mathbf{Q}(\sqrt{r})$ are*

1. $a + b\sqrt{r}$ *for a and b integers.*
2. *If $r \equiv 1$ (mod 4), $a + b\sqrt{r}$ are also algebraic integers if a and b are halves of odd integers.*

There are no other algebraic integers in quadratic fields.

Proof: We showed that if $2a$ and $a^2 - rb^2$ are integers, the conditions of the theorem are met; that is, if $a + b\sqrt{r}$, with a and b rational, is to be an algebraic integer, it must have one of the forms of the theorem. To complete the proof, we must show that if a and b are as in the theorem, then $2a$ and $a^2 - rb^2$ are integers. This is obvious if a and b are integers. If a and b are halves of odd integers, then $2a$ is an integer. To show that $a^2 - rb^2$ is an integer, for $r \equiv 1$ (mod 4), write $a = d/2$, $b = e/2$ for d and e odd and perform the following calculation:

$$a^2 - rb^2 = \left(\frac{d}{2}\right)^2 - r\left(\frac{e}{2}\right)^2 = \frac{d^2 - re^2}{4},$$

$$d^2 - re^2 \equiv 1 - r \equiv 0 \ (\text{mod } 4).$$

The proofs of the following two corollaries are left as exercises.

Corollary 1.   Let r be a square-free integer; then the algebraic integers in $Q(\sqrt{r})$ are the numbers $x + y\alpha$, where x and y are integers and $\alpha = \sqrt{r}$ if $r \not\equiv 1 \pmod 4$ and $\alpha = \frac{1}{2}(1 + \sqrt{r})$ if $r \equiv 1 \pmod 4$.

Corollary 2.   The units of $Q(\sqrt{r})$ are $x + y\sqrt{r}$, with $x^2 - ry^2 = \pm 1$, x and y integers if $r \not\equiv 1 \pmod 4$; x and y can also be halves of odd integers if $r \equiv 1 \pmod 4$.

Now that we have defined what we mean by algebraic integers, we are faced with a problem that must be disposed of before we can get on with the main concern of this chapter. The question is: In any algebraic extension of the field of rational numbers, are the sum and product of two algebraic integers both algebraic integers? The answer must certainly be "yes" if we are to make any progress. For this we need a theorem about symmetric functions which we shall use again and again in this chapter and which has many other uses. We shall state and prove this theorem in Section 3.

3.   Symmetric Functions

You will recall from college algebra that if

$$f(x) = (x - x_1)(x - x_2) \cdots (x - x_n)$$
$$= x^n - c_1 x^{n-1} + c_2 x^{n-2} - \cdots + (-1)^n c_n,$$

then

(3.1) $$c_1 = \sum_i x_i, \quad c_2 = \sum_{i<j} x_i x_j, \ldots, \quad c_n = x_1 x_2 \cdots x_n.$$

That is, c_k is the sum of all products of k of the x's. Each of the c_k is called an *elementary symmetric function* of the x_i. More generally, a function of the x's is called a *symmetric function* if it is unchanged by every permutation of the x's. There are other symmetric functions beside the elementary ones. For instance,

$$x_1^3 + x_2^3 + \cdots + x_n^3$$

is also a symmetric function of the x's. The fundamental theorem says that every symmetric function which is a polynomial in the x's can be expressed as a polynomial in the elementary symmetric functions. Furthermore, if the polynomial in the x's has integral coefficients, it can be expressed as a polynomial in the elementary symmetric functions with integral coefficients. Note that here the c's are functions of the n variables x_i. In the corollary of Theorem 3.1 we shall consider them in another light.

To see how this comes about in a specific case, we take $n = 2$ and compute $x_1^3 + x_2^3$ as a polynomial in c_1 and c_2:

$$c_1^3 = (x_1 + x_2)^3 = x_1^3 + 3x_1^2 x_2 + 3x_1 x_2^2 + x_2^3$$
$$= x_1^3 + x_2^3 + 3x_1 x_2 (x_1 + x_2)$$
$$= x_1^3 + x_2^3 + 3c_2 c_1.$$

This shows that $x_1^3 + x_2^3 = c_1^3 - 3c_2 c_1$.

Before stating and proving the theorem on symmetric functions, let us see in a specific example how it may be used to prove that the sum of two algebraic integers is an algebraic integer. Suppose that $f(d) = 0$, where $f(x) = x^3 + x + 1$, and that $g(e) = 0$, where $g(x) = x^2 + 2x + 3$. We want to show that $d + e$ is an algebraic integer. Let $d = d_1, d_2$, and d_3 be the zeros of $f(x)$, that is, the conjugates of d, and let $e = e_1$ and e_2 be the zeros of $g(x)$. Then

$$F(x) = (x - e_1 - d_1)(x - e_1 - d_2)(x - e_1 - d_3)(x - e_2 - d_1)$$
$$\cdot (x - e_2 - d_2)(x - e_2 - d_3)$$

is a polynomial with leading coefficient 1 that has $d + e$ as a zero. We need to prove that the coefficients of $F(x)$ are integers. Now

$$F(x) = f(x - e_1)f(x - e_2)$$
$$= [(x - e_1)^3 + (x - e_1) + 1][(x - e_2)^3 + (x - e_2) + 1].$$

But $F(x)$ is unchanged if we interchange e_1 and e_2. Hence when the indicated product is computed, it has the property that the coefficients of the various powers of x are symmetric functions of e_1 and e_2. For instance, the constant term of $F(x)$ is $(e_1^3 + e_1 + 1)(e_2^3 + e_2 + 1)$. The theorem on symmetric functions would inform us that the coefficients of $F(x)$ are polynomials with integral coefficients in the coefficients of $g(x)$, that is, are integers. This would complete the proof. Now we state and prove the theorem.

Theorem 3.1. (Fundamental theorem on symmetric functions.) Let $h(x_1, x_2, \ldots, x_n)$ be a polynomial in the x's with integer coefficients and suppose that h is a symmetric function of the x's, that is, is unchanged by every permutation of the x's; then h can be expressed as a polynomial $g(c_1, c_2, \ldots, c_n)$ with integral coefficients, where the c_i are defined by (3.1), that is, they are the elementary symmetric functions of the x's. (After the proof of the theorem we shall indicate a somewhat more general form, which is also valid.)

Proof: Now h is a sum of products of powers of the x's. Given any such product, since h is symmetric it must contain the sum of all products obtainable from it by permutations of the subscripts. Thus define

$$(3.2) \qquad\qquad h_0 = \sum_P x_1^{a_1} x_2^{a_2} \cdots x_n^{a_n},$$

where the exponents are nonnegative integers and the subscript P on the summation sign means that the sum is over all products obtained by permuting the x's. For instance, if $n = 3$,

$$\sum_P x_1^3 x_2^3 x_3 = x_1^3 x_2^3 x_3 + x_1^3 x_2 x_3^3 + x_1 x_2^3 x_3^3.$$

Notice that the sums of the exponents in all terms of h_0 are the same.

We call h_0 a *transitive symmetric polynomial*, since every product of the sum can be obtained from every other by permuting the subscripts. Now h must be the sum of a finite number of such polynomials. Hence, to prove the theorem, we need only prove it for transitive symmetric polynomials.

Since h_0 is transitive, it is determined by any product. Thus, in writing the typical term, we may assume that

(3.3) $a_1 \geq a_2 \geq \cdots \geq a_n \geq 0.$

We call this the *canonical term* of h_0. Under these conditions we call a_1 the *maximum exponent* of h_0. The sum of the a's is the *degree* of h_0.

First, if the maximum exponent of h_0 is 1, then h_0 is one of the elementary symmetric functions and no further proof is needed. Now we want to set up a proof by induction. To do this, we need to define some kind of measure of a transitive symmetric polynomial by which we can compare such polynomials. Call the sequence of a_i subject to the condition (3.3) the *exponent sequence* of h_0. Suppose that g is another symmetric transitive polynomiol with exponents b_i in nonincreasing order. We call the exponent sequence of g *less than* that of h_0 if the first b_i that is not equal to the corresponding a_i is less, that is, if

$$b_1 < a_1 \quad \text{or for some} \quad t < n, \ b_i = a_i \quad \text{for} \quad 1 \leq i \leq t$$
$$\text{and} \qquad b_{t+1} < a_{t+1}.$$

If all the exponents are the same, then $g = h_0$.

Before going on with the proof, let us look at the exponent sequence of the product of two transitive symmetric polynomials. Suppose that h_0 is defined by (3.2) and

$$g_0 = \sum_P x_1^{b_1} x_2^{b_2} \cdots x_r^{b_r}.$$

Assume that the term shown after $\sum$ is the canonical term of g_0. Then the maximum exponent sequence of the product $h_0 g_0$ will occur when the canonical terms "work together," so to speak. That is, one transitive symmetric polynomial in the product will have the exponent sequence

$$a_1 + b_1, a_2 + b_2, \ldots, a_n + b_n,$$

where we assume that $n \geq r$, and if $i > r$, it is understood that $b_i = 0$; all other transitive symmetric polynomials in the product $h_0 g_0$ will have lesser or equal exponent sequences.

Now we assume as an induction hypothesis that the theorem holds for all transitive symmetric polynomials g whose exponent sequences are less than that of h_0 and prove the theorem for h_0. If $a = a_i$ for $1 \leq i \leq n$, then $h_0 = c_n^a$, and the theorem holds for $h = h_0$. Otherwise we have for some k less than n,

$$a = a_1 = a_2 = \cdots = a_k > a_{k+1}.$$

Let $b = a - a_{k+1}$; notice that b is positive and define

(3.4) $$G' = \sum_P (x_1 x_2 \cdots x_k)^{a-b} x_{k+1}^{a_{k+1}} \cdots x_n^{a_n},$$

(3.5) $$G = h_0 - (c_k)^b G'.$$

Now the maximum exponent of G' is less than a; hence by the induction hypothesis G' can be expressed as a polynomial in the c's with integer coefficients. The induction hypothesis will result in a proof if we can show that

(3.6) $$(c_k)^b G' = h_0 + H,$$

where H is a sum of transitive symmetric polynomials whose exponent sequences are less than that of h_0. Now consider

$$(c_k)^b G' = \left[\sum_P x_1 x_2 \cdots x_k \right]^b \left[\sum_P (x_1 x_2 \cdots x_k)^{a-b} x_{k+1}^{a_{k+1}} \cdots x_n^{a_n} \right].$$

Note that the exponents of the second summation are in nonincreasing order.

The expression $(c_k)^b$ is expressible as a sum of transitive symmetric polynomials. Of all of these, the one whose exponent sequence is a maximum is

$$\sum_P x_1^b x_2^b \cdots x_k^b = \sum_P (x_1 x_2 \cdots x_k)^b.$$

The exponents for each product in G' will be some permutation of

$$a - b, a - b, \ldots, a - b, a_{k+1}, \ldots, a_n.$$

Thus the maximum sequence of exponents for $(c_k)^b G'$ is

$$a, a, \ldots, a, a_{k+1}, \ldots, a_n,$$

which is just the sequence of exponents for h_0. Thus we have shown in (3.6) that the maximum exponent sequence in H is less than that of h_0. Hence H, as well as G', is expressible as a polynomial in the c_i with integral coefficients. Then (3.6) establishes the theorem for h_0.

Let us illustrate the proof for $n = 4$ and

$$h_0 = \sum_P x_1^3 x_2^3 x_3.$$

Then

$$G = h_0 - \left[\sum_P x_1 x_2 \right]^2 \left[\sum_P x_1 x_2 x_3 \right] = h_0 - c_2^2 c_3.$$

For this example $k = 2$ and $b = 3 - 1 = 2$. To find h_0, we calculate G, which has a lesser exponent sequence. Now

$$(3.7) \qquad (c_2)^2 = \left[\sum_P x_1 x_2 \right]^2 = \sum_P x_1^2 x_2^2 + 2 \sum_P x_1^2 x_2 x_3 + 6 x_1 x_2 x_3 x_4.$$

We can check this calculation by counting the number of terms before combining. On the left the number is $6^2 = 36$. On the right the first sum has 6 terms, the second $2 \cdot 12 = 24$ terms, and the last 6 terms. This checks because $6 + 24 + 6 = 36$. Then

$$G = \sum_P x_1^3 x_2^3 x_3 - \left[\sum_P x_1^2 x_2^2 + 2 \sum_P x_1^2 x_2 x_3 + 6 x_1 x_2 x_3 x_4 \right] \left[\sum_P x_1 x_2 x_3 \right]$$

$$= \sum_P x_1^3 x_2^3 x_3 - \sum_P x_1^3 x_2^3 x_3 - \sum_P x_1^3 x_2^2 x_3 x_4$$

$$- 2 \sum_P x_1^3 x_2^2 x_3^2 - 4 \sum_P x_1^2 x_2^2 x_3 x_4 - 6 \sum_P x_1^2 x_2^2 x_3^2 x_4 - 6 \sum_P x_1^2 x_2^2 x_3^2 x_4.$$

The first two terms cancel each other. From the first term of (3.7) comes the exponent sequence 3, 2, 1, 1. From the second term of (3.7) come three exponent sequences, 3, 2, 2; 3, 2, 1, 1; and 2, 2, 2, 1. From the third term of (3.7) comes the exponent sequence 2, 2, 2, 1. Notice that the sequences of exponents that include 3, 3, 1 represent all the sets of four or less integers whose sum is 7, the degree of h_0. Thus

$$h_0 = c_2^2 c_3 + G = c_2^2 c_3 - 5 \sum_P x_1^3 x_2^2 x_3 x_4 - 2 \sum_P x_1^3 x_2^2 x_3^2 - 12 \sum_P x_1^2 x_2^2 x_3^2 x_4.$$

Each of the sums on the right has lower maximum exponent sequence than h_0 and can be evaluated by the same process. If one were to calculate h_0, one would need to evaluate each of these in terms of the coefficients. Fortunately, we are not really concerned with such evaluations, since we only need to know that they exist. So, we leave the example at this point. Our chief use of the theorem will be for the result expressed in the following corollary.

Corollary. If $\theta_1, \theta_2, \ldots, \theta_n$ are the zeros of the irreducible polynomial

$$f(x) = x^n - c_1 x^{n-1} + c_2 x^{n-2} - \cdots + (-1)^n c_n,$$

where the c_i are complex numbers, then every symmetric polynomial in the θ's with integer coefficients is a polynomial in the c_i with integer coefficients. In particular, if the θ's are algebraic integers, that is, the c_i are integers, then every symmetric polynomial in the θ's with integer coefficients is an integer.

If we look back at the proof of theorem, we can see that the hypothesis and conclusion can be somewhat generalized. Suppose that $h(x_1, x_2, \ldots, x_n)$ is written as a polynomial

$$h = d_1 h_1 + d_2 h_2 + \cdots + d_s h_s,$$

where each h_i is a transitive symmetric polynomial in the x's in the form (3.2) and the d's are in some field. We showed in the course of the proof that each h_i is expressible as a polynomial $f_i(c_1, c_2, \ldots, c_n)$ with integer coefficients. Then

$$h = d_1 f_1 + d_2 f_2 + \cdots + d_s f_s.$$

However, we shall be using the corollary for the most part in the form given.

Now we apply Theorem 3.1 to show that the sum of two algebraic integers is an algebraic integer.

Theorem 3.2. The sum of two algebraic integers is an algebraic integer.

Proof: Earlier in this section we outlined a proof for a special case. Our proof follows the same pattern. Let $f(x)$ and $g(x)$ be two polynomials in $\mathbf{Z}[x]$ with leading coefficient 1. Let the zeros of the former be $d_1, d_2, \ldots, d_n$ and those of the latter be $e_1, e_2, \ldots, e_s$. The d's and e's are algebraic integers. Certainly, $d_1 + e_1$ is a zero of

$$F(x) = \prod (x - d_i - e_j),$$

where the product is over $1 \leq i \leq n$ and $1 \leq j \leq s$. The leading coefficient of $F(x)$ is 1. We want to show that the coefficients are integers. We write $z_j = x - e_j$ and see that

$$f(z_j) = \prod_i (z_j - d_i) \qquad \text{and} \qquad F(x) = \prod_j f(z_j) = \prod_j f(x - e_j).$$

The coefficients of $F(x)$ are symmetric polynomials in the e_j with integer coefficients and hence, by the corollary of Theorem 3.1, are integers. This completes the proof.

Corollary. If d and e are two algebraic integers in $\mathbf{Q}(\theta)$, then $d + e$ is an algebraic integer in $\mathbf{Q}(\theta)$.

The corollary follows, since the sum of any two numbers of $\mathbf{Q}(\theta)$ is in $\mathbf{Q}(\theta)$, in view of the fact that $\mathbf{Q}(\theta)$ is a field.

We leave as an exercise the proof of the companion theorem.

Theorem 3.3. The product of two algebraic integers is an algebraic integer.

Corollary. The product of two algebraic integers in $\mathbf{Q}(\theta)$ is an algebraic integer in $\mathbf{Q}(\theta)$.

Let S be either the set of all algebraic integers or the set of algebraic integers in $\mathbf{Q}(\theta)$ for θ an algebraic number over $\mathbf{Q}$. Since S is a subset of the complex numbers, it contains no divisors of zero, addition and multiplication are commutative and associative, and the distributive property holds. We have just shown that S is closed under addition and multiplication. Certainly, S contains the additive and multiplicative identities and the additive inverse of each element. Thus we have proved the following theorem and its corollary.

Theorem 3.4.   The set of algebraic integers forms an integral domain.

Corollary.   The set of algebraic integers in $\mathbf{Q}(\theta)$, any extension of the rational field, forms an integral domain.

Somewhat aside from our main path is the following theorem, which is not hard to prove. We give a sketch of the proof.

Theorem 3.5.   The set S of all algebraic numbers is a field.

Sketch of the proof: We need to show that S is closed under addition, multiplication, and division. (The other properties follow from the fact that S is a subset of the complex numbers.) To prove closure under division, let c be a nonzero root of

$$f(x) = a_n x^n + a_{n-1} x^{n-1} + \cdots + a_1 x + a_0 = 0.$$

Then $1/c$ is a zero of the polynomial $x^n f(1/x)$, which has coefficients in $\mathbf{Q}$. Thus the reciprocal of an algebraic number is an algebraic number. Let α and β be two algebraic numbers. Then by Theorem 2.2 there are integers a and b such that $a\alpha$ and $b\beta$ are algebraic integers; by Theorem 3.3 $ab\alpha\beta$ is an algebraic integer and hence $\alpha\beta$ is an algebraic number. Thus the set of algebraic numbers is closed under multiplication. To prove closure under addition, use Theorem 3.2 to show that $a\alpha + b\beta$ and $a\alpha - b\beta$ are algebraic integers. Hence

$$(a + b)(\alpha + b\beta) + (b - a)(a\alpha - b\beta) = 2ab(\alpha + \beta)$$

is an algebraic integer, which shows that $\alpha + \beta$ is an algebraic number. Alternatively, closure under addition and multiplication could be shown by the method we used to prove Theorems 3.2 and 3.3.

Exercises

1. Show that every algebraic number is a zero of a polynomial in $\mathbf{Z}[x]$.

2. Prove that in the integral domain $\{x + y\sqrt{-5}\}$, for x and y integers, the number 3 is a prime.

3. Prove the corollaries of Theorem 2.4.

4. Find the algebraic integers in $\mathbf{Q}(\sqrt{-11})$. Notice that $9 = 3 \cdot 3 = \frac{1}{2}(5 + \sqrt{-11})\frac{1}{2}(5 - \sqrt{-11})$ and prove that 3 and $\frac{1}{2}(5 + \sqrt{-11})$ are not prime algebraic integers.

5. Prove that 3 is a prime algebraic integer in $Q(i)$ but that 5 is not.

6. Show that if p is a prime number such that $p \equiv 3 \pmod 4$, then p is a prime algebraic integer in $Q(i)$.

7. The numbers -1 and $+1$ are units in every algebraic field. For what negative integers r are 1 and -1 the only units in $Q(\sqrt r)$?

8. Find six units of $Q(\sqrt 2)$.

9. Let r be a positive integer that is not a square and suppose that for integers a and b, the number $a + b\sqrt r$ is a unit of $Q(\sqrt r)$. Show that $(a + b\sqrt r)^k$ is a unit for all positive integers k. Hence show that if $Q(\sqrt r)$ has units different from 1 and -1, it has infinitely many units.

10. Let $f(x)$ have coefficients in $Q(\sqrt r)$ for r a square-free integer. Show that all the zeros of $f(x)$ are algebraic numbers.

11. Prove Theorem 3.3.

12. Complete the proof of Theorem 3.5.

13. (a) Express $\sum_i x_i^3$ as a polynomial in the elementary symmetric polynomials in x_1, x_2, and x_3.
    (b) Express $\sum_i x_i^4$ as a polynomial in the elementary symmetric polynomials in x_1, x_2, and x_3.

14. Repeat Exercise 13 for x_1, x_2, x_3, and x_4.

15. Illustrate the proof of Theorem 3.1 for $n > 4$ and

    (a) $h_0 = \sum_P x_1^3 x_2 x_3$.   (b) $\sum_P x_1^2 x_2^2 x_3$.

16. Let $n > 4$ and

    $$h_0 = \sum_P x_1^3 x_2^3 x_3^2 x_4.$$

    Find G as defined in the proof of Theorem 3.1.

17. Show that $\sum_P x_1^r - [\sum_P x_i]^r$ has maximum exponent less than r.

18. Let c be a zero of a polynomial $f(x)$ whose leading coefficient is 1 and whose other coefficients are Gaussian integers. Prove that c is an algebraic integer. (Recall that the Gaussian integers are the numbers $a + bi$, where a and b are integers.)

19. Let $s_k = \sum_P x_1^k$; then the following formula is called *Newton's formula*:

    $$s_k - c_1 s_{k-1} + c_2 s_{k-2} - \cdots + (-1)^k k c_k = 0,$$

    where the c_i are defined by (3.1) and $c_j = 0$ if $j > n$, the number of x_i. By means of this formula one can evaluate successively the various s_k in terms of the c's. Prove Newton's formula for $k = 1, 2, 3, 4$.

20. Let $p(x)$ be an irreducible polynomial over $\mathbf{Q}$ and $g(x) = x^n p(1/x)$, where n is the degree of $p(x)$. Prove that $g(x)$ is also irreducible over $\mathbf{Q}$.

21. Let r be a square-free integer and rb^2 an integer, where b is a rational number. Prove that b is an integer.

★22. Without assuming the theorem on symmetric functions but assuming Newton's formula, prove by induction or otherwise that s_k, defined in Exercise 19, can be expressed as a polynomial in the c_i with integral coefficients.

★23. Let θ be an algebraic integer and $g(x)$ a polynomial in $\mathbf{Z}[x]$ such that $g(\theta) = 0$. *Must* the leading coefficient of $g(x)$ be 1? Why or why not?

4.   The Concept of an Ideal

Now we can begin to deal directly with the problem proposed at the beginning of this chapter. We shall explore informally the way to a solution in $\mathbf{Q}(\sqrt{-5})$ and gradually evolve useful concepts and results. Consider the set of algebraic integers in $\mathbf{Q}(\sqrt{-5})$, that is, from Theorem 2.4, the set $\{x + y\sqrt{-5}\}$, where x and y are integers. We designate this set by $\mathbf{Z}[\sqrt{-5}]$, using square brackets. Recall that in $\mathbf{Z}[\sqrt{-5}]$

$$9 = 3 \cdot 3 = (2 + \sqrt{-5})(2 - \sqrt{-5}),$$

where 3 and $2 + \sqrt{-5}$ are primes in $\mathbf{Z}[\sqrt{-5}]$. Thus 9 can be expressed in two different ways as a product of prime algebraic integers.

Faced with this phenomenon, it was Richard Dedekind† who had the brilliant idea that instead of considering the single number 3, one could consider the set $\mathscr{A} = \{3x\}$, where x is in $\mathbf{Z}[\sqrt{-5}]$, and $\mathscr{B} = \{(2 + \sqrt{-5})y\}$, where y is in $\mathbf{Z}[\sqrt{-5}]$. The sets $\mathscr{A}$ and $\mathscr{B}$ are examples of what are called *ideals*. The integer 3 determines $\mathscr{A}$ and the number $2 + \sqrt{-5}$ determines $\mathscr{B}$, and vice versa. The utility of this way of looking at things becomes apparent if we consider also

(4.1) $$\mathscr{C} = \{3x + (2 + \sqrt{-5})y\},$$

where x and y are in $\mathbf{Z}[\sqrt{-5}]$. For sets $\mathscr{A}$, $\mathscr{B}$, and $\mathscr{C}$ we start with the same ring $\mathbf{Z}[\sqrt{-5}]$, and they each have the property of being closed under addition and multiplication by all elements of the ring. Any nonempty set over a ring that has these two closure properties is called an *ideal*. (We define this more

---

† Actually, a concept equivalent to an ideal was developed before Dedekind by E. Kummer in connection with Fermat's Last Theorem (Section 14 of Chapter II). But the modern concept of an ideal and its application to the problem of this chapter is due to Dedekind.

formally later in the section.) We now proceed to explore in terms of this example what properties ideals have.

Now, $\mathscr{C}$ contains $\mathscr{A}$ and $\mathscr{B}$, and any ideal over $\mathbf{Z}[\sqrt{-5}]$ that contains $\mathscr{A}$ and $\mathscr{B}$ must contain $\mathscr{C}$. If we were to replace "contains" by "divides," $\mathscr{C}$ would satisfy the requirements of a greatest common divisor, as we now show.

We use the customary notation of sets and write $\mathscr{D} \supseteq \mathscr{E}$ if every element of $\mathscr{E}$ is in $\mathscr{D}$, or $\mathscr{D} \supset \mathscr{E}$ if $\mathscr{E}$ is a proper subset of $\mathscr{D}$, that is, if $\mathscr{D}$ contains $\mathscr{E}$ but not all the elements of $\mathscr{D}$ are in $\mathscr{E}$. So we have

$$\mathscr{C} \supseteq \mathscr{A} \quad \text{and} \quad \mathscr{C} \supseteq \mathscr{B},$$

and if $\mathscr{D}$ is an ideal over $\mathbf{Z}[\sqrt{-5}]$ that contains $\mathscr{A}$ and $\mathscr{B}$, it contains $\mathscr{C}$, since $\mathscr{D}$ is closed under addition and multiplication by elements of the ring $\mathbf{Z}[\sqrt{-5}]$. We call $\mathscr{C}$ the "least common container" of $\mathscr{A}$ and $\mathscr{B}$.

In a more natural sense $\mathscr{C}$ defined in (4.1) is also a greatest common divisor of $\mathscr{A}$ and $\mathscr{B}$. For suppose that $\mathscr{D}' = \{3x + (2 - \sqrt{-5})y\}$ for x and y in $\mathbf{Z}[\sqrt{-5}]$. Define $\mathscr{C}\mathscr{D}'$ to be all sums of products cd, where c is in $\mathscr{C}$ and d is in $\mathscr{D}'$. Then

$$\mathscr{C}\mathscr{D}' = \{9r' + (6 + 3\sqrt{-5})s + (6 - 3\sqrt{-5})t + 9r''\}$$
$$= \{9r + (6 + 3\sqrt{-5})s + (6 - 3\sqrt{-5})t\},$$

where r, s, and t are in $\mathbf{Z}[\sqrt{-5}]$. If we let $r = -w$ and $s = t = w$ for w in $\mathbf{Z}[\sqrt{-5}]$, we see that $\mathscr{C}\mathscr{D}'$ contains $3w$ for all w in $\mathbf{Z}[\sqrt{-5}]$. Hence $\mathscr{C}\mathscr{D}' \supseteq \mathscr{A}$. But $\mathscr{A} \supseteq \mathscr{C}\mathscr{D}'$, since for all r, s, and t,

$$9r + (6 + 3\sqrt{-5})s + (6 - 3\sqrt{-5})t = 3u$$

for some u in $\mathbf{Z}[\sqrt{-5}]$. Thus $\mathscr{C}\mathscr{D}' = \mathscr{A}$. Hence in this case $\mathscr{C}$ is a divisor of $\mathscr{A}$. Similarly, it can be shown that $\mathscr{C}$ is a divisor of $\mathscr{B}$. Now it would seem natural to define the sum of two ideals $\mathscr{A} + \mathscr{B}$ to be the set $\{a + b\}$, where a is in $\mathscr{A}$ and b is in $\mathscr{B}$. Then for $\mathscr{C}$, $\mathscr{A}$, and $\mathscr{B}$ defined as above, $\mathscr{C} = \mathscr{A} + \mathscr{B}$. Furthermore, if $\mathscr{E}$ is a common divisor of $\mathscr{A}$ and $\mathscr{B}$, we can write $\mathscr{E}\mathscr{E}_1 = \mathscr{A}$ and $\mathscr{E}\mathscr{E}_2 = \mathscr{B}$, which gives

$$\mathscr{C} = \mathscr{A} + \mathscr{B} = \mathscr{E}\mathscr{E}_1 + \mathscr{E}\mathscr{E}_2.$$

Formally, this would imply that $\mathscr{C} = \mathscr{E}(\mathscr{E}_1 + \mathscr{E}_2)$, assuming the distributive property for ideals. Thus, intuitively, it would seem that $\mathscr{C} = \mathscr{A} + \mathscr{B}$ is a divisor of both $\mathscr{A}$ and $\mathscr{B}$ and any common divisor of $\mathscr{A}$ and $\mathscr{B}$ is a divisor of $\mathscr{C}$. In other words, $\mathscr{C}$ would be a greatest common divisor of $\mathscr{A}$ and $\mathscr{B}$.

We shall see that for ideals over rings of algebraic integers, the property of containment and divisor are equivalent. It is clear that much of the above exploratory material needs justification. So, the first thing to do is to evolve a definition of an ideal by successive generalizations of the above example.

First, instead of ideals over $\mathbf{Z}[\sqrt{-5}]$, let R be the set of all algebraic integers in some algebraic extension of $\mathbf{Q}$ and $\alpha_1, \alpha_2, \ldots, \alpha_k$ a set of k elements of R. Then the set of all numbers

$$(4.2) \qquad \mathscr{A} = \{x_1\alpha_1 + x_2\alpha_2 + \cdots + x_k\alpha_k\},$$

where the x_i are in R is called an *ideal* over R. By its definition this set is closed under addition and multiplication by elements of R.

With this as a background, we might better define an ideal formally in terms of how it behaves rather than how it looks. Then, after the definition we can explore more how it looks. In Section 9 we shall define and consider ideals over noncommutative rings, but since all our applications up to that point are to commutative rings, we make this restriction in the following definition.

Definition.   *If R is a commutative ring with identity element, any nonempty set $\mathscr{A}$ of elements of R is called an* ideal *if the following conditions hold:*

1. *If a_1 and a_2 are in $\mathscr{A}$, then $a_1 + a_2$ is in $\mathscr{A}$.*
2. *If a is in $\mathscr{A}$ and r in R, then ra is in $\mathscr{A}$.*

In other words, a set of elements of R is called an ideal if it is closed under addition as well as multiplication by an element of R. (This definition is extended to noncommutative rings in Section 9.) Notice that an ideal is also closed under subtraction, since -1 is in R.

It should be emphasized that although most of our illustrations and examples have to do with rings of algebraic integers, we make no such restriction until we come to Section 6. In general, we shall use script letters for ideals. In fact, we could also use a script letter for the basic ring R, since R is an ideal over itself. (A boldface $\mathbf{R}$ denotes something quite different—the field of real numbers. Of course, this is trivially an ideal over itself, too.)

It happens that most of the ideals we shall be considering are like $\mathscr{A}$ in (4.2). In that case we borrow terminology from vector spaces and say that the α's *span* $\mathscr{A}$. Then we use the notation

$$\mathscr{A} - (\alpha_1, \alpha_2, \ldots, \alpha_k)$$

If it happens that all elements of $\mathscr{A}$ in (4.2) are obtained by restricting the x's to be integers, we say that the α's *integrally span* $\mathscr{A}$ and use square brackets:

$$\mathscr{A} = [\alpha_1, \alpha_2, \ldots, \alpha_k].$$

Before considering the properties of ideals, let us point out two special ideals. First, there is the trivial ideal consisting of the zero element of R alone; it satisfies the requirements imposed in the definition. We call it the *zero ideal* and designate it by (0). The next simplest ideal is one spanned by a single element alone. That is, if a is a nonzero element of R, then the set $\{xa\}$ as x ranges over all elements of R is called a *principal ideal*. It can be written in accordance with the above as (*a*). The ideals $\mathscr{A} = (3)$ and $\mathscr{B} = (2 + \sqrt{-5})$ dealt with earlier in this section are principal ideals. Notice that $\mathscr{A}$ contains the number $3\sqrt{-5}$, for instance, which is not a *rational* multiple of 3.

The ring **Z** of integers has the property that every ideal over **Z** is a principal ideal, as we now show. Suppose that $\mathscr{A}$ is any set of integers which is closed under addition and multiplication by an integer, that is, which is an ideal over **Z**. Since **Z** is well-ordered (see Section 6 of Chapter III), the set $\mathscr{A}$ has a least positive element; call it *m*. The following argument, like that in the proof of Theorem 9.3 of Chapter III, shows that every element of $\mathscr{A}$ is a multiple of *m*. Suppose that *a* is an element of $\mathscr{A}$; then from Theorem 9.1 of Chapter III, there are integers *q* and *r* such that

$$a = mq + r, \qquad 0 \le r < m.$$

But $r = a - mq$ is in $\mathscr{A}$, since *a* and *m* are in $\mathscr{A}$ and $\mathscr{A}$ is an ideal. Since *m* is the least positive element of $\mathscr{A}$, it follows that $r = 0$ and *m* is a factor of *a*. Thus we have shown the following theorem.

Theorem 4.1. Every ideal over **Z** except the zero ideal is a principal ideal.

There are other rings that have this property. Now we define a term.

Definition. *A ring R is said to be a* principal ideal ring *if every nonzero ideal over R is a principal ideal.*

Another example of a principal ideal ring is the set of Gaussian integers, as you are asked to show in Exercise 19 after Section 5. The set **Q**[*x*] of polynomials in *x* with rational coefficients is also a principal ideal ring, as you are asked to show in Exercise 20 after Section 5.

Before proceeding with the theory, let us revert to the example with which this section began and show that over $\mathbf{Z}[\sqrt{-5}]$ the ideal

$$\mathscr{C} = (3, 2 + \sqrt{-5})$$

is not a principal ideal. Suppose that $\mathscr{C} = (a + b\sqrt{-5})$, for *a* and *b* integers, and seek a contradiction. If this were true, then for some α and β in $\mathbf{Z}[\sqrt{-5}]$ the following equalities would hold:

$$(a + b\sqrt{-5})\alpha = 3 \qquad \text{and} \qquad (a + b\sqrt{-5})\beta = 2 + \sqrt{-5}.$$

But we showed in Section 1 of this chapter that 3 and $2 + \sqrt{-5}$ are distinct prime algebraic integers. Hence $a + b\sqrt{-5}$ would have to be a unit. Thus $\mathscr{C}$ would contain the number 1. This means that the following would have to be solvable for integers *x*, *y*, *z*, and *t* :

$$3(x + y\sqrt{-5}) + (2 + \sqrt{-5})(z + t\sqrt{-5}) = 1.$$

That is,

$$3x + 2z - 5t = 1 \quad \text{and} \quad 3y + z + 2t = 0.$$

If we multiply the second equation by 2 and subtract from the first, we get

$$3x - 6y - 9t = 1,$$

which is not solvable in integers. Thus we have a contradiction. This shows that $\mathscr{C} = (3, 2 + \sqrt{-5})$ is not a principal ideal.

A fundamental property of a principal ideal ring is given in the following theorem.

Theorem 4.2.   If R is a principal ideal ring, then every pair of elements r and r' of R not both zero have a greatest common divisor g expressible in the form $g = rx + r'y$, where x and y are in R.

Proof: First, we should explain that by "greatest common divisor" in the above theorem, we of course mean an element g of R such that $gu = r$ and $gu' = r'$ for u and u' in R and that if h is in this sense a divisor of both r and r', it is a divisor of g.

For the proof see that the set $\{rx + r'y\}$, where x and y range over R, is an ideal $\mathscr{A}$. Since R is a principal ideal ring, $\mathscr{A}$ is a principal ideal and we can write $\mathscr{A} = (c)$ for c in R. Since r and r' are in (c), then $cv = r$ for some v in R and $cv' = r'$ for some v' in R. Therefore, in this sense, c is a common divisor of r and r'. Now if $gu = r$ and $gu' = r'$ for u and u' in R, we note that $c = rx_0 + r'y_0$ for some x_0 and y_0 in R and hence that $c = g(ux_0 + u'y_0)$, which shows that g is a factor of c. This completes the proof.

Once one has such a greatest common divisor, he has made some progress toward unique decomposition into prime factors, but we shall see that there is at least one other fundamental difficulty further on.

The next step is to define the sum and product of two ideals. This is done in the most natural way as follows.

Definition.   *If $\mathscr{A}$ and $\mathscr{B}$ are two ideals over R, then $\mathscr{A} + \mathscr{B}$ is the set of all sums $a + b$, where a and b are in $\mathscr{A}$ and $\mathscr{B}$, respectively. The product $\mathscr{A}\mathscr{B}$ is the sum of all products ab, where a is in $\mathscr{A}$ and b is in $\mathscr{B}$.*

Notice that for multiplication we need to consider sums of products in order to have closure of $\mathscr{A}\mathscr{B}$ under addition. The set of ideals over R has most of the properties of a ring. We list them and give a brief indication of the reasons why they have these properties.†

---

† Details of some of these properties are left as exercises.

1. If $\mathscr{A}$ and $\mathscr{B}$ are ideals, then $\mathscr{A} + \mathscr{B}$ is an ideal, for $(a_1 + b_1) + (a_2 + b_2)$ $= (a_1 + a_2) + (b_1 + b_2)$; and $r(a + b) = ra + rb$, where the a's are in $\mathscr{A}$ and the b's in $\mathscr{B}$.
2. $\mathscr{A}\mathscr{B}$ is an ideal, since $a_1 b_1 + a_2 b_2$ is in $\mathscr{A}\mathscr{B}$ as well as $(ra)b = r(ab)$.
3. The addition and multiplication of ideals are associative and commutative, since these properties hold for R.
4. The identity for multiplication is R, since $R\mathscr{A}$ consists of all the sums of products ra, where r is in R and a is in $\mathscr{A}$. Since all such sums are the ideal $\mathscr{A}$, it follows that $R\mathscr{A} = \mathscr{A}$.
5. The distributive property holds: $\mathscr{A}(\mathscr{B} + \mathscr{C}) = \mathscr{A}\mathscr{B} + \mathscr{A}\mathscr{C}$.

For ideals the additive and multiplicative inverses do not usually exist; this may be seen as follows. First, since every ideal $\mathscr{X}$ contains 0, $\mathscr{A} + \mathscr{X} \supseteq \mathscr{A}$ for all ideals $\mathscr{X}$, and hence $\mathscr{A} + \mathscr{X} = (0)$ has no solutions unless $\mathscr{A} = (0)$. Second, if $\mathscr{A}\mathscr{X} = R$, the identity element for multiplication, then $\mathscr{A} \supseteq R$. Since $\mathscr{A} \subseteq R$, it follows that $\mathscr{A} = R$. This means that $\mathscr{A}\mathscr{X} = R$ is not solvable unless $\mathscr{A} = R$, and then the solution is $\mathscr{X} = R$. So if $\mathscr{A} \neq R$, it has no multiplicative inverse. Notice that for ideals $\mathscr{A} + \mathscr{A} = \mathscr{A}$. In fact, $\mathscr{A} + \mathscr{B} = \mathscr{A}$ if $\mathscr{B} \subseteq \mathscr{A}$. But $\mathscr{A} \cdot \mathscr{A}$ is not always $\mathscr{A}$. Why?

Now let us point out some general properties arising from the special case at the beginning of this section.

Theorem 4.3. If $\mathscr{A}$ and $\mathscr{B}$ are two ideals over R, then $\mathscr{A} + \mathscr{B}$ contains $\mathscr{A}$ and $\mathscr{B}$, and if $\mathscr{D}$ is any ideal that contains $\mathscr{A}$ and $\mathscr{B}$, then $\mathscr{D}$ contains $\mathscr{A} + \mathscr{B}$.

Proof: Since every ideal contains the element 0 of R, $a = a + 0$ and $b = 0 + b$ are in $\mathscr{A} + \mathscr{B}$. If $\mathscr{D}$ contains $\mathscr{A}$ and $\mathscr{B}$, then since $\mathscr{D}$ is closed under addition it must contain $\mathscr{A} + \mathscr{B}$. This completes the proof.

This theorem means that $\mathscr{A} + \mathscr{B}$ behaves in a sense like a greatest common divisor, for if in the theorem we replace "contains" by "divides" we would have the definition of greatest common divisor. As noted previously, we could call $\mathscr{A} + \mathscr{B}$ the "least common container" of $\mathscr{A}$ and $\mathscr{B}$.

Now the relation "$\mathscr{A}$ divides $\mathscr{C}$" has a perfectly natural meaning, as contained in the following definition.

Definition. *If $\mathscr{A}$ and $\mathscr{C}$ are two ideals over R, $\mathscr{A}$ is a divisor of $\mathscr{C}$ or $\mathscr{C}$ is a multiple of $\mathscr{A}$ if there is an ideal $\mathscr{B}$ over R such that $\mathscr{A}\mathscr{B} = \mathscr{C}$. If $\mathscr{A}$ is a divisor of $\mathscr{C}$, we use the notation corresponding to that for integers: $\mathscr{A} \mid \mathscr{C}$.*

Here we have a peculiar phenomenon. If $\mathscr{A}\mathscr{B} = \mathscr{C}$, then $\mathscr{A} = \mathscr{A}R \supseteq \mathscr{A}\mathscr{B} = \mathscr{C}$, so that *if $\mathscr{A}$ divides $\mathscr{C}$, it contains $\mathscr{C}$*. This does violence to our intuition, since we are accustomed to a divisor's being smaller than what it divides. But this is a penalty we have to pay for defining ideals as we have done.

Now we have two concepts: $\mathscr{A} \supseteq \mathscr{B}$ and $\mathscr{A} \mid \mathscr{B}$. We know that if $\mathscr{A} \mid \mathscr{B}$, then $\mathscr{A} \supseteq \mathscr{B}$. But the question is:

(4.3)   If $\mathscr{A} \supseteq \mathscr{B}$, must $\mathscr{A}$ be a divisor of $\mathscr{B}$?

We shall show that for ideals over a ring of algebraic integers the answer is "yes," but it is not easy. Once this is done and an additional basic property proved (Theorem 7.3), the path to unique decomposition is easy. At the close of Section 5 and in an exercise following that section, examples are given to show that the answer is not always "yes."

However, there is a concept that is equivalent whether we take the point of view of divisor or container, that is, the property of being relatively prime. Recall that two integers a and b are relatively prime if $ax + by = 1$ is solvable in integers x and y. In term of ideals this means that $(a) + (b)$ contains 1, that is, is the ideal **Z**. So by analogy we have the following definition.

Definition.   *If $\mathscr{A}$ and $\mathscr{B}$ are two ideals over R and $\mathscr{A} + \mathscr{B} = R$, we call $\mathscr{A}$ and $\mathscr{B}$ relatively prime.*

This has two connotations. In terms of containing, Theorem 4.3 shows us that any ideal containing $\mathscr{A}$ and $\mathscr{B}$ under these conditions contains R. Hence if $\mathscr{A}$ and $\mathscr{B}$ are relatively prime, then R is the least common container of $\mathscr{A}$ and $\mathscr{B}$. Conversely, if R is the least common container of $\mathscr{A}$ and $\mathscr{B}$, then $R = \mathscr{A} + \mathscr{B}$, for $\mathscr{C} = \mathscr{A} + \mathscr{B}$ contains both $\mathscr{A}$ and $\mathscr{B}$ and is contained in R.

In terms of dividing, note that $R\mathscr{A} = \mathscr{A}$ and $R\mathscr{B} = \mathscr{B}$ imply that R divides $\mathscr{A}$ and $\mathscr{B}$. Suppose that an ideal $\mathscr{C}$ divides both $\mathscr{A}$ and $\mathscr{B}$; that is, $\mathscr{A} = \mathscr{C}\mathscr{A}'$ and $\mathscr{B} = \mathscr{C}\mathscr{B}'$. Then

$$R = \mathscr{A} + \mathscr{B} = \mathscr{C}\mathscr{A}' + \mathscr{C}\mathscr{B}' = \mathscr{C}(\mathscr{A}' + \mathscr{B}')$$

implies that $\mathscr{C} \mid R$. But we showed just before (4.3) that $\mathscr{C} \mid R$ implies that $\mathscr{C} \supseteq R$ and hence that $\mathscr{C} = R$. Thus if $\mathscr{A}$ and $\mathscr{B}$ are relatively prime, R is the only common divisor of $\mathscr{A}$ and $\mathscr{B}$.

Thus if $\mathscr{A} + \mathscr{B} = R$, then R is the minimum ideal containing both $\mathscr{A}$ and $\mathscr{B}$ and R is the only common factor of $\mathscr{A}$ and $\mathscr{B}$.

For example, suppose that $\mathscr{A} = (2 + \sqrt{-5})$ and $\mathscr{B} = (2 - \sqrt{-5})$, where $R = \mathbf{Z}[\sqrt{-5}]$. Then $\mathscr{A} + \mathscr{B}$ contains $2 + \sqrt{-5} + 2 - \sqrt{-5} = 4$ and also $(2 - \sqrt{-5})(2 + \sqrt{-5}) = 9$. Hence $\mathscr{A} + \mathscr{B}$ contains $9 - 4 \cdot 2 = 1$, which shows that $\mathscr{A} + \mathscr{B} = R$. Therefore, $\mathscr{A}$ and $\mathscr{B}$ are relatively prime.

5.   *Progress Toward Unique Decomposition*

Now, with the background of the first four sections of this chapter we can define more clearly the problem with which we are faced. Starting with a commutative ring R with identity element, we want to develop a concept of a prime ideal so that every ideal can be expressed as a product of prime ideals uniquely except for the order of the factors. We shall find in this section that we can make some progress toward our goal without restricting the ring. Our first task thus should be to define prime ideals. Two allied concepts play this role.

Definition. *An ideal $\mathscr{P}$ over R is called a* prime ideal *if it is neither R nor the zero ideal and if the only ideals that divide $\mathscr{P}$ are $\mathscr{P}$ and R.*

NOTE: This definition is not the usual one nowadays for prime ideals over a ring. But for ideals over rings of algebraic integers, it is equivalent to the usual definition. It seems better to use the natural definition here. In Section 8 we give the other definition (with a slight modification) and show it to be equivalent to this one.

Definition. *An ideal $\mathscr{M}$, with $(0) \neq \mathscr{M} \subset R$, over R is called* maximal *over R if R is the only ideal over R that properly contains $\mathscr{M}$.*

Above (4.3) we noted that if $\mathscr{A} \mid \mathscr{B}$, then $\mathscr{A} \supseteq \mathscr{B}$. Hence if $\mathscr{B}$ is not a prime ideal, there is an ideal $\mathscr{S}$ neither $\mathscr{B}$ nor R such that $\mathscr{S} \mid \mathscr{B}$. Thus $\mathscr{S} \supseteq \mathscr{B}$ and $\mathscr{B}$ is not maximal. Therefore, if $\mathscr{B}$ *is maximal, then it is prime.*

We want to show that for algebraic integer rings the prime ideals and the maximal ideals are the same. As one might expect, for a general commutative ring they behave the same in the following two theorems. Since the proofs are much like those for integers, we leave the proofs as exercises.

Theorem 5.1a. *If $\mathscr{A} \mid \mathscr{B}\mathscr{C}$ and $\mathscr{A}$ and $\mathscr{B}$ are relatively prime, then $\mathscr{A} \mid \mathscr{C}$.*

Theorem 5.1b. *If $\mathscr{A} \supseteq \mathscr{B}\mathscr{C}$ and $\mathscr{A}$ and $\mathscr{B}$ are relatively prime, then $\mathscr{A} \supseteq \mathscr{C}$.*

Now let us see how close we can come to the unique decomposition theorem, first, for prime ideals and, second, for maximal ideals. Let $\mathscr{N}$ be any ideal over R. If $\mathscr{N}$ is a prime ideal, there is no problem. If not, $\mathscr{N} = \mathscr{A}\mathscr{B}$, where neither $\mathscr{A}$ nor $\mathscr{B}$ is $\mathscr{N}$. If $\mathscr{A}$ and $\mathscr{B}$ are prime ideals, stop the process. If not, express each as a product of prime ideals. So, it would seem that $\mathscr{N}$ can be expressed as a product of prime ideals. *But* here appears the first difficulty. We need to know that the process stops after a finite number of steps, that is, that $\mathscr{N}$ has only a finite number of distinct divisors. Furthermore, even if this obstacle were to be surmounted we would have difficulty in proving the decomposition unique. For suppose that

$$\mathscr{P}_1\mathscr{P}_2 \cdots \mathscr{P}_k = \mathscr{Q}_1\mathscr{Q}_2 \cdots \mathscr{Q}_s.$$

If $\mathscr{P}_1$ divides $\mathscr{Q}_1$, then $\mathscr{P}_1 = \mathscr{Q}_1$, since $\mathscr{Q}_1$ is a prime ideal and we can presumably divide both sides by $\mathscr{P}_1$. But if $\mathscr{P}_1$ does not divide $\mathscr{Q}_1$, the question is:

Does this imply that $\mathscr{P}_1$ and $\mathscr{Q}_1$ are relatively prime; that is, is $\mathscr{P}_1 + \mathscr{Q}_1 = R$? We are not sure of this, since we do not know that a prime ideal is maximal. Hence for prime ideals we are blocked in two places.

Suppose that we take the maximal ideal approach. That is, suppose that we have first proved that maximal ideals and prime ideals are the same; then how far could we progress toward unique decomposition of ideals? We would still have the first difficulty above, that is, we would have to show that every ideal $\mathscr{N}$ over R is properly contained in only a finite number of ideals. But the second difficulty is surmountable, as we now show. That is, we need to prove that if $\mathscr{P}_1$ and $\mathscr{Q}_1$ are distinct maximal ideals, then they are relatively prime. To this end, let $\mathscr{P}_1 + \mathscr{Q}_1 = D$. Since $\mathscr{P}_1$ and $\mathscr{Q}_1$ are distinct and maximal, neither is contained in the other. Hence $D \neq \mathscr{P}_1$ and $D \supset \mathscr{P}_1$. Since $\mathscr{P}_1$ is maximal, $D = \mathscr{R}$, and $\mathscr{P}_1$ and $\mathscr{Q}_1$ are relatively prime. Actually, we have proved a somewhat stronger result.

Theorem 5.2.   If $\mathscr{P}$ is a maximal ideal over R and $\mathscr{Q}$ an ideal over R not contained in $\mathscr{P}$, then $\mathscr{P}$ and $\mathscr{Q}$ are relatively prime.

In summary, we have shown that there are two paths toward the unique decomposition theorem for ideals. The first is to show that (1) every ideal has a finite number of divisors, and (2) two distinct prime ideals are relatively prime. The second is to show that (3) maximal ideals and prime ideals are the same, and (4) every ideal is properly contained in only a finite number of ideals. Our plan of action will be to restrict ourselves to rings of algebraic integers and, after proving some auxiliary results, follow the second path.

You might be interested to know that property 4 plays a crucial role in generalizing the theorem of unique decomposition to rings other than those of algebraic integers. In fact, a commutative ring R is called a *Noetherian ring* (see Section 10) if it has the property that any sequence of ideals $\mathscr{A}_i$ in R such that

$$\mathscr{A}_1 \subset \mathscr{A}_2 \subset \mathscr{A}_3 \subset \cdots$$

has a finite number of terms. Over a Noetherian ring a kind of unique decomposition theorem is possible. (See N. H. McCoy, *Rings and Ideals*, and the second volume of van der Waerden's *Modern Algebra*, listed in the Bibliography.) A concept closely allied to this is that of a *Dedekind ring*, which is defined to be an integral domain in which every ideal is a product (not necessarily unique) of prime ideals.

We close this section with an example to show that the answer to the question (4.3) is not always "yes." That is, we give an example of two ideals $\mathscr{A}$ and $\mathscr{B}$ over a ring R such that $\mathscr{A} \supset \mathscr{B}$ but $\mathscr{A}$ is not a divisor of $\mathscr{B}$.

EXAMPLE. $R = \mathbf{Z}[x]$, $\mathscr{A} = \{2f(x) + xg(x)\}$, and $\mathscr{B} = \{4f(x) + xg(x)\}$, where f and g range over all polynomials in $\mathbf{Z}[x]$. Using the notation in

Section 4 after the definition of an ideal, we can write $\mathscr{A} = (2, x)$ and $\mathscr{B} = (4, x)$. Now $\mathscr{A}$ consists of all polynomials in $\mathbf{Z}[x]$ of the form $2k + xh(x)$, where k is an integer, for first we can choose $f(x) = k$ and $g(x) = h(x)$ to see that $2k + xh(x)$ is in $\mathscr{A}$; second, writing $f(x) = k + xf_0(x)$ we see that every element of $\mathscr{A}$ can be written

$$2k + x[2f_0(x) + g(x)],$$

which is of the form desired with $h(x) = 2f_0(x) + g(x)$. Similarly, $\mathscr{B}$ is the set of all polynomials in $\mathbf{Z}[x]$ whose constant term is a multiple of 4. Thus $\mathscr{A} \supset \mathscr{B}$ follows from noting that $\mathscr{A}$ contains 2, which is not in $\mathscr{B}$.

Now we want to show that $\mathscr{A}$ does not divide $\mathscr{B}$. This is done by showing that the assumption $\mathscr{A}\mathscr{C} = \mathscr{B}$, for some ideal $\mathscr{C}$, leads to a contradiction. Now $\mathscr{A}\mathscr{C} = \mathscr{B}$ implies that $\mathscr{C} \supseteq \mathscr{B}$. If $\mathscr{C} = \mathscr{B}$, then $\mathscr{B} = \mathscr{A}\mathscr{B} = (8, 2x, 4x, x^2) = (8, 2x, x^2)$, which cannot be $\mathscr{B}$ because it does not contain x. Therefore, $\mathscr{C} \supset \mathscr{B}$. This means that $\mathscr{C}$ must contain a polynomial whose constant term c is not divisible by 4. There are two possibilities.

1. Every constant term in polynomials of $\mathscr{C}$ is even. In that case some $c \equiv 2 \pmod 4$, 2 is in $\mathscr{C}$, and $\mathscr{C} = \mathscr{A}$. This cannot be, since $\mathscr{A}\mathscr{C} = \mathscr{A}^2 = (4, 2x, x^2)$, which does not contain x and hence is not $\mathscr{B}$.

2. If some constant term c is odd, then $c \equiv \pm 1 \pmod 4$, $\mathscr{C}$ contains 1, and hence is R. But in that case we would have $\mathscr{A}\mathscr{C} = \mathscr{A}R = \mathscr{A}$, which is not $\mathscr{B}$.

In all cases we have arrived at a contradiction and have proved that, for this example, $\mathscr{A}$ contains $\mathscr{B}$ but is not a divisor of $\mathscr{B}$.

In the next two sections we shall restrict the ring R to be a ring of algebraic integers in $\mathbf{Q}(\theta)$, where θ is an algebraic number of degree n. For this case we shall find, among other things, that if an ideal $\mathscr{A}$ contains an ideal $\mathscr{B}$, then $\mathscr{A}$ is a divisor of $\mathscr{B}$.

Exercises

1. Let $\mathscr{D}$ be the ideal $\{3x + (2 - \sqrt{-5})y\}$ and $\mathscr{C} = \{3x + (2 + \sqrt{-5})y\}$, where x and y are in $\mathbf{Z}$. Show that $\mathscr{D} \neq \mathscr{C}$. Find $\mathscr{D}\mathscr{C}$ and $\mathscr{D} + \mathscr{C}$.

2. Suppose that x and y in Exercise 1 are taken to be in $\mathbf{Z}[\sqrt{-5}]$. Would the conclusion be the same?

3. Is an ideal closed under subtraction? Why or why not?

4. Let $\mathscr{D}$ be the ideal $(7, 6 + \sqrt{-3})$ and $\mathscr{C}$ the ideal $(7, 6 - \sqrt{-3})$ over $\mathbf{Z}[\sqrt{-3}]$. Show that $\mathscr{D} \neq \mathscr{C}$ and that $\mathscr{D}$ and $\mathscr{C}$ are relatively prime. Find $\mathscr{D}\mathscr{C}$ and $\mathscr{D} + \mathscr{C}$.

5. Let R be the ring $\mathbf{Z}[\sqrt{-5}]$ and $\mathscr{C} = (3, 2 + \sqrt{-5})$ over $\mathbf{Z}[\sqrt{-5}]$. Find an ideal $\mathscr{E}$ such that $\mathscr{C}\mathscr{E} = (2 + \sqrt{-5})$.

6. Let $\mathbf{Q}[x, y]$ be the set of all polynomials in x and y with rational coefficients. Show that $\mathbf{Q}[x, y]$ is a ring.

7. Which of the following are ideals over $\mathbf{Q}[x, y]$ as defined in Exercise 6?
   (a) All $f(x, y)$ such that $f(0, 0) = 0$.
   (b) All $f(x, y)$ that are functions of y alone.
   (c) All $f(x, y)$ such that $f(1, 1) = 0$.
   (d) All $f(x, y)$ having no terms of the first degree, that is, having the form $a + cx^2 + c'xy + c''y^2 +$ terms of higher degree.

8. Let $\mathbf{Q}[x, y]$ be as defined in Exercise 6. Show that the principal ideal (x) is a prime ideal. Show that (x) is a proper subset of $\mathscr{T} = \{f(x, y)\}$, where $f(x, y)$ is as defined in Exercise 7a and hence (x) is not maximal in $\mathbf{Q}[x, y]$.

9. Show that, in the notation of Exercise 8, $\mathscr{T} \supseteq (x)$ but $\mathscr{T}$ is not a divisor of (x).

10. If $\mathscr{A}$ and $\mathscr{B}$ are two ideals over a ring, show that $\mathscr{A}\mathscr{B} \subseteq (\mathscr{A} \cap \mathscr{B})$, and give an example to show that $\mathscr{A}\mathscr{B} \subset (\mathscr{A} \cap \mathscr{B})$ is possible.

11. Prove that if $\mathscr{A}$ and $\mathscr{B}$ are two ideals over a ring R, then $\mathscr{A} + \mathscr{B}$ and $\mathscr{A}\mathscr{B}$ are also ideals over R.

12. Prove Theorem 5.1a.

13. Prove Theorem 5.1b.

14. Show that in the example at the close of the section, $\mathscr{B}$ is a prime ideal even though it is not maximal in R.

15. Prove that if R is a field, it contains exactly two ideals, R itself and the zero ideal.

16. Let U be a set and S all subsets of U. If $\mathscr{A}$ and $\mathscr{B}$ are in S, define $\mathscr{A} + \mathscr{B} = \mathscr{A} \cup \mathscr{B}$ and $\mathscr{A} \cdot \mathscr{B} = \mathscr{A} \cap \mathscr{B}$. Is S a ring? Is so, why; if not, what properties does it lack?

17. If $\mathscr{A}$ and $\mathscr{B}$ are two ideals over a ring of algebraic integers whose product is the zero ideal, show that at least one of $\mathscr{A}$ and $\mathscr{B}$ is the zero ideal.

18. Show that if $\mathscr{A}$ is an ideal, $\mathscr{A} \cdot \mathscr{A}$ is not necessarily $\mathscr{A}$.

19. Prove that the ring of Gaussian integers is a principal ideal ring.

20. Prove that the ring of polynomials with rational coefficients is a principal ideal ring.

21. Prove that the distributive property holds for ideals over a commutative ring R with identity element.

22. Prove that in the example at the close of Section 5,
    (a) $\mathscr{B}$ is maximal in $\mathscr{A}$.
    (b) $\mathscr{A}$ is maximal in R.
    (c) The ideals $\mathscr{A}$ and R are the only ideals over R that properly contain $\mathscr{B}$.

23. Which of the following are principal ideals over R?
    (a) $R = \mathbf{Z}[\sqrt{-3}], \mathscr{A} = (2, 1 + \sqrt{-3})$.
    (b) $R = \mathbf{Z}[\sqrt{-6}], \mathscr{A} = (2, \sqrt{-6})$.
    (c) $R = \mathbf{Z}[\omega], \omega = \frac{1}{2}(-1 + \sqrt{-3}), \mathscr{A} = (2, 1 + \sqrt{-3})$.
    (d) $R = \mathbf{Z}[\sqrt{-6}], \mathscr{A} = (10, 12 + \sqrt{-6})$.

*24. Prove that if R is a Euclidean domain, then every ideal $\mathscr{A}$ over R is a principal ideal.

*25. Does the conclusion of Exercise 17 necessarily hold if R is not restricted to be a ring of algebraic integers? Why?

*26. Let $\mathscr{A}$ and $\mathscr{B}$ be two ideals over a ring R. Prove that $\mathscr{A} \cap \mathscr{B}$ is an ideal over R. In what respects does $\mathscr{A} \cap \mathscr{B}$ behave like the least common multiple of $\mathscr{A}$ and $\mathscr{B}$?

6.   Ideals over the Ring of Algebraic Integers

In this section, when we refer to an ideal, we mean a nonzero ideal. We have just seen in the summary after Theorem 5.2 that two key results are needed to prove a theorem of unique decomposition for ideals. We have already noted that unless we impose some restriction on the ring, we encounter trouble; therefore, in this section and the next we restrict ourselves to rings of algebraic integers. The big advantage of this restriction is that we can develop relationships with the (rational) integers, where unique decomposition does hold, to prove the theorem for the set of ideals over algebraic integers as well. The proofs of Theorems 6.2 and 6.4 are rather difficult. You may prefer merely to follow through the proofs for the case $n = 2$, or, indeed, assume the theorems outright.

Recall Section 14 of Chapter III, and let $\mathbf{Q}(\theta)$ be some algebraic extension of the field of rational numbers, where θ is a zero of an irreducible polynomial of degree n over $\mathbf{Q}$. Let S be the ring of algebraic integers in $\mathbf{Q}(\theta)$. Every element of S can be expressed as a linear combination of $1, \theta, \theta^2, \ldots, \theta^{n-1}$ with rational coefficients, that is, in the form

(6.1) $$a_0 + a_1\theta + \cdots + a_{n-1}\theta^{n-1},$$

where the a_i are rational numbers.

As we noted in Section 4 of Chapter IV, $\mathbf{Q}(\theta)$ is a vector space of dimension n over $\mathbf{Q}$ and has the set $1, \theta, \theta^2, \ldots, \theta^{n-1}$ as a basis. In fact, by the corollary of Theorem 4.4 of Chapter IV, every set of n elements of $\mathbf{Q}(\theta)$ that are linearly independent over $\mathbf{Q}$ is a basis of $\mathbf{Q}(\theta)$. Let $\mathscr{A}$ be an ideal over S, the set of algebraic integers of $\mathbf{Q}(\theta)$. Since $\mathscr{A} \subseteq \mathbf{Q}(\theta)$, every element of $\mathscr{A}$ is expressible as a linear combination of a set of basis elements of $\mathbf{Q}(\theta)$, even though $\mathscr{A}$ may not be a vector space. One might guess that since $\mathscr{A}$ may be a proper subset of $\mathbf{Q}(\theta)$, it might be spanned by fewer than n elements of $\mathbf{Q}(\theta)$. This is certainly true, since $\mathscr{A}$ may be a principal ideal and hence spanned by a single element a in $\mathbf{Q}(\theta)$; that is, $\mathscr{A}$ would then be the set of all elements $\{as\}$, where s is in S. But if every element of $\mathscr{A}$ is to be a linear combination of a set W of elements of $\mathscr{A}$ with *rational* coefficients, it turns out, somewhat surprisingly, that W must have n elements.

For example, let $\theta = \sqrt[3]{2}$ and $\mathscr{A}$ the principal ideal $(\sqrt[3]{2})$. Now θ is an algebraic integer because it is a root of $x^3 - 2 = 0$, and hence $\mathscr{A}$ must contain

$$\theta, \ \theta \cdot \theta = \theta^2 \qquad \text{and} \qquad \theta^2 \cdot \theta = 2.$$

These three elements are linearly independent over $\mathbf{Q}$ and

$$W = \{2a_1 + a_2\theta + a_3\theta^2\} \subseteq \mathscr{A} \subseteq \{2x_1 + x_2\theta + x_3\theta^2\} = Q(\theta),$$

where the a_i are integers and the x_i are rational numbers. Now we have no assurance that $W = \mathscr{A}$; in fact, we shall shortly give an example that sheds light on this. But first we show that every ideal over S, the set of algebraic integers of $\mathbf{Q}(\theta)$, contains n elements that are linearly independent over $\mathbf{Q}$.

Theorem 6.1. Let $\mathscr{A}$ be an ideal over S, the set of algebraic integers of $\mathbf{Q}(\theta)$, where θ is of degree n over $\mathbf{Q}$. Then $\mathscr{A}$ contains a set of n elements that are linearly independent over $\mathbf{Q}(\theta)$. (Remember that we are excluding zero ideals.)

Proof: Let $\alpha \neq 0$ be an element of $\mathscr{A}$. We know from Theorem ?.? that there is an integer b such that $b\theta = \varphi$ is an algebraic integer. Hence $\mathscr{A}$ contains the following set of algebraic integers:

$$(6.2) \qquad\qquad a, a\varphi, a\varphi^2, \ldots, a\varphi^{n-1}.$$

The elements of this set are linearly independent over $\mathbf{Q}$, since the powers of θ, and hence the powers of φ, are linearly independent. This completes the proof.

Since the elements of (6.2) are linearly independent, they are a basis of $\mathbf{Q}(\theta)$ over $\mathbf{Q}$, and since $\mathscr{A} \subseteq \mathbf{Q}(\theta)$, we have the following corollary.

Corollary. Each ideal $\mathscr{A}$ over S contains a set W of n elements such that every element of $\mathscr{A}$ is a linear combination of the elements of W with *rational* coefficients.

Notice that unless $\mathscr{A} = \mathbf{Q}(\theta)$, not every linear combination over $\mathbf{Q}$ of the elements of (6.2) is in $\mathscr{A}$. Furthermore, we do not know that every element of $\mathscr{A}$ is a linear combination of the elements of (6.2) with integral coefficients. Our next step is to show that *n linearly independent elements of $\mathscr{A}$ can be so chosen* that $\mathscr{A}$ consists of all linear combinations of these elements with *integral* coefficients.

Definition. *If $\mathscr{A}$ is an ideal over S, the set of algebraic integers of $\mathbf{Q}(\theta)$, and if the linearly independent set $\alpha_1, \alpha_2, \ldots, \alpha_n$ has the property that*

$$\mathscr{A} = \{x_1\alpha_1 + x_2\alpha_2 + \cdots + x_n\alpha_n\},$$

where the x's assume all integer values, then the set of α's is called an integral basis *of $\mathscr{A}$.*

For instance, in the example immediately above, 2, θ, θ^2 is an integral basis of W. The next theorem asserts that every ideal over S has an integral basis. As a preview of the method of proof we consider the following problem.

EXAMPLE. Let S be the set of algebraic integers over $\mathbf{Q}(\sqrt{-11})$. Find an integral basis for S.
 We have already solved this problem for $\mathbf{Q}(\sqrt{r})$ in Theorem 2.4. The method used there is more effective than that which we now use. But it is still worthwhile to look at another method of proof as a preview of one that applies to any ring of algebraic integers. We know that S contains all numbers $x + y\sqrt{-11}$, for x and y integers. But it also contains $\alpha = \frac{1}{2}(1 + \sqrt{-11})$, since α is a zero of

$$x^2 - x + 3.$$

We shall show that 1 and α constitute an integral basis of S. So we suppose this is not the case and reach a contradiction.
 Suppose that S contains an element $a + b\alpha$, where a and b are rational numbers but not both integers. Write $\beta = a + b\alpha$, suppose that b is not an integer and write $b = b_0 + c$ for $0 < c < 1$ and b_0 an integer. Then if

$$\gamma = \beta - b_0\alpha = a + b\alpha - b_0\alpha = a + c\alpha,$$

γ is in S, since β and $-b_0\alpha$ are. Let α' be the conjugate of α; that is, $\alpha' = \frac{1}{2}(1 - \sqrt{-11})$. Let $\gamma' = a + c\alpha'$. Then $\gamma - \gamma' = c(\alpha - \alpha') = c\sqrt{-11}$ and

(6.3) $(\gamma - \gamma')^2 = c^2(\alpha - \alpha')^2 = -11c^2.$

Now since γ and γ' are algebraic integers, so is $(\gamma - \gamma')^2$. But (6.3) shows that $(\gamma - \gamma')^2$ is rational. Hence for (6.3) to hold, $11c^2$ must be an integer. To show that this is impossible, let $c = s/t$, where s and t are relatively prime integers. Then $-11c^2 = -11s^2/t^2$. But s and t relatively prime implies that s^2 and t^2 are also and hence that t^2 is a factor of 11; thus $t^2 = 1$ and c is an integer, contrary to supposition. Thus we have shown that if $a + b\alpha$ is an algebraic integer, b is an integer. To prove that a is an integer, one can start with $a = a_0 + c_1$, where a_0 is an integer and $0 < c_1 < 1$, then use the same argument.

The proof in general consists of noting that (6.3) implies

$$|(\gamma - \gamma')^2| < |(\alpha - \alpha')^2|,$$

repeating the process to give a sequence of positive integers, and noting that therefore the process must stop. Now we state and prove the general theorem.

Theorem 6.2.   Let $\mathscr{A}$ be an ideal over S, the algebraic integers of $\mathbf{Q}(\theta)$. Then $\mathscr{A}$ has an integral basis.

Proof: To give a feeling for how the proof goes, we first prove the theorem for the case $n = 2$. On this model we later construct the general proof. Using Theorem 6.1, we may let α_1 and α_2 be two linearly independent (over $\mathbf{Q}$) elements of $\mathscr{A}$. Every element of $\mathscr{A}$ is a linear combination of these α's with rational coefficients. Since α_i are both in $\mathbf{Q}(\theta)$, we have

(6.4)
$$\alpha_1 = a_{11} + a_{12}\theta,$$
$$\alpha_2 = a_{21} + a_{22}\theta,$$

where the a_{ij} are rational numbers. Since the α's are linearly independent, $a_{11}a_{22} - a_{12}a_{21} \neq 0$. We want to modify this basis into an integral basis. We suppose that $\beta = b_1\alpha_1 + b_2\alpha_2$ is an element of $\mathscr{A}$ for which b_1 is not an integer. (The proof would be the same if b_2 were not an integer.) Then write $h_1 = b_0 + c$, where $0 < c < 1$ and b_0 is an integer; that is, b_0 is the greatest integer less than b_1. Then $\beta_1 = \beta - b_0\alpha_1$ is in $\mathscr{A}$, and every element of $\mathscr{A}$ is a linear combination of the following, with rational coefficients:

(6.5)
$$\beta_1 = c\alpha_1 + b_2\alpha_2,$$
$$\beta_2 = \qquad\quad \alpha_2,$$

for β_1 and β_2 are linearly independent and hence a basis of $\mathbf{Q}(\theta)$ over $\mathbf{Q}$. The reaction to this might well be: "Fine, but what progress are we making?"

To show that we are making progress, define $\alpha_i' = a_{i1} + a_{i2}\theta'$, where θ' is the conjugate of θ and consider the matrix product

(6.6)
$$M = \begin{bmatrix} \alpha_1 & \alpha_1' \\ \alpha_2 & \alpha_2' \end{bmatrix} = \begin{bmatrix} a_{11} & a_{12} \\ a_{21} & a_{22} \end{bmatrix} \cdot \begin{bmatrix} 1 & 1 \\ \theta & \theta' \end{bmatrix}.$$

Since $\theta' - \theta \neq 0 \neq a_{11}a_{22} - a_{12}a_{21}$, the multiplicative property of determinants shows that $\det(M) \neq 0$. Since $\det^2(M)$ is a symmetric polynomial in θ and θ', Theorem 3.1 (the fundamental theorem on symmetric functions) shows that $\det^2(M)$ is a rational number. If we can show that the elements of M are algebraic integers, it will follow that $\det^2(M)$, being both an algebraic integer and rational, must be an integer. To do this, we need to show that α_i' is the conjugate of α_i for $i = 1$ and 2. Hence we suppress the subscripts for the time being and let $\alpha = a + b\theta$ and $\alpha' = a + b\theta'$. Then $\alpha + \alpha' = 2a + b(\theta + \theta')$ and $\alpha\alpha' = a^2 + b^2\theta\theta' + ab(\theta + \theta')$. Then α and α' are zeros of the polynomial $q(x) = x^2 - (\alpha + \alpha')x + \alpha\alpha'$ with rational coefficients. Since α is an algebraic integer, the coefficients must be integers and α' is its conjugate. Thus we have shown that $\det^2(M)$ is an integer.

Now we perform a similar computation for the β's, by defining $\beta_1' = c\alpha_1' + b_2\alpha_2'$, $\beta_2' = \alpha_2'$, and see that

$$\begin{bmatrix} \beta_1 & \beta_1' \\ \beta_2 & \beta_2' \end{bmatrix} = \begin{bmatrix} c & b_2 \\ 0 & 1 \end{bmatrix} \cdot \begin{bmatrix} \alpha_1 & \alpha_1' \\ \alpha_2 & \alpha_2' \end{bmatrix}.$$

Taking the squares of determinants of the matrices, we have

$$(\beta_1\beta_2' - \beta_1'\beta_2)^2 = (\alpha_1\alpha_2' - \alpha_1'\alpha_2)^2 c^2.$$

Now $(\beta_1\beta_2' - \beta_1'\beta_2)^2$ is an integer by the same reasoning that we used for $\det^2(M)$ in (6.6). But $c^2 < 1$, showing that

$$|(\beta_1\beta_2' - \beta_1'\beta_2)^2| < |(\alpha_1\alpha_2' - \alpha_1'\alpha_2)^2|.$$

What we have done is to decrease the absolute value of the square of the determinant of M.

If every element of $\mathscr{A}$ is a linear combination of β_1 and β_2 with integral coefficients, the β's form an integral basis. If this is not the case, we can repeat the process with the α's replaced by the β's. As this process is continued, the absolute values of the squares of the determinants of the M's form a decreasing sequence of positive integers. (The square of the determinant of M can be a negative integer; see Exercise 15 after Section 7.) Since a decreasing set of positive integers has a least integer, the process must cease after a finite number of steps. This completes the proof for the case $n = 2$.

Before embarking on the proof for $n > 2$, it will be simpler to state and prove as a lemma a result that we derived for $n = 2$.

Lemma 1. Let $\sigma \neq 0$ be an algebraic number in $\mathbf{Q}(\theta)$ for θ an algebraic number of degree $n > 1$. Let $\theta_1, \theta_2, \theta_3, \ldots, \theta_n$ be the conjugates of θ, where $\theta_1 = \theta$; $g(x)$ be the irreducible polynomial of degree r with rational coefficients such that $g(\sigma) = 0$; and σ_i be defined by

$$\sigma = \sigma_1, \quad \sigma_i = h(\theta_i), \quad h(x) = c_0 + c_1 x + \cdots + c_{n-1}x^{n-1},$$

where the c_i are in $\mathbf{Q}$. Then each σ_i is a conjugate of σ, r is a factor of n, and

$$F(x) = \prod_{i=1}^{n} (x - \sigma_i) = [g(x)]^{n/r}.$$

To begin, let us see what this lemma means. One starts with an algebraic number θ and some algebraic number σ in $\mathbf{Q}(\theta)$. Now there are two sorts of conjugates of σ that can be identified. First, we can look at the irreducible polynomial $g(x)$ such that $g(\sigma) = 0$ and then see that the conjugates of σ are the other zeros of $g(x)$. Second, one can look at the expression of σ as a linear combination $h(\theta)$ of powers of θ and hope that the conjugates of σ can be obtained by replacing θ in $h(\theta)$ by *its* conjugates. The lemma asserts that both calculations yield the same conjugates, but the second may yield some duplicates.

For example, suppose that θ is a zero of $x^6 + 4x^3 + 2$. (This is irreducible by Eisenstein's criterion, Theorem 13.6 of Chapter III.) Let $\sigma = \theta^3$. Then σ is a zero of the irreducible polynomial $g(x) = x^2 + 4x + 2$. The other zero of $g(x)$ is $-4 - \sigma = -4 - \theta^3$, since the sum of the zeros of $g(x)$ is -4. (Also, $-4 - \sigma = 2/\sigma$, since 2 is the product of σ and its conjugate.) Now if θ_i are the conjugates of θ, we define $\sigma_i = \theta_i^3$. The lemma asserts that each σ_i is either σ or $-4 - \sigma$, and that

$$F(x) = \prod_{i=1}^{6} (x - \sigma_i) = [g(x)]^3.$$

This means that three of θ_i^3 are σ and three are $-4 - \sigma$. This can be verified for this example by noting that the conjugates of θ can be written

$$\theta, \quad \theta_2 = \omega\theta, \quad \theta_3 = \omega^2\theta, \quad \theta_4 = \frac{\sqrt[3]{2}}{\theta}, \quad \theta_5 = \omega\theta_4, \quad \theta_6 = \omega^2\theta_4,$$

where ω is an imaginary cube root of 1; that is, $\omega = \frac{1}{2}(-1 + \sqrt{-3})$. Then

$$\sigma = \theta^3 = \theta_2^3 = \theta_3^3 \quad \text{and} \quad -4 - \sigma = \frac{2}{\sigma} = \theta_4^3 = \theta_5^3 = \theta_6^3.$$

Proof of Lemma 1: Set $g[h(x)] = G(x)$ and see that $g[h(\theta)] = g(\sigma) = 0$ implies that $G(\theta) = 0$. Hence, by Theorem 14.7 of Chapter III, the irreducible polynomial of θ is a factor of $G(x)$ and thus $G(\theta_i) = 0$ for all conjugates of θ. Therefore, $0 = g[h(\theta_i)] = g(\sigma_i) = 0$ for $1 \le i \le n$. This shows that each σ_i is a conjugate of σ.

Now, $F(\sigma) = 0$ and $g(\sigma) = 0$ imply that $g(x) \mid F(x)$ and $F(x)/g(x) = F_2(x)$ is a polynomial with rational coefficients. But every zero of $F_2(x)$ is a conjugate of σ, and hence $g(x) \mid F_2(x)$. Continuing this process yields the rest of the lemma.

Now we begin the proof of the theorem for $n > 2$. Let $\mathscr{A}$ be any ideal over the algebraic integers of $\mathbf{Q}(\theta)$ and $\alpha_1, \alpha_2, \ldots, \alpha_n$ be a linearly independent set of elements of $\mathscr{A}$ such that every element of $\mathscr{A}$ is a linear combination of these α's with rational coefficients. Then we have

(6.7)
$$\begin{aligned}
\alpha_1 &= a_{11} + a_{12}\theta + a_{13}\theta^2 + \cdots + a_{1n}\theta^{n-1}, \\
\alpha_2 &= a_{21} + a_{22}\theta + a_{23}\theta^2 + \cdots + a_{2n}\theta^{n-1}, \\
&\cdots\cdots\cdots\cdots\cdots\cdots\cdots\cdots\cdots\cdots\cdots\cdots\cdots\cdots \\
\alpha_n &= a_{n1} + a_{n2}\theta + a_{n3}\theta^2 + \cdots + a_{nn}\theta^{n-1}.
\end{aligned}$$

We want to modify the set of α's into an integral basis. Suppose that (6.7) is not an integral basis and there is some β in $\mathscr{A}$ such that

$$\beta = b_1\alpha_1 + b_2\alpha_2 + \cdots + b_n\alpha_n,$$

where not all the b's are integers. By permuting the subscripts of the α's if necessary, we may assume that b_1 is not an integer and write $b_1 = b_0 + c$, where $0 < c < 1$ and b_0 is an integer. Then define

(6.8)
$$\begin{aligned}
\beta_1 &= c\alpha_1 + b_2\alpha_2 + \cdots + b_n\alpha_n \\
\beta_i &= \alpha_i \quad \text{for } 2 \le i \le n.
\end{aligned}$$

The β_i span $\mathscr{A}$, since

$$\begin{aligned}
\sum_{i=1}^{n} d_i\alpha_i &= \frac{d_1}{c}(\beta_1 - b_2\alpha_2 - \cdots - b_n\alpha_n) + \sum_{i=2}^{n} d_i\alpha_i \\
&= \frac{d_1}{c}(\beta_1 - b_2\beta_2 - \cdots - b_n\beta_n) + \sum_{i=2}^{n} d_i\beta_i.
\end{aligned}$$

Denote by θ_i the conjugates of θ as in the lemma, define $\theta = \theta_1$, and

(6.9) $$\alpha_{ij} = a_{i1} + a_{i2}\theta_j + a_{i3}\theta_j^2 + \cdots + a_{in}\theta_j^{n-1} \quad \text{for } 1 \le j \le n.$$

Note that $\alpha_{i1} = \alpha_i$ and define

$$\Delta(\alpha) = \begin{bmatrix} \alpha_{11} & \alpha_{12} & \cdots & \alpha_{1n} \\ \alpha_{21} & \alpha_{22} & \cdots & \alpha_{2n} \\ \cdots & \cdots & \cdots & \cdots \\ \alpha_{n1} & \alpha_{n2} & \cdots & \alpha_{nn} \end{bmatrix}, \quad D(\theta) = \begin{bmatrix} 1 & 1 & \cdots & 1 \\ \theta_1 & \theta_2 & \cdots & \theta_n \\ \cdots & \cdots & \cdots & \cdots \\ \theta_1^{n-1} & \theta_2^{n-1} & \cdots & \theta_n^{n-1} \end{bmatrix}.$$

Then multiplication of matrices shows that

(6.10) $$\Delta(\alpha) = A D(\theta),$$

where $A = (a_{ij})$.

Now, by Lemma 1, the elements of $\Delta(\alpha)$ are conjugates of α_i for $(1 \le i \le n)$ and hence are algebraic integers. Also, $\det^2[\Delta(\alpha)]$ is a symmetric polynomial in θ and its conjugates. Hence $\det^2[\Delta(\alpha)]$ is a rational number and thus is an integer. Since the α's are linearly independent, $\det(A) \ne 0$, and we showed in Exercise 18 of Section 9 in Chapter IV that $\det[D(\theta)]$ is ± 1 times the product of differences $\theta_i - \theta_j$, $j < i$. Since θ and its conjugates are zeros of the same irreducible polynomial over $\mathbf{Q}$, the corollary of Theorem 16.2 of Chapter III shows that the conjugates of θ are all distinct. Hence the determinant of $D(\theta)$ is not zero. Thus the determinant of $\Delta^2(\alpha)$ is an integer different from zero.

Now if we define $\beta_{1j} = c\alpha_{1j} + b_2\alpha_{2j} + \cdots + b_n\alpha_{nj}$, $\beta_{ij} = \alpha_{ij}$ for $2 \le i \le n$, $1 \le j \le n$, $\Delta(\beta) = (\beta_{ij})$ and

$$
C = \begin{bmatrix}
c & b_2 & b_3 & \cdots & b_n \\
0 & 1 & 0 & \cdots & 0 \\
0 & 0 & 1 & \cdots & 0 \\
\cdots & \cdots & \cdots & \cdots & \cdots \\
0 & 0 & 0 & \cdots & 1
\end{bmatrix},
$$

it follows that $\Delta(\beta) = C\,\Delta(\alpha)$ and

$$
\det{}^2[\Delta(\beta)] = c^2 \det{}^2[\Delta(\alpha)],
$$

which implies that the square of the determinant of $\Delta(\beta)$ is less in absolute value than the square of the determinant of $\Delta(\alpha)$. Since both squares of determinants are integers, it follows that after a finite number of steps the above process yields an integral basis. Note that the squares of the determinants need not be *positive* integers, but this does not matter. This completes the proof of the theorem.

Corollary 1.   Given a nonzero ideal $\mathscr{A}$ over a ring of algebraic integers, the linearly independent set $\alpha_1, \alpha_2, \ldots, \alpha_n$ of elements of $\mathscr{A}$ form an integral basis if and only if the square of the determinant of $\Delta(\alpha)$ is less than or equal in absolute value to that of the square of the determinant of $\Delta(\beta)$ for every other set of n linearly independent elements of $\mathscr{A}$.

Corollary 2.   The set of algebraic integers of $\mathbf{Q}(\theta)$ has an integral basis.

It should be noted that while the theorem and corollary give a process that leads to a basis, in any particular case it might be very difficult to determine when one had arrived at an integral basis.

For ideals over $\mathbf{Q}(\sqrt{r})$, however, where r is a square-free integer, it is not hard to find an integral basis for all ideals over S, the set of algebraic integers in $\mathbf{Q}(\sqrt{r})$. We do this for $r \equiv 2$ or $3 \pmod 4$, leaving the case $r \equiv 1 \pmod 4$ as Exercise 19 after Section 7. From Theorem 2.4, under the conditions we have imposed on r, the algebraic integers in $\mathbf{Q}(\sqrt{r})$ are just the elements of $\mathbf{Z}[\sqrt{r}]$, all linear combinations of 1 and $\sqrt{r}$ with integer coefficients. Let $\mathscr{A}$ be an ideal over $S = \mathbf{Z}[\sqrt{r}]$. It consists of certain numbers of the form $x + y\sqrt{r}$, where x and y are integers. We seek an integral basis for $\mathscr{A}$.

Now, $\mathscr{A}$ contains an integer, since if $a + b\sqrt{r}$ is in $\mathscr{A}$, so is $(a + b\sqrt{r}) \cdot (a - b\sqrt{r}) = a^2 - b^2r$. Therefore, let u be the least positive integer in $\mathscr{A}$. If

$\mathscr{A}$ contains only integers, then, from Theorem 4.1, $\mathscr{A}$ consists of the multiples of u. If $\mathscr{A}$ contains an element that is not an integer, let T be the set of integers y such that for some x, $x + y\sqrt{r}$ is in $\mathscr{A}$. Let m be the least positive element in T. Then it must divide all coefficients of $\sqrt{r}$ in elements of $\mathscr{A}$. Since if $a + b\sqrt{r}$ is in $\mathscr{A}$, so is $(a + b\sqrt{r})\sqrt{r} = br + a\sqrt{r}$, and it follows that m divides a. Also, since $u\sqrt{r}$ is in $\mathscr{A}$, m divides u. So, $\mathscr{A} = (m)\mathscr{B}$, where $\mathscr{B}$ contains the following two elements:

$$v = \frac{u}{m} \quad \text{and} \quad c + \sqrt{r}$$

for some integer c. Now every integer in $\mathscr{B}$ is divisible by v, and hence $(c + \sqrt{r})(c - \sqrt{r}) = c^2 - r$ is divisible by v. Thus we have shown all but the last sentence of the following theorem.

Theorem 6.3. If $r \equiv 2$ or $3 \pmod 4$ and is square-free and $\mathscr{A}$ is an ideal over $S = \mathbf{Z}[\sqrt{r}]$, then either $\mathscr{A}$ consists of all integral multiples of an integer or $\mathscr{A} = (m)\mathscr{B}$, where $\mathscr{B}$ is an ideal spanned by

$$(6.11) \qquad\qquad v \quad \text{and} \quad c + \sqrt{r};$$

v is the least positive integer in $\mathscr{B}$; and c is a positive integer satisfying the condition $v \mid (c^2 - r)$. Furthermore, (6.11) is an integral basis of $\mathscr{B}$.

Proof: All that remains to prove is that (6.11) is an integral basis. To do this, we need to show that for all integers x, y, z, and t, the following equation is solvable for integers s and w:

$$v(x + y\sqrt{r}) + (c + \sqrt{r})(z + t\sqrt{r}) = sv + w(c + \sqrt{r}).$$

This is equivalent to the solvability of the following two equations:

$$vx + cz + tr = sv + wc,$$
$$vy + z + tc = w.$$

If we subtract c times the second equation from the first, we have

$$v(x - cy) + t(r - c^2) = sv.$$

Since $r - c^2$ is divisible by v, we can solve for an integer s and choose w so that $vy + z + tc = w$ to complete the proof.

In the course of the proof we have actually proved a little more, which we embody in the following corollary.

Corollary. If r is as in Theorem 6.3 and if $\mathscr{B}$ is an ideal over $\mathbf{Z}[\sqrt{r}]$ spanned by v and $c + \sqrt{r}$, where v and c are integers, $v \neq 0$, and $v \mid c^2 - r$, then v and $c + \sqrt{r}$ form an integral basis of $\mathscr{B}$.

It can be shown that if the α_i for $1 \leq i \leq n$ are an integral basis of $\mathscr{A}$, then the linearly independent β_i form an integral basis of $\mathscr{A}$ if and only if the determinants of $\Delta^2(\alpha)$ and $\Delta^2(\beta)$ are equal in absolute value. You are asked to show this in Exercise 17 after Section 7.

A much more important theorem for our purpose is Theorem 6.4. The proof of this theorem is difficult. It is possible to avoid this by another path to the theorem of unique decomposition. (See Harry Pollard, *The Theory of Algebraic Integers*, Chapter VIII, listed in the Bibliography.) But the alternative path seems at this stage harder to motivate. In fact, there is some justice in the contention that the classical approach, which is what we give, is the chief motivation for the modifications that lead to the other path and generalizations.

Theorem 6.4.   If $\mathscr{A}$ is an ideal over the algebraic integers of $\mathbf{Q}(\theta)$, then there is an ideal $\mathscr{B}$ such that for some integer c, $\mathscr{A}\mathscr{B}$ is the principal ideal (c).

Proof: To help motivate the proof, you should look at the beginning of Section 4. There we showed that if $\mathscr{C} = \{3x + (2 + \sqrt{-5})y\}$, where x and y are in $\mathbf{Z}[\sqrt{-5}]$ and the ideal $\mathscr{D}'$ is defined by $\mathscr{D}' = \{3x + (2 - \sqrt{-5})y\}$, then $\mathscr{C}\mathscr{D}' = (3)$. One might think of $\mathscr{D}'$ as a kind of conjugate ideal.

To see what is involved in the proof, we first consider the case $n = 2$. Let α_1, α_2 be an integral basis for $\mathscr{A}$. Since we want to have (c) the product of $\mathscr{A}$ and $\mathscr{B}$, it should seem natural to choose $\mathscr{B}$ as a kind of conjugate of $\mathscr{A}$. So let $\mathscr{B} = [\alpha_1', \alpha_2']$, where the α_i' are defined as in (6.6), and see that

$$(6.12) \qquad \mathscr{A}\mathscr{B} = \{x_1\alpha_1\alpha_1' + x_2\alpha_1\alpha_2' + x_3\alpha_1'\alpha_2 + x_4\alpha_2\alpha_2'\},$$

where the x_i range over the algebraic integers of $\mathbf{Q}(\theta)$. Now each of the following is symmetrical in θ and θ', is an algebraic integer, and hence is an integer,

$$(6.13) \qquad \alpha_1\alpha_1',\ \alpha_2\alpha_2',\ \alpha_1\alpha_2' + \alpha_1'\alpha_2.$$

If $\mathscr{A}\mathscr{B}$ is to be equal to (c), c must divide each number of (6.13). Hence we let c be the greatest common divisor of the numbers of (6.13). It remains to show that c is a divisor of every element of (6.12) in the sense that each number of (6.12) divided by c is an algebraic integer. Since, by the choice of c, it is a factor of $\alpha_1\alpha_1'$ and $\alpha_2\alpha_2'$, we must show that each of $\alpha_1\alpha_2'$ and $\alpha_1'\alpha_2$ is of the form $c\omega$, where ω is an algebraic integer. To this end, the first step is to notice that the numbers of (6.13) are the coefficients of the powers of x in

$$ch(x) = (\alpha_1 + \alpha_2 x)(\alpha_1' + \alpha_2' x).$$

Write $h(x) = d_0 + d_1 x + d_2 x^2$, where the d_i are integers. Then $h(\sigma) = 0$, where $\sigma = -\alpha_1/\alpha_2$. By Theorem 2.2, $d_2\sigma$ is an algebraic integer. Thus

$$d_2\sigma = \frac{\alpha_2\alpha_2' - \alpha_1}{c} = \frac{-\alpha_1\alpha_2'}{c}$$

is an algebraic integer. Since $(\alpha_1\alpha_2' + \alpha_1'\alpha_2)/c$ is an algebraic integer, so is $\alpha_1'\alpha_2/c$, and the proof for this case is complete. In the general case, what corresponds to the previous few lines is the result of a made-to-order lemma that is stated and proved after the rest of the proof is complete.

To use the above pattern for $n > 2$, let $\alpha_1, \alpha_2, \ldots, \alpha_n$ be an integral basis for an ideal $\mathcal{A}$ and define the conjugates α_{ij} of α_i as in (6.7) and (6.9). Then we write

$$f(x) = \alpha_1 + \alpha_2 x + \alpha_3 x^2 + \cdots + \alpha_n x^{n-1},$$

(6.14)
$$g(x) = \prod_{j=2}^{n} (\alpha_{1j} + \alpha_{2j}x + \cdots + \alpha_{nj}x^{n-1})$$

$$= \gamma_1 + \gamma_2 x + \gamma_3 x^2 + \cdots + \gamma_t x^{t-1}, \qquad t = (n-1)^2 + 1,$$

$$h_0(x) = f(x)g(x).$$

Now since the coefficients of $h_0(x)$ are symmetric polynomials in θ and its conjugates, and since the coefficients are algebraic integers by Lemma 1, the coefficients are integers. Let c be the greatest common divisor of the coefficients of $h_0(x)$ and $h_0(x) = ch(x)$. We define the ideal $\mathcal{B}$ to be the set of all linear combinations of the γ's with algebraic integers as coefficients. Now from the definition of c, it is a factor of each coefficient of the product $f(x)g(x)$. We want to show that each product $\alpha_i\gamma_j$ is c multiplied by an algebraic integer, and hence that $\mathcal{A}\mathcal{B} = (c)$. This results from the made-to-order lemma, which we now state and prove.

Lemma 2.   Let $f(x)$ and $g(x)$ be defined as in the first and third lines of (6.14), where the coefficients are algebraic integers, and suppose that each coefficient of $f(x)g(x)$ is, for some algebraic integer δ, a product of δ by an algebraic integer. Then every product $\alpha_i\gamma_j$ is δ multiplied by an algebraic integer. (In the theorem $c = \delta$.)

Proof: Define $h(x)$ by $\delta h(x) = f(x)g(x)$ and

$$h(x) = \delta_0 + \delta_1 x + \delta_2 x^2 + \cdots + \delta_u x^u,$$

where the δ's are algebraic integers.

The following auxiliary result turns out to be useful: If $p(x)$ is a polynomial with algebraic integers as coefficients and if $p(\lambda) = 0$, then $p(x)/(x - \lambda)$ has coefficients that are algebraic integers. (Notice that the factor theorem does not give us enough information, namely, that the coefficients are algebraic integers.) This can be proved using Taylor's series, as indicated in Exercise 12 after Section 17 of Chapter III. But we give a direct proof by induction on the degree of $p(x)$. If $p(x) = \alpha + \beta x$, then $p(x) = \alpha + \beta(x - \lambda) + \lambda\beta$ and $p(\lambda) = 0$ implies $\alpha + \lambda\beta = 0$ and hence $p(x)/(x - \lambda) = \beta$, an algebraic integer. Assume that $u > 1$ is the degree of $p(x)$ and that the result desired holds for all polynomials of degree less than u. Define $q(x) = p(x) - \alpha_u x^{u-1}(x - \lambda)$, where α_u is the leading coefficient of $p(x)$. Then $q(x)$ is of degree less than u and hence by the induction hypothesis may be written $p'(x)(x - \lambda)$, where the coefficients of $p'(x)$ are algebraic integers. Hence

$$p(x) = (x - \lambda)p'(x) + \alpha_u x^{u-1}(x - \lambda)$$

and

$$\frac{p(x)}{x - \lambda} = p'(x) + \alpha_u x^{u-1},$$

which establishes the auxiliary result.

To return to the proof of Lemma 2, let $f(x)$ and $g(x)$ be factored as follows:

$$f(x) = \alpha_n(x - \lambda_1)(x - \lambda_2) \cdots (x - \lambda_{n-1}),$$
$$g(x) = \gamma_t(x - \mu_1)(x - \mu_2) \cdots (x - \mu_{t-1}).$$

Then $\pm \alpha_i/\alpha_n$ for $1 \le i \le n - 1$ are the elementary symmetric functions of $f(x)$ and $\pm \gamma_j/\gamma_t$ for $1 \le j \le t - 1$ are the elementary symmetric functions of $g(x)$. If we take $p(x)$ in the auxiliary result to be $h(x)$ and divide successively by

$$(x - \lambda_1), (x - \lambda_3), \ldots, (x - \lambda_{n-1}), (x - \mu_2), (x - \mu_3), \ldots, (x - \mu_{t-1}),$$

we see that the polynomial

$$\left(\frac{\alpha_n \gamma_t}{\delta}\right)(x - \lambda_2)(x - \mu_1)$$

has coefficients that are algebraic integers. Thus $\alpha_n \gamma_t/\delta$ is an algebraic integer, call it ϵ, and $\epsilon \lambda_2 \mu_1$ is also an algebraic integer. Similarly, one can show that $\epsilon \lambda_i \mu_j$ is an algebraic integer for all i and j. This implies that

$$\epsilon(\lambda_1 + \lambda_2 + \cdots + \lambda_{n-1})(\mu_1 + \mu_2 + \cdots + \mu_{t-1}) = \epsilon\left(\frac{\alpha_{n-1}}{\alpha_n}\right)\left(\frac{\gamma_{t-1}}{\gamma_t}\right)$$

is an algebraic integer.

One can apply the same process to show first that $\epsilon \lambda_2 \mu_1 \mu_2$ is an algebraic integer, then that $\epsilon \lambda_i \mu_j \mu_k$ for $j \ne k$ are algebraic integers, and hence that

$$\frac{\epsilon \gamma_{t-2} \alpha_{n-1}}{\alpha_n \gamma_t}$$

is an algebraic integer. Thus it will follow from the same procedure that $\epsilon \alpha_i \gamma_j/\alpha_n \gamma_t$ is an algebraic integer for all i and j. But

$$\frac{\epsilon \alpha_i \gamma_j}{\alpha_n \gamma_t} = \frac{\alpha_n \gamma_t}{\delta} \frac{\alpha_i}{\alpha_n} \frac{\gamma_j}{\gamma_t} = \frac{\alpha_i \gamma_j}{\delta}.$$

This completes the proof of the lemma and establishes Theorem 6.4.

Note that Lemma 2 is very close to Gauss's lemma, Theorem 13.4 of Chapter III. The important difference is that there the coefficients are in a Euclidean ring, whereas here they are algebraic integers that may not form a Euclidean ring. The proofs are very different. Actually, as you are asked to do in Exercise 20 after Section 7, it is possible to derive Gauss's lemma from Lemma 2. As we shall see in the next section, Theorem 6.4 is, as far as our problem goes, the theorem from which all blessings flow.

7.   The Unique Decomposition Theorem

Recall that immediately after Theorem 5.2 we outlined what is necessary for unique decomposition for ideals over algebraic integers. There were two paths mentioned. The first is to show that (1) every ideal has a finite number of divisors and (2) two distinct prime ideals are relatively prime. The second is to show that (3) maximal ideals and prime ideals are the same and (4) every ideal is properly contained in only a finite number of ideals. We choose the second path, and incidentally show the results of the first.

The first theorem answers in the affirmative the question of (4.3) and has as a corollary result 3 of the previous paragraph. The proofs of Theorems 7.1 and 7.3 follow roughly the same pattern: first a proof for principal ideals and then an application of Theorem 6.4.

Theorem 7.1. If $\mathscr{A}$ and $\mathscr{D}$ are two ideals over the set S of algebraic integers of $\mathbf{Q}(\theta)$ and if $\mathscr{A} \supseteq \mathscr{D}$ and $\mathscr{A} \neq (0)$, then there is an ideal $\mathscr{C}$ over S such that $\mathscr{A}\mathscr{C} = \mathscr{D}$. That is, if $\mathscr{A}$ contains $\mathscr{D}$, then $\mathscr{A}$ is a divisor of $\mathscr{D}$.

Proof: First, we show the theorem for $\mathscr{A}$ a principal ideal, $\mathscr{A} = (g)$, where g is an algebraic integer. Now $(g) \supseteq \mathscr{D}$ means that every element of $\mathscr{D}$ is g multiplied by an algebraic integer. Let E be the set $\{e\}$ of elements of S such that ge is in $\mathscr{D}$. E is not an empty set, since $\mathscr{D}$ is not. First, to show that E is an ideal, note that ge_1 and ge_2 in $\mathscr{D}$ imply that $ge_1 + ge_2 = g(e_1 + e_2)$ is in $\mathscr{D}$ and hence that $e_1 + e_2$ is in E. Similarly, rge_1 is in $\mathscr{D}$ for every algebraic integer r, and hence re_1 is in E. Second, to show that $\mathscr{A}E = \mathscr{D}$, note that since every element of $\mathscr{A}$ is of the form sg with s in S, it follows that every element of $\mathscr{A}E$ is of the form sge for e an element of E; since $\mathscr{D}$ is an ideal, $sge = s(ge)$ is in $\mathscr{D}$ and hence $\mathscr{A}E \subseteq \mathscr{D}$. On the other hand, if d is an element of $\mathscr{D}$, it follows from $\mathscr{D} \subseteq \mathscr{A}$ that there is an element s of S such that $d = gs$, and by the definition of E, this s is in E. Hence $\mathscr{D} \subseteq \mathscr{A}E$. Thus for $\mathscr{A} = (g)$ we have shown that E as defined is an ideal and $\mathscr{D} = \mathscr{A}E$.

To show the theorem in general, use Theorem 6.4 to justify the existence of an ideal $\mathscr{B}$ such that $\mathscr{A}\mathscr{B} = (g)$, some principal ideal with g an integer. Then

$\mathscr{A} \supseteq \mathscr{D}$ implies that $\mathscr{AB} \supseteq \mathscr{DB}$. Hence $(g) \supseteq \mathscr{DB}$. Then, by the previous paragraph, we know the existence of an ideal $\mathscr{E}$ such that

$$\mathscr{DB} = (g)\mathscr{E} = \mathscr{ABE} = \mathscr{AEB}.$$

To complete the proof, we want to "cancel the $\mathscr{B}$." The next theorem shows that this can be done and justifies the conclusion $\mathscr{AE} = \mathscr{D}$, which is what we want.

Corollary 1. If $\mathscr{A}$ and $\mathscr{B}$ are two ideals over S, then $\mathscr{A} \supseteq \mathscr{B}$ if and only if $\mathscr{A} \mid \mathscr{B}$. [See (4.3).]

Corollary 2. An ideal $\mathscr{A}$ over S is maximal if and only if it is a prime ideal.

The second corollary follows since if $\mathscr{A}$ is maximal, the only ideals that contain it are $\mathscr{A}$ and S, and hence its only divisors are $\mathscr{A}$ and S. Conversely, if $\mathscr{A}$ is prime its only divisors are $\mathscr{A}$ and S, and hence the only ideals that contain it are $\mathscr{A}$ and S. This corollary is property 3 in the first paragraph of this section.

Theorem 7.2. If $\mathscr{P}, \mathscr{R}$, and $\mathscr{T}$ are three ideals over S, the ring of algebraic integers of $\mathbf{Q}(\theta)$; if $\mathscr{P}$ is not the zero ideal; and if $\mathscr{RP} = \mathscr{TP}$, then $\mathscr{R} = \mathscr{T}$.

Proof: By Theorem 6.4 there is an ideal $\mathscr{W}$ such that $\mathscr{PW} = (w)$, a principal ideal. Hence $\mathscr{RP} = \mathscr{TP}$ implies that $\mathscr{RPW} = \mathscr{TPW}$ and $\mathscr{R}(w) = \mathscr{T}(w)$. This means that for every r in $\mathscr{R}$, $rw = tuw$ for some t in $\mathscr{T}$ and u in S. Since the lowercase letters stand for complex numbers, $rw = tuw$ implies that $r = tu$. This shows that $\mathscr{R} \subseteq \mathscr{T}$. Similarly, we can show that $\mathscr{T} \subseteq \mathscr{R}$ and hence that $\mathscr{T} = \mathscr{R}$. This completes the proof of Theorem 7.2 and establishes the truth of Theorem 7.1.

The other property (property 4) that we need for unique decomposition is stated in the following theorem.

Theorem 7.3. Every ideal over the ring of algebraic integers in $\mathbf{Q}(\theta)$ is contained in only a finite number of ideals over the same ring.

Proof: First, we prove the theorem for principal ideals. Let $\beta_1, \beta_2, \ldots, \beta_n$ be an integral basis for the set of algebraic integers in $\mathbf{Q}(\theta)$. Suppose that $\mathscr{A}$ is an ideal that contains the integer c; that is, $\mathscr{A} \supseteq (c)$. Since $(c) = (-c)$, we may assume c to be positive. Let $\alpha_1, \alpha_2, \ldots, \alpha_n$ be an integral basis of $\mathscr{A}$. We want to show that the number of ideals $\mathscr{A}$ containing (c) is finite. Now every element of $\mathscr{A}$ is expressible as a linear combination of the α_i and c with integer coefficients. [Of course, the statement is true without including c, but we must make use of the assumption that $\mathscr{A}$ contains (c).] Let

$$\alpha_i = \sum_{j=1}^{n} b_{ij}\beta_j,$$

where the b_{ij} are integers. There are integers q_{ij} and r_{ij} such that

$$b_{ij} = cq_{ij} + r_{ij}, \qquad 0 \le r_{ij} < c.$$

Then

$$\alpha_i = \sum_{j=1}^{n} (cq_{ij} + r_{ij})\beta_j = c \sum_{j=1}^{n} q_{ij}\beta_j + \sum_{j=1}^{n} r_{ij}\beta_j.$$

So the α_i and c span integrally the same ideal as

$$(7.1) \qquad \left[\sum_{j=1}^{n} r_{1j}\beta_j, \ \sum_{j=1}^{n} r_{2j}\beta_j, \ \ldots, \ \sum_{j=1}^{n} r_{nj}\beta_j, \ c \right].$$

That is, the elements of $\mathscr{A}$ are the linear combinations of (7.1) with integer coefficients. Since each r_{ij} is an integer that satisfies the condition $0 \le r_{ij} < c$, there are c different possibilities for each r_{ij}. Hence the number of different sets (7.1) is c^m, where $m = n^2$. Thus there is only a finite number of ideals $\mathscr{A}$ containing (c) and the theorem is proved for principal ideals, in fact, for any ideal containing a principal ideal.

Now, for the general case, let $\mathscr{B}$ be any ideal over the algebraic integers of $\mathbf{Q}(\theta)$. Then by Theorem 6.4 there is an ideal $\mathscr{D}$ such that $\mathscr{B}\mathscr{D} = (c)$ for some integer c. Then if $\mathscr{A}$ is an ideal containing $\mathscr{B}$, we have

$$\mathscr{A} \supseteq \mathscr{B} \supseteq \mathscr{B}\mathscr{D} = (c).$$

Thus $\mathscr{A}$ is an ideal that contains (c). We know by the previous paragraph that there are only finitely many such ideals $\mathscr{A}$. Thus the proof is complete. [We have just shown (see Section 5) that the set of algebraic integers of $\mathbf{Q}(\theta)$ is a Noetherian ring.]

Corollary 1. Every ideal over S has a finite number of divisors.

Corollary 2. Two distinct prime ideals over S are relatively prime.

The first corollary is property 1 at the beginning of this section. It follows from Theorems 7.1 and 7.3. The second corollary is property 2 in the first paragraph of this section. It holds because if $\mathscr{P}$ and $\mathscr{Q}$ are distinct prime ideals, then $\mathscr{Q} \subseteq \mathscr{P}$ is false, since this would imply that $\mathscr{P} \mid \mathscr{Q}$ and hence that $\mathscr{P} = \mathscr{Q}$ or $\mathscr{P} = S$. Thus $\mathscr{Q}$ is not contained in $\mathscr{P}$ and, by Theorem 5.2, $\mathscr{P}$ and $\mathscr{Q}$ are relatively prime.

The theorem on unique decomposition can now be proved somewhat along the lines of the proof of Theorem 11.1 in Chapter III.

Theorem 7.4.   (Unique decomposition theorem for ideals over rings of algebraic integers.)   Let $\mathscr{A}$ be any ideal over the ring of integers of $\mathbf{Q}(\theta)$. Then

$$(7.2) \qquad\qquad \mathscr{A} = \mathscr{P}_1\mathscr{P}_2\mathscr{P}_3\cdots\mathscr{P}_k,$$

where the $\mathscr{P}_i$ are prime ideals. The factorization is unique except for the order of the $\mathscr{P}$'s. (Note that no units occur as in Theorem 11.1 of Chapter III.)

Proof: If $\mathscr{A}$ is a prime ideal, there is nothing to prove. If not, then $\mathscr{A} = \mathscr{B}\mathscr{C}$, where $\mathscr{B}$ and $\mathscr{C}$ are ideals, neither of which is $\mathscr{A}$. If $\mathscr{B}$ is not a prime ideal, it can be written $\mathscr{B} = \mathscr{B}_1\mathscr{B}_2$, and similarly for $\mathscr{C}$. We may continue this process until the divisors are all primes, since, by Corollary 1 of Theorem 7.3, the ideal $\mathscr{A}$ has only a finite number of divisors. Thus

$$\mathscr{A} = \mathscr{P}_1\mathscr{P}_2\cdots\mathscr{P}_k$$

for $\mathscr{P}_i$ prime ideals not necessarily distinct. To prove uniqueness, suppose that

$$(7.3) \qquad\qquad \mathscr{P}_1\mathscr{P}_2\cdots\mathscr{P}_k = \mathscr{Q}_1\mathscr{Q}_2\cdots\mathscr{Q}_u,$$

where the $\mathscr{Q}$'s are prime ideals. If $\mathscr{P}_1 \neq \mathscr{Q}_1$, then, by Corollary 2 of Theorem 7.3, $\mathscr{P}_1$ and $\mathscr{Q}_1$ are relatively prime. Then, by Theorem 5.1a, $\mathscr{P}_1 \mid \mathscr{Q}_2\mathscr{Q}_3\cdots\mathscr{Q}_u$ follows from (7.3). We continue this process to show that $\mathscr{P}_1$ is equal to one of the $\mathscr{Q}_i$. We permute the subscripts of the $\mathscr{Q}$'s and arrive at $\mathscr{P}_1 = \mathscr{Q}_1$. Then (7.3) and Theorem 7.2 shows that $\mathscr{P}_2\mathscr{P}_3\cdots\mathscr{P}_k = \mathscr{Q}_2\mathscr{Q}_3\cdots\mathscr{Q}_u$. Continuing this process establishes the theorem.

Let us see how all this applies to the example with which we started this chapter (see Section 4):

$$\mathscr{R} = [\sqrt{-5}], \quad \mathscr{A} = (3), \quad \mathscr{B} = (2 + \sqrt{-5}), \quad \mathscr{C} = (3, 2 + \sqrt{-5}) = \mathscr{A} + \mathscr{B}.$$

From the corollary of Theorem 6.3 or Exercise 2 after Section 5, $\mathscr{C} = [3, 2 + \sqrt{-5}]$; that is, 3 and $2 + \sqrt{-5}$ form an *integral* basis of $\mathscr{C}$. Now $\mathscr{C} = \mathscr{A} + \mathscr{B}$ implies that $\mathscr{C}$ is the greatest common divisor of $\mathscr{A}$ and $\mathscr{B}$, from Theorem 4.3 and Corollary 1 of Theorem 7.1. It would be convenient if $\mathscr{C}$ turned out to be a prime ideal, that is, a maximal ideal. Hence our first task is to show that if $\mathscr{D}$ is an ideal for which $\mathscr{R} \supseteq \mathscr{D} \supseteq \mathscr{C}$, then $\mathscr{D} = \mathscr{R}$ or $\mathscr{D} = \mathscr{C}$. From Theorem 6.3 we can write

$$\mathscr{D} = (m)\mathscr{E}, \qquad \text{where } \mathscr{E} = [v, c + \sqrt{-5}] \text{ for some integer } c,$$
$$\text{and} \quad v \mid (c^2 + 5).$$

Since $2 + \sqrt{-5}$, being in $\mathscr{C}$, is in $\mathscr{D}$ and m divides the coefficient of $\sqrt{-5}$ in all elements of $\mathscr{D}$, it follows that $m = 1$ and $\mathscr{D} = \mathscr{E}$. Since v divides all integers in $\mathscr{D}$ and $\mathscr{D}$ contains 3, it is a divisor of 3. Hence $v = 1$ or 3. In the former case $\mathscr{D} = \mathscr{R}$. In the latter case we need to consider only two possibilities for c, namely, $c = 1$ and $c = 2$, since if c is in $\mathscr{E}$ with $v = 3$, so is $c - 3x$ for all integers x. If $c = 1$, then $\mathscr{D}$ contains

$$2 + \sqrt{-5} - (1 + \sqrt{-5}) = 1,$$

and hence $\mathscr{C} = \mathscr{R}$. If $c = 2$, then $\mathscr{D} = \mathscr{C}$. This completes the proof that $\mathscr{C}$ is a prime ideal.

Now, $\mathscr{C}$ contains $\mathscr{B}$ and hence is a divisor of $\mathscr{B}$. Let us find an ideal $\mathscr{F}$ such that $\mathscr{C}\mathscr{F} = \mathscr{B}$. To find $\mathscr{F}$, we assume that it exists and by this means narrow the possibilities before testing it. $\mathscr{F}$ also divides $\mathscr{B}$ and hence contains $\mathscr{B}$. Therefore, $2 + \sqrt{-5}$ is an element of $\mathscr{F}$. By Theorem 6.3 and its corollary there is a positive integer e such that

$$\mathscr{F} = [e, 2 + \sqrt{-5}].$$

Then

$$\mathscr{C}\mathscr{F} = [3e, e(2 + \sqrt{-5}), 3(2 + \sqrt{-5}), (2 + \sqrt{-5})^2].$$

Now if $\mathscr{C}\mathscr{F}$ is to be a subset of $\mathscr{B}$, then every element of $\mathscr{C}\mathscr{F}$ must be $2 + \sqrt{-5}$ multiplied by an element of $\mathscr{R}$. Thus $3e$ must be such a multiple and we see that

$$(2 + \sqrt{-5})(x + y\sqrt{-5}) = 3e$$

for some integers x and y. This is equivalent to

(7.4) $\qquad\qquad 2x - 5y = 3e \qquad$ and $\qquad x + 2y = 0.$

From the second equation of (7.4), $x = -2y$. We replace x by $-2y$ in the first equation of (7.4) and get $-9y = 3e$, which implies that $3 \mid e$.

This exploration would indicate that $\mathscr{F} = [3, 2 + \sqrt{-5}] = \mathscr{C}$ might work. We wish to show that $\mathscr{C}\mathscr{F} = \mathscr{C}^2 = \mathscr{B}$. Now

$$\mathscr{C}^2 = [9, 3(2 + \sqrt{-5}), (2 + \sqrt{-5})^2].$$

This shows that every element of $\mathscr{C}^2$ is $2 + \sqrt{-5}$ multiplied by an algebraic integer. Hence $\mathscr{C}^2 \subseteq \mathscr{B}$. But $(2 + \sqrt{-5})^2 = -1 + 4\sqrt{-5}$ and

$$9 - (6 + 3\sqrt{-5}) + (-1 + 4\sqrt{-5}) = 2 + \sqrt{-5},$$

which shows that $\mathscr{B} \subseteq \mathscr{C}^2$ and thus that $\mathscr{B} = \mathscr{C}^2$. We have expressed the ideal $\mathscr{B}$ as a product of prime ideals.

Exercises

1. Express 3 as a product of prime ideals in $\mathbf{Z}[\sqrt{-5}]$.

2. Show that $\mathscr{A} = [5, 4 + \sqrt{11}]$ is a prime ideal over $\mathbf{Z}[\sqrt{11}]$.

3. Express the principal ideal (15) as a product of prime ideals over $\mathbf{Z}[\sqrt{11}]$.

4. Why do units not appear in Theorem 7.4?

5. In $\mathbf{Z}[\sqrt{-5}]$, express (3) as a product of prime ideals.

6. In $\mathbf{Z}[\sqrt{-5}]$, express (6) as a product of prime ideals.

7. In the proof of Theorem 7.2, show that $\mathscr{T} \subseteq \mathscr{R}$.

8. In $\mathbf{Z}[\sqrt{-5}]$, prove that the principal ideal $\mathscr{A} = (m)$ is not a prime ideal if there is an integer c such that $m \mid c^2 + 5$.

9. Prove that in $\mathbf{Z}[\sqrt{11}]$, the principal ideal $\mathscr{A} = (m)$ is not a prime ideal if there is an integer c such that $m \mid c^2 - 11$.

10. Let r be a square-free integer with $r \equiv 2$ or $3 \pmod 4$. Prove that in $\mathbf{Z}[\sqrt{r}]$, $\mathscr{A} = (m)$ is not a prime ideal if there is an integer c such that $m \mid c^2 - r$.

11. Show that if in Theorem 6.3, the ideal $\mathscr{B}$ has an integral basis $[v, c + r]$ with $v \mid c^2 - r$, then $\mathscr{B}$ also has the integral basis $[v, d + r]$, where $0 \le d < v$ and $c \equiv d \pmod v$.

12. What is wrong with the following "proof" of Theorem 7.2? The equation $\mathscr{R}\mathscr{P} = \mathscr{T}\mathscr{P}$ implies that $\mathscr{R}\mathscr{P} - \mathscr{T}\mathscr{P} = (0)$ and hence that $(\mathscr{R} - \mathscr{T})\mathscr{P} = (0)$. Since $\mathscr{P}$ is not the zero ideal, it follows that $\mathscr{R} - \mathscr{T} = (0)$ and hence that $\mathscr{R} = \mathscr{T}$.

13. Use Taylor's series from elementary calculus (see Exercise 12 after Section 17 in Chapter III) to prove the auxiliary result in the proof of Lemma 2.

14. If α is in an ideal over $\mathbf{Q}(\theta)$, are the conjugates of α necessarily in the same ideal? Prove that this is true or give an example to show that it is a not true.

15. Give an example for which $\Delta^2(\alpha)$ in Section 6 has a determinant that is a negative integer.

16. Prove Corollaries 1 and 2 of Theorem 7.1.

17. Let $\alpha_1, \alpha_2, \ldots, \alpha_n$ be an integral basis of an ideal $\mathscr{A}$ over the algebraic integers of $\mathbf{Q}(\theta)$ and $\beta_1, \beta_2, \ldots, \beta_n$ a linearly independent (over $\mathbf{Q}$) set of elements of $\mathscr{A}$. Prove that the β's form an integral basis of $\mathscr{A}$ if and only if the determinants of $\Delta^2(\alpha)$ and $\Delta^2(\beta)$ are equal in absolute value, where $\Delta(\alpha)$ and $\Delta(\beta)$ are defined in the proof of Theorem 6.1.

18. Prove that if $\mathscr{A}$ is not the zero ideal and $\mathscr{A}$ and $\mathscr{B} \subseteq \mathscr{A}$ are ideals over S, a ring of algebraic integers, then $\mathscr{A}\mathscr{X} = \mathscr{B}$ holds for a unique ideal $\mathscr{X}$.

★19. Find an integral basis for any ideal $\mathscr{A}$ over the ring of algebraic integers of $\mathbf{Q}(\sqrt{r})$, where r is a square-free integer and $r \equiv 1 \pmod 4$.

★20. Show how Gauss's lemma may be derived from Lemma 2.

8. Quotient Rings

Although we have solved the problem that we set out to deal with at the beginning of the chapter, there are a few ramifications worth exploring which in some respects are more basic than the problem itself. We have found for ideals over rings of algebraic integers a unique decomposition theorem. There is also for more general rings a result analogous to that of Theorem 2.1 of Chapter III, namely, that the set of integers modulo m is a field if and only if m is a prime. We can prove a corresponding result in a much more general setting by use of the ideas of this chapter and Sections 9 and 10 of Chapter II.

In this section we take R to be a commutative ring with identity element but not necessarily over the algebraic integers. Let $\mathscr{A}$ be an ideal over R. Since R is an Abelian additive group and $\mathscr{A}$ is an additive subgroup, a "quotient group" exists just as the cosets of $\mathscr{A}$ form an additive group in the sense of Sections 4 and 10 of Chapter II. That is, a coset of $\mathscr{A}$ is the set $\{r + x\}$, where x ranges over the elements of $\mathscr{A}$. Just as previously, we designate the coset by $r + \mathscr{A}$, and we define addition of cosets by

$$(r + \mathscr{A}) + (r' + \mathscr{A}) = (r + r') + \mathscr{A}.$$

But now we are concerned with a ring that has *two* operations so that we should consider *products* of cosets as well. We define a product as follows:

$$(r + \mathscr{A})(r' + \mathscr{A}) = rr' + \mathscr{A}.$$

Notice that if $r + a$ and $r' + a'$ are elements of $r + \mathscr{A}$ and $r' + \mathscr{A}$, respectively, then $(r + a)(r' + a') = rr' + ar' + ra' + aa'$ is in $rr' + \mathscr{A}$, since $\mathscr{A}$, being an ideal, implies that ar' and ra' are in $\mathscr{A}$. It is conceivable, to be sure, that there might be some element of $rr' + \mathscr{A}$ which is not the product of an element of $r + \mathscr{A}$ with one of $r' + \mathscr{A}$, but that does not affect our definition. We denote the set of cosets modulo $\mathscr{A}$ by $R/\mathscr{A}$, and we see that we have already proved part of the following theorem.

Theorem 8.1. If $\mathscr{A}$ is an ideal over a commutative ring R with identity element, the set of cosets $R/\mathscr{A}$ is a commutative ring with identity element.

Proof: We have already proved that the set $R/\mathscr{A}$ is an additive group and is closed under multiplication. That multiplication is associative follows directly from the definition. The (multiplicative) identity of $R/\mathscr{A}$ is the coset $1 + \mathscr{A}$. The distributive property is shown by the following:

$$
\begin{aligned}
(r + \mathscr{A})[(r' + \mathscr{A}) + (r'' + \mathscr{A})] &= (r + A)(r' + r'' + \mathscr{A}) \\
&= r(r' + r'') + \mathscr{A} \\
&= (rr' + \mathscr{A}) + (rr'' + \mathscr{A}).
\end{aligned}
$$

Also the cosets are commutative. This completes the proof.

In Section 9 we shall see that this result is easy to extend to noncommutative rings.

Now, suppose that $R = \mathbf{Z}$ and $\mathscr{A}$ is the principal ideal (m) for m an integer. Then, referring to the first paragraph of this section and the concepts we have just been considering, we see that $R/\mathscr{A}$ is a field if and only if m is a prime number. It is this result that we shall now generalize. Notice that since prime ideals and maximal ideals can be different over general rings, we must distinguish between them in the theorems that follow.

Theorem 8.2. If R is a commutative ring with identity element and if $\mathscr{A}$ is a nonzero ideal over R such that $R/\mathscr{A}$ is an integral domain, then $\mathscr{A}$ is a prime ideal.

Proof: We show that if $\mathscr{A}$ is not a prime ideal, then $R/\mathscr{A}$ has divisors of zero. Let $\mathscr{A} = \mathscr{B}\mathscr{C}$, where neither $\mathscr{B}$ nor $\mathscr{C}$ is $\mathscr{A}$. Then $\mathscr{B} \supset \mathscr{A}$ and $\mathscr{C} \supset \mathscr{A}$ imply that there are elements b and c in $\mathscr{B}$ and $\mathscr{C}$, respectively, but not in $\mathscr{A}$. Also, $\mathscr{B}\mathscr{C} = \mathscr{A}$ implies that bc is in $\mathscr{A}$ and hence that $(b + \mathscr{A})(c + \mathscr{A}) = bc + \mathscr{A} = \mathscr{A}$. Thus the product of two cosets, neither of which is $\mathscr{A}$, is contained in $\mathscr{A}$. This shows that $R/\mathscr{A}$ has divisors of zero.

Unfortunately, the converse of this theorem is not true. To show this, consider $R = \mathbf{Z}[x]$ and $\mathscr{B} = [4, x]$. (Compare the example at the end of Section 5.) Then $(2 + \mathscr{B})(2 + \mathscr{B}) \subseteq \mathscr{B}$, and hence $R/\mathscr{B}$ has divisors of zero. But we shall now show that $\mathscr{B}$ is a prime ideal. Here $\mathscr{B}$ is the set of all polynomials in $\mathbf{Z}[x]$ whose constant term is divisible by 4. Then if $\mathscr{B} = \mathscr{A}\mathscr{C}$, where neither $\mathscr{A}$ nor $\mathscr{C}$ is $\mathscr{B}$, we have, as previously, $\mathscr{A} \supset \mathscr{B}$ and $\mathscr{C} \supset \mathscr{B}$. If the constant term of any polynomial in $\mathscr{A}$ is an odd integer, $\mathscr{A}$ contains 1 and hence is R. This is impossible because $R\mathscr{C} = \mathscr{C} \neq \mathscr{B}$. So, $\mathscr{A}$ must be $[2, x]$ and this is the only possibility for $\mathscr{C}$ as well. But $\mathscr{A}\mathscr{C} = \mathscr{A}^2 = [2, x][2, x] = [4, 2x, x^2]$ implies that the coefficient of x in all elements of $\mathscr{A}^2$ is even. Hence x is in $\mathscr{B}$ but not in $\mathscr{A}^2 = \mathscr{A}\mathscr{C}$; this denies that $\mathscr{B} = \mathscr{A}\mathscr{C}$. Thus we have shown that $\mathscr{B}$ is a prime ideal and that $R/\mathscr{B}$ has divisors of zero. This is the example we wanted.

It is partly for this reason that a prime ideal over a general ring is usually defined differently from our definition. To avoid confusion, we define a property P that is more restrictive than our definition of prime ideal. It is the property that usually is taken for primality (the property of being a prime).

Definition. *An ideal $\mathscr{A}$ over a ring R has property P if $\mathscr{A} \neq (0)$† and if bc an element of $\mathscr{A}$ implies that b or c is in $\mathscr{A}$.*

---

† In the usual definition of a prime ideal the restriction $\mathscr{A} \neq (0)$ is not present. But if we are to show property P to be equivalent to our (the classical) definition of a prime ideal, we must keep the restriction, since for us the zero ideal is not a prime.

If $\mathscr{A}$ is not a prime ideal, it does not have property *P*, for $\mathscr{A} = \mathscr{B}\mathscr{C}$ implies, as above, that *b* and *c* may be chosen in $\mathscr{B}$ and $\mathscr{C}$ so that *bc* is in $\mathscr{A}$ without either *b* or *c* being in $\mathscr{A}$. But the above example shows that if $\mathscr{A}$ is a prime ideal, it need not have property *P*. Thus property *P* implies primality but not conversely. We shall show at the end of this section that for ideals over rings of algebraic integers, property *P* is equivalent to primality. Meanwhile we have the following theorem.

Theorem 8.3. If *R* is a commutative ring with identity element and if $\mathscr{A}$ is an ideal over *R*, then $R/\mathscr{A}$ is an integral domain if and only if $\mathscr{A}$ has property *P*.

Proof: If $\mathscr{A}$ does not have property *P*, then there are elements *b* and *c* of *R*, neither of which is in $\mathscr{A}$, such that *bc* is in $\mathscr{A}$. Thus $b + \mathscr{A}$ and $c + \mathscr{A}$ are two cosets, neither of which is $\mathscr{A}$ but whose product is $\mathscr{A}$. Thus $R/\mathscr{A}$ is not an integral domain.

Conversely, if $R/\mathscr{A}$ is not an integral domain, then cosets $b + \mathscr{A}$ and $c + \mathscr{A}$ exist, neither of which is $\mathscr{A}$ but whose product is $\mathscr{A}$. This shows that *bc* is in $\mathscr{A}$ and hence that property *P* does not hold. This completes the proof.

There is a corresponding stronger theorem for maximal ideals.

Theorem 8.4. If *R* is a commutative ring with identity element and $\mathscr{A}$ is an ideal over *R*, then $R/\mathscr{A}$ is a field if and only if $\mathscr{A}$ is maximal in *R*.

Proof: Suppose that $\mathscr{A}$ is maximal in *R* and let *c* be an element of *R* that is not in $\mathscr{A}$. Then form the ideal $(c) + \mathscr{A} = \mathscr{B}'$. Now $\mathscr{B}' \supset \mathscr{A}$, and since $\mathscr{A}$ is maximal, it follows that $\mathscr{B}' = R$. Hence $\mathscr{B}'$ contains the identity element 1 of *R*. Thus for some *x* in *R*, $1 - cx$ is in $\mathscr{A}$. This implies that $(c + \mathscr{A})(x + \mathscr{A})$ $- 1$ is in $\mathscr{A}$. Hence if $\mathscr{B} = c + \mathscr{A}$, then $\mathscr{B}(x + \mathscr{A}) = cx + \mathscr{A}$, which contains $cx + (1 - cx) = 1$ and hence is $1 + \mathscr{A}$, the identity coset. Thus we have shown that any coset different from the zero coset has a multiplicative inverse, and hence $R/\mathscr{A}$ is a field.

Conversely, suppose that $R/\mathscr{A}$ is a field. We want to prove that $\mathscr{A}$ is maximal. Then (see Exercise 15 after Section 5), $R/\mathscr{A}$, being a field, contains just two ideals, the zero coset and $R/\mathscr{A}$ itself. We suppose that there is an ideal $\mathscr{B}$ such that $R \supset \mathscr{B} \supset \mathscr{A}$ and seek a contradiction. Let *b* be some element of $\mathscr{B}$ not in $\mathscr{A}$ and let $\mathscr{C}$ be the ideal $(b) + \mathscr{A}$. Then $\mathscr{C}/\mathscr{A}$ consists of the coset $\mathscr{A}$ and cosets $\beta + \mathscr{A}$, where β is in (b) but is not in $\mathscr{A}$. Now $\mathscr{C} \subseteq \mathscr{B}$ implies that $\mathscr{C} \subset R$ and, $\mathscr{C}/\mathscr{A}$ is an ideal in $R/\mathscr{A}$. (Why?) Since $R/\mathscr{A}$ contains only two ideals, it follows that either $\mathscr{C}/\mathscr{A} = (0)$, the zero coset, or $\mathscr{C}/\mathscr{A} = R/\mathscr{A}$. The former is impossible, since $\mathscr{C} \supset \mathscr{A}$. The latter is impossible, since $\mathscr{C}/\mathscr{A} = R/\mathscr{A}$ implies that for every *r* in *R*, $r + \mathscr{A} = c + \mathscr{A}$ for some *c* in $\mathscr{C}$.

This would imply that $R \subseteq \mathscr{C}$, which denies $\mathscr{C} \subset R$. This is the contradiction which shows that $R \supset \mathscr{B} \supset \mathscr{A}$ is impossible and hence that $\mathscr{A}$ is maximal.

You are asked to prove, as an exercise, the following theorem.

Theorem 8.5. If $\mathscr{A}$ is a maximal ideal over a commutative ring R with identity element, then $\mathscr{A}$ has property P.

The converse of this theorem is false, as you are asked to show in Exercise 16 at the end of Section 9. However, if we add another condition on R, we have the following theorem.

Theorem 8.6. If an ideal $\mathscr{A}$ over R has property P and if $R/\mathscr{A}$ has a finite number of elements, then $\mathscr{A}$ is maximal.

Partial proof: If an integral domain has a finite number of elements, it is a field, by Theorem 3.2 in Chapter III. One can apply Theorems 8.3 and 8.4 to complete the proof.

So far in this section we have only specified that R is a commutative ring with identity element. In the rest of this section we restrict R to be the ring of integers of an algebraic field $\mathbf{Q}(\theta)$. Corollary 2 of Theorem 7.1 shows that an ideal $\mathscr{A}$ over R is maximal if and only if it is prime. The next theorem is one that we promised when we gave our definition of prime ideal.

Theorem 8.7. If R is the ring of algebraic integers over $\mathbf{Q}(\theta)$, an ideal $\mathscr{A}$ over R is a prime ideal if and only if it has property P.

Proof: If $\mathscr{A}$ is a prime ideal, then it is maximal by Corollary 2 of Theorem 7.1, and by Theorem 8.5 it has property P.

To prove that if $\mathscr{A}$ has property P, then it is a prime ideal, we need only show that property P implies maximality; for, from Section 5, if $\mathscr{A}$ is maximal, it is a prime ideal. Now Theorem 8.6 shows that if $\mathscr{A}$ has property P, it is maximal *if $R/\mathscr{A}$ has a finite number of elements*. To complete the proof of Theorem 8.7, we need only establish the truth of the following theorem.

Theorem 8.8. Let R be the ring of algebraic integers in $\mathbf{Q}(\theta)$ and $\mathscr{A}$ an ideal over R. Then the number of elements in $R/\mathscr{A}$ is finite.

Proof: The proof is quite similar to that of Theorem 7.3. First, we prove the theorem for $\mathscr{A} = (a)$, a principal ideal. Let $\beta_1, \beta_2, \ldots, \beta_n$ be an integral basis of R. Consider any element

$$\beta = \sum_{i=1}^{n} b_i \beta_i$$

of R, where the b_i are integers. For each b_i there exist integers q_i and r_i such that $b_i = q_i a + r_i$ with $0 \le r_i < a$ and

$$\beta = a \sum_{i=1}^{n} q_i \beta_i + \sum_{i=1}^{n} r_i \beta_i.$$

Hence β is in the same coset as $\sum r_i \beta_i$. There are a^n distinct elements $\sum_{i=1}^{n} r_i \beta_i$ and hence a^n cosets of $\mathscr{A}$.

Suppose that $\mathscr{B}$ is any ideal over R. We know from Theorem 6.4 that there is an ideal $\mathscr{C}$ such that $\mathscr{B}\mathscr{C} = (d)$, a principal ideal. Now $\mathscr{B} \supseteq (d)$. This shows that if $r - r'$ is in (d), then it is in $\mathscr{B}$; that is, if $r - r'$ is not in $\mathscr{B}$, then it is not in (d). Hence the number of different cosets of $\mathscr{B}$ is not greater than the number of different cosets of (d), which is finite. This completes the proof.

Let us summarize our findings on prime and maximal ideals $\mathscr{A}$ and ideals with property P over commutative rings with identity elements.

1. If $\mathscr{A}$ is a maximal ideal, it is prime. (See after the definition of maximal in Section 5.)
2. If $\mathscr{A}$ has property P, it is prime. (Before Theorem 8.3)
3. If $\mathscr{A}$ is maximal, it has property P. (Theorem 8.5)

We could condense these three findings into the following statement:

(8.1)   Maximal implies property P implies primality.

The converse of 3 holds only with the additional condition that $R/\mathscr{A}$ has a finite number of elements. There was given a counterexample to the converse of statement 1. You are asked in Exercise 20 after Section 9 to give a counterexample to the converse of the statement 2.

If the ring is a ring of algebraic integers, then we know from Corollary 2 of Theorem 7.1 that an ideal is maximal if and only if it is prime. Thus (8.1) shows that for a ring of algebraic integers, all three concepts are equivalent.

9.   Homomorphisms of Rings

At the end of Chapter III we defined a ring and indicated how the ideas of homomorphisms and isomorphisms could be carried over to rings from the context of groups. In particular, we shall see that the concept of an ideal, which was developed to solve the problem of unique decomposition over algebraic integers, enters in a basic way into the discussion of homomorphisms of rings. In fact, an ideal is to a ring homomorphism what a normal subgroup is to a group homomorphism.

Here, since we want to establish connection with homomorphisms of groups, we allow the ring to be noncommutative but still require it to have a (multiplicative) identity element. You may want to look back at the definition in Section 9 of Chapter II of a homomorphism for a group and in Section 2 of Chapter III, where an isomorphism for fields is defined.

Definition.   *Let R and R′ be two rings with identity elements. We call a linear transformation σ of R onto R′ a homomorphism if it preserves structure with respect to both addition and multiplication; that is,*

$$(9.1) \quad (r_1 + r_2)\sigma = r_1\sigma + r_2\sigma \quad and \quad (r_1 r_2)\sigma = (r_1\sigma)(r_2\sigma)$$
$$for\ all\ r_1\ and\ r_2\ in\ R.$$

From the definition of homomorphism of sets in Section 9 of Chapter II, this means that σ is a homomorphism of R onto R′ if it is a homomorphism for the additive groups of elements of R and R′ and for the sets R and R′ with respect to multiplication. (The nonzero elements of R do not necessarily form a multiplicative group.)

In dealing with groups, we found that the kernel of a homomorphism was important. Therefore, we have the corresponding definition for rings.

Definition.   *If σ is a homomorphism of a ring R onto R′, then the* kernel *of σ is the set of elements of R whose image is the additive identity, zero, of R′.*

What does the kernel of a homomorphism look like? It follows immediately from Theorem 9.1 of Chapter II that the kernel is an additive group. But what is it with respect to multiplication? Let K be the kernel of σ. Then K consists of those elements k of R such that $k\sigma = 0'$, the zero element of R′. Since in a ring $0 \cdot r = 0$ for all elements r (why?),

$$(0 \cdot r)\sigma = (0\sigma)(r\sigma) = 0'(r\sigma) = 0',$$
$$(r \cdot 0)\sigma = (r\sigma)(0\sigma) = (r\sigma)0' = 0',$$

where $0'$ is the zero of R′. This means that if k is in K, then rk and kr are both in K for r any element in R. This looks like a normal subgroup except that it is not a multiplicative group. We have proved the following theorem.

Theorem 9.1.   Let K be the kernel of a homomorphism of a ring R onto a ring R′. Then K is an additive (Abelian) group and has the property that if k is in K, so is rk and kr for all elements r in R.

Recall from Section 4 that K just described is an ideal. However, in Section 4, ideals were defined only over commutative rings. We now state the more general definition.

Definition.   *If R is a ring with identity element, any set $\mathscr{A}$ of elements of R is called a* right ideal *over R if*

1. *Whenever a_1 and a_2 are in $\mathscr{A}$, then $a_1 + a_2$ is in $\mathscr{A}$.*
2. *If a is in $\mathscr{A}$, then ar is in $\mathscr{A}$ for all r in R.*

A right ideal is a set of R that is closed under addition and under multiplication on the right by elements of R. If in condition 2, ar is replaced by ra, the set A is called a *left ideal*. The set A is called a *two-sided ideal* or *ideal* if for every a in A and r in R, $ar = r'a$ and $ra = ar''$ for elements r' and r'' in R, that is, if A is both a left ideal and a right ideal.

If a ring is commutative, all ideals are two-sided and we omit the adjectives "left" or "right." If the ring is not commutative, there are ideals that are not two-sided; but even in such a ring some ideals (for example, the kernel of a homomorphism) are two-sided. With this terminology, Theorem 9.1 affirms that the kernel K of the homomorphism is a (two-sided) ideal over R.

In Section 8 we defined a quotient ring for commutative rings. This can be carried over completely for noncommutative rings for two-sided ideals $\mathscr{A}$. The proof of Theorem 8.1 also applies for such rings if we omit the condition of commutativity so that we have the following theorem.

Theorem 9.2.   If $\mathscr{A}$ is an ideal over a ring R with an identity element, the set of cosets $R/\mathscr{A}$ is a ring with identity element.

The next two theorems follow directly from Theorems 9.2 and 9.3 of Chapter II. Their proofs are left as exercises.

Theorem 9.3.   If σ is a homomorphism of a ring R onto a ring R' and K is the kernel of the homomorphism, then two elements of R have the same image in R' if and only if they are in the same coset $c + K$ of K. That is $r_1\sigma = r_2\sigma$ if and only if $r_1 + K = r_2 + K$.

Theorem 9.4.   A homomorphism of a ring R onto a ring R' is an isomorphism if and only if its kernel is the element 0 of R.

We have the following theorem, the first sentence of which is a restatement of Theorem 9.1. The proof of the isomorphism is similar to that for groups in the proof of Theorem 10.6 in Chapter II. The proof is left as an exercise.

Theorem 9.5.   The kernel of a homomorphism of a ring R onto a ring R' is an ideal K over R. Also, R' is isomorphic to R/K.

Corresponding to the diagram in Section 10 of Chapter II, we have the following, where φ and σ are homomorphisms and $\bar{\sigma}$ is an isomorphism:

Just as for groups in Section 10 of Chapter II, every two-sided ideal $\mathscr{K}$ over R determines a homomorphism whose kernel is $\mathscr{K}$, and every homomorphism determines a two-sided ideal, its kernel. The latter we have already shown in the proof of Theorem 9.1. The truth of the former is affirmed in the following theorem. This could be proved by citing Theorem 10.4 of Chapter II, but perhaps we can better understand what is happening by proving it directly.

Theorem 9.6.   Let $\mathscr{A}$ be an ideal over a ring R. Then there is a homomorphism σ of R onto a ring R' such that $\mathscr{A}$ is the kernel of σ.

Proof: Let S be the set of cosets of $\mathscr{A}$ and, as in the proof of Theorem 10.4 of Chapter II, let a correspondence σ be set up between the elements of R and cosets of $\mathscr{A}$, that is,

$$r \overset{\sigma}{\to} r + \mathscr{A} \qquad \text{or} \qquad r\sigma = r + \mathscr{A}.$$

To show that this is a homomorphism, suppose that $r\sigma = r + \mathscr{A}$ and $r'\sigma = r' + \mathscr{A}$. Then

$$(r\sigma)(r'\sigma) = (r + \mathscr{A})(r' + \mathscr{A}) = rr' + \mathscr{A} = (rr')\sigma,$$
$$r\sigma + r'\sigma = (r + \mathscr{A})(r' + \mathscr{A}) = r + r' + \mathscr{A} = (r + r')\sigma.$$

This shows the homomorphism. Since $r\sigma = \mathscr{A}$ if and only if r is in $\mathscr{A}$ and $\mathscr{A}$ is the additive identity for cosets modulo $\mathscr{A}$, we see that $\mathscr{A}$ is the kernel of the homomorphism. This completes the proof.

One other result is interesting and not hard to prove. Recall that in Exercise 15 after Section 5 we showed that the only ideals in a field are the zero ideal and the field itself. Here is a generalization of this result.

Theorem 9.7.   Let R be a division ring, that is, a noncommutative field. Its only ideals are itself and the zero ideal.

Proof: Let $\mathscr{A}$ be an ideal over R that contains an element a not zero. Then since $\mathscr{A}$ is an ideal, $a^{-1}a$ is in $\mathscr{A}$ and thus 1 is in $\mathscr{A}$. This implies that every element of R is in $\mathscr{A}$ and $\mathscr{A} = R$.

Exercises

1. Which of the following are ring homomorphisms?
   (a) $b \to 3b$, where the b are integers in **Z**.
   (b) $f(x) \to f(i)$, where $f(x)$ is a polynomial with rational coefficients and $i^2 = -1$.

2. Let R be the ring of matrices of order 2 over a field and S be the subset of R consisting of all matrices whose first row has all its elements zero. Show that S is a right ideal. Is it a left ideal also? Why or why not?

3. Let R be the ring defined in Exercise 2 and S the subset of singular matrices. Is S a left ideal, a right ideal, or both? Why or why not?

4. An element r of a ring is called *nilpotent* if $r^m = 0$ for some positive integer m. Prove that the set N of nilpotent elements of a commutative ring R form an ideal over R.

5. If R and N are as in Exercise 4, prove that the quotient ring R/N has no nilpotent elements except the zero coset.

6. Prove that the intersection of two left ideals over a ring R is a left ideal.

7. Let $\mathscr{A}$ be a left ideal over a ring R and $\mathscr{B}$ a right ideal. Is $\mathscr{A} \cap \mathscr{B}$ a left ideal, a right ideal, both, or neither?

8. Prove Theorem 9.3.

9. Prove Theorem 9.4.

10. Prove Theorem 9.5.

11. Prove Theorem 8.5.

12. Complete the proof of Theorem 8.6.

13. Let R be a ring with no divisors of zero and $\mathscr{A}$ an ideal of R. Show by an example that $R/\mathscr{A}$ may have divisors of zero.

14. Show that for each positive integer m, there is up to an isomorphism, exactly one ring R' with an identity element with m elements which is the homomorphic image of the ring of integers.

15. Let R be a commutative ring with identity element. Let c be an element of R with a multiplicative inverse and d a nilpotent element of R. Note that if $d^2 = 0$, then the inverse of $c + d$ is $(c^{-1} - c^{-2}d)$. Prove that $c + d$ has a multiplicative inverse for every nilpotent element d.

16. Give an example to show that the converse of Theorem 8.5 is false.

17. Let R be the ring of polynomials in x with real coefficients. Let A be the subset of polynomials $f(x)$ in R such that $f(1) = 0$. Show that A is an ideal.

18. Let $\mathscr{A}$ be defined as in Exercise 17. Fill in the details of the following sketch of a proof that $\mathscr{A}$ is maximal. Let $\mathscr{B}$ be an ideal over R such that $\mathscr{B} \supset \mathscr{A}$ and let $g(x)$ be in $\mathscr{B}$ with $g(1) = c \neq 0$. Then $g(x)$ and $h(x) = g(x) - c$ are both in $\mathscr{B}$ and hence c is in $\mathscr{B}$. Thus $cc^{-1} = 1$ is in $\mathscr{B}$ and $\mathscr{B} = R$.

19. For R and $\mathscr{A}$ as in Exercises 17 and 18, show that $R/\mathscr{A}$ is isomorphic to the field of real numbers.

20. Use the example at the end of Section 5 or by other means give an example of a prime ideal that does not have property P.

21. In proving Theorem 8.3, we used the fact that there are just two ideals over a field. After Theorem 9.6 it was noted that this same fact follows from Theorem 9.6. Is this circular reasoning? Why or why not?

22. In defining cosets of a ring, we used sums: $r + \mathscr{A}$. Could we have used products instead, that is, defined a coset of $\mathscr{A}$ as the set of elements rx where x ranges over $\mathscr{A}$? Why or why not?

23. Define $a \oplus b = a + b - 1$ and $a \odot b = a + b - ab$, where a and b are in **Z**. Prove that there is a ring R with the operations $\oplus$ and $\odot$ instead of addition and multiplication. Prove that **Z** and R are isomorphic.

24. Given a ring R, $\mathscr{A}$ a set of elements of R, and σ a homomorphism of R onto R'.
    (a) Prove that if $\mathscr{A}$ is a right ideal, then $\mathscr{A}\sigma$, the image of $\mathscr{A}$, is also a right ideal.
    (b) If R is commutative and $\mathscr{A}$ is a prime ideal, does it follow that $\mathscr{A}\sigma$ is a prime ideal? Explain.

25. Answer the question "why?" in the proof of Theorem 8.4.

*26. Let R be a ring and $\mathscr{A}$ and $\mathscr{B}$ two ideals over R, where $\mathscr{A} \supset \mathscr{B}$. Prove:
    (a) $\mathscr{A}/\mathscr{B}$ is an ideal in $R/\mathscr{B}$.
    (b) $R/\mathscr{A}$ is isomorphic to $(R/\mathscr{B})/(\mathscr{A}/\mathscr{B})$.

*27. Let R be the ring of matrices of order 2 over a field. Prove that this ring has no ideals except the zero ideal and R itself.

*28. In Exercise 4e after Section 4 of Chapter I we defined quaternions over the field of real numbers. Suppose that Q is the ring of quaternions over Z_p for some odd prime number p. Prove that Q is a ring whose only ideals are Q and the zero ideal. Prove that this ring is not a division ring. (See Theorem 9.6.)

*29. Let R be a ring in which $x^3 = x$ for every x in R. Prove that R is a commutative ring.

*30. Let $\mathscr{S}$ and $\mathscr{T}$ be ideals in a ring R. Show that $\mathscr{S}$ is an ideal in $\mathscr{S} + \mathscr{T}$ and that $\mathscr{S} \cap \mathscr{T}$ is an ideal in $\mathscr{T}$. Prove that the quotient ring $\mathscr{T}/(\mathscr{S} \cap \mathscr{T})$ is isomorphic to the quotient ring $(\mathscr{S} + \mathscr{T})/\mathscr{S}$.

10. *Richard Dedekind and Emmy Noether*

Richard Dedekind (1831–1916) began life in Brunswick, Germany, where Gauss was born. He was Gauss's last student, receiving his doctorate from Göttingen University in 1852. He became "private docent" there and in 1857 gave what were probably the first university lectures on Galois theory. He taught for five years at Zurich Polytechnic Institute and in 1862 returned to Brunswick, where he taught the rest of his active life in the technical high school. It was he who first defined groups by means of postulates. His most notable mathematical contributions were the development of the theory of ideals to achieve unique decomposition for algebraic integer ideals and the definition of irrational numbers by a process that is now referred to as a Dedekind cut.

In his commemorative address to the Royal Society of Göttingen in 1917, published in the *Göttingen Nachrichten*, Edmund Landau said "Richard Dedekind was not only a great mathematician, but one of the wholly great in the history of mathematics, now and in the past, the last hero of a great epoch, the last pupil of Gauss, for four decades himself a classic from whose works not only we, but our teachers and the teachers of our teachers, have drawn."

Although her name does not often appear in texts in modern algebra, the spirit of Emmy Noether (1882–1935) pervades this subject. Her father, Max Noether, himself a distinguished mathematician, was a member of the mathematics department at the University of Erlangen, Germany, and she grew up and received her early mathematical training there. She lectured occasionally at Erlangen, substituting for her father when he was ill. While there she visited, from time to time, the prestigious Mathematical Institute at Göttingen and in 1916 moved to that institution. Because she was a woman she did not have the position in German universities that was her due, in spite of the great efforts of her mathematician colleagues, but her position in the mathematical world was unquestioned. Hermann Weyl (see below) said that when he was at Göttingen in 1930–1933, "she was without doubt the strongest center of mathematical activity there, considering both the fertility of her scientific research program and her influence upon a large circle of pupils."

Her genius was to find the abstraction that would unify and be fruitful. For instance, she freed the notion of an ideal promulgated by Dedekind from the restrictions of algebraic integers so that her approach gave a unique decomposition over more general ideals, especially for polynomials. A key concept which she recognised was that of a ring with "ascending chain condition," named, after her, a Noetherian ring. (See Section 5 of this chapter.) Her publications were few, and most of her results appeared in the works of her students and associates. Measured by the inspiration and far-reaching consequences of her thought, she is considered the greatest woman mathematician in the Western World.

When political pressures scattered the members of the Mathematical Institute of Göttingen in 1933, Emmy Noether came to America, and while at Bryn Mawr was a frequent visitor at the Institute for Advanced Study in Princeton. She died suddenly of a brain tumor. Hermann Weyl (himself a great mathematician) in a memorial address at Bryn Mawr shortly after her death said, "Two traits determined all her nature: first, the native productive power of her mathematical genius. She was not clay, pressed by the artistic hands of God into harmonic form, but rather a chunk of human primary rock into which he had blown his creative breath of life. Second, her heart knew no malice; she did not believe in evil—indeed it never entered her mind that it could play a role among men." You should read his address in *Scripta Matematica*, vol. 3, 1935, pp. 201–220. In his description of her work is the whole depth and sweep of modern algebra and the warmth of an admiring associate.

VI

Constructions and Galois Theory

1. Introduction

This chapter is devoted largely to the solution of two classical problems: constructions with straightedge and compass, and solutions of equations by radicals. The first of these is indubitably classical, having been considered by the Greeks, and the second is a problem that was not solved until the early part of the nineteenth century. We consider them together, since some of the theory that we must develop in order to solve the first problem also applies to the second. The crucial tool is Galois theory, which is a very important part of modern algebra. In fact, the development of the mathematics needed to solve these problems is more important than the problems themselves.

In high school geometry you dealt with various constructions: to construct through a given point a line parallel to a given line, to construct an equilateral triangle, to construct a line perpendicular to a given line at a given point, to construct the bisector of a given angle, and so forth. In all of these, the only instruments allowed by the rules of the game were a straightedge for drawing straight lines and a compass for drawing circles or as dividers. By using a compass as a divider, we mean adjusting the compass so that the point and lead are at the ends of a given segment and are thereby measuring off on any line a segment of the same length.

If we are very particular here, we can get into all sorts of logical difficulties.

As we know, the concept of a point is undefined; but we think of it to be "as small a dot as possible." A line is undefined, but we think of it as having "no width," whatever that means. We postulate that two distinct lines can intersect in at most one point. We cannot draw a line, but we think that we can draw something that "represents" a line, whatever that means. The more we try to explain, the more complex it becomes. Hence without shame we speak of "drawing a line or a circle" and using the compass as a divider.

We use the term "straightedge" to emphasize the fact that the use of any marks like those occurring on a ruler is barred. The use of such aids as T-squares and triangles is also barred. These restrictions, of course, are very artificial, for there is no reason why a straightedge or compass need be "more accurate" than a T-square or the marks on a ruler. Hence one must view the problem as a kind of game with prescribed rules. Now, a game has little interest except as one develops proficiency in playing it against opposition. Therefore, the method of solution is of chief importance here. In the interest of brevity we shall use the word *constructible* as an abbreviation for *constructible with straightedge and compass*, as described above. This will become more definite in the discussion that follows.

2.   *The Algebraic Equivalent to Constructibility*

To make the problem more precise and tractable, we now proceed to express it in numerical terms. Starting with any line segment, we can construct the line that contains it and, using a compass along that line, a line segment whose length is any integral multiple of the given segment. That is, if the given segment is of length m, we can construct on that line segments of lengths mr, where r is any positive integer. In fact, by using coordinates on the line, the multiple r could also be negative. Then, given two segments of lengths a and b, Figure VI.1 shows a construction of a line segment of length ab. (You are asked in an exercise to show why this works.) You are also asked in an exercise to show how, given line segments of lengths b and c, one can construct a line segment of length b/c.

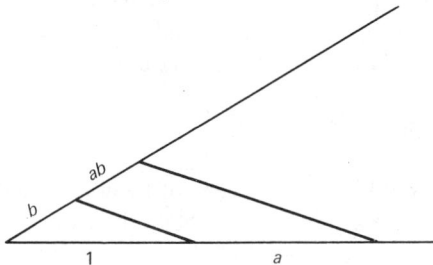

Figure VI-1

Step 1.   Given line segments of lengths b and c, line segments of lengths

$$\frac{bx + cy}{bz + ct}$$

can be constructed for all integers x, y, z, and t for which the fraction is positive. In other words, given a line segment of unit length, a segment whose length is any positive rational number is constructible. In fact, by the use of a coordinate system on a line, a point whose coordinate is any rational number can be constructed. Therefore, we say more briefly that *any rational number is constructible*.

So far we have used a compass only to lay off distances on a line. We can construct some irrational numbers by intersecting circles with lines. Assuming a coordinate system, we know that every line has an equation of the form $ax + by + c = 0$, where not both a and b are zero. Every circle has an equation of the form

$$x^2 + y^2 + dx + ey + f = 0.$$

If $b \neq 0$, we can solve the linear equation for y and substitute it in the quadratic one to get a quadratic equation in x. If $b = 0$, we can solve the linear equation for x and, by substituting in the equation of the circle, get a quadratic equation in y. So, we can construct solutions of some quadratic equations.

To consider the intersection of two circles, we write their equations as follows:

$$x^2 + y^2 + d_i x + e_i y + f_i = 0, \qquad i = 1, 2.$$

Then, by subtraction, we see that the coordinates of all points common to the two circles satisfy the equation

$$g(x, y) = (d_1 - d_2)x + (e_1 - e_2)y + (f_1 - f_2) = 0.$$

If the coefficients of g are all zero, the circles coincide. If the coordinates of x and y are both zero and $f_1 \neq f_2$, there are no points in common. If not both coefficients of x and y are zero, then the points of intersection of the two circles are the same as the points of intersection of the line $g(x, y) = 0$ with either of the circles. Hence the intersection of two circles yields no numbers that cannot be obtained by intersecting a line with a circle.

A line can be constructed if its slope and a point through which it passes can be constructed; and a circle is constructible if its center and radius are.

Step 2.   If a line and a circle can be constructed, their intersection can be constructed, also.

We have shown that if a line segment of length c is constructible starting with a segment of unit length, then either c is a rational number or is an element of a field F in a sequence:

$$F_0 = Q, F_0(c_1) = F_1, F_2 = F_1(c_2), \ldots, F = F_k = F_{k-1}(c_k),$$

where (see Section 14 of Chapter III) each c_i is a solution of a quadratic equation with coefficients in F_{i-1}. It saves verbiage if we have the understanding that whenever we write of "constructing a number," we mean constructing a line segment whose length is that number. So we can express the result more briefly in the statement of the following theorem.

Theorem 2.1.   The only numbers that can be constructed, starting with a unit, are numbers of a field F which arises from the field of rational numbers by a finite number of adjunctions of solutions of quadratic equations, that is, by quadratic adjunctions.

Conversely, suppose that $ax^2 + bx + c = 0$ is a quadratic equation such that the numbers a, b, and c are constructible. The solutions are $(-b \pm \sqrt{b^2 - 4ac})/2$, and they can be constructed if $\sqrt{b^2 - 4ac}$ can be. But there is a construction given in high school for the square root of a given positive number. Incorporating this result with that of Theorem 2.1, we have the following theorem.

Theorem 2.2.   If we start with a unit segment, a number k can be constructed if and only if k is an element of some field F obtained from the rational field by a finite number of quadratic adjunctions.

Up to this point we have been tacitly assuming that the numbers which we construct are real and positive. By introducing a coordinate system, we can construct negative numbers as well. Constructions can also be extended to imaginary numbers. In fact, it is almost necessary to do this, since it is conceivable that a sequence of adjunctions that in the end yield a real number could include along the way some adjunctions of quadratic imaginaries. To make this generalization, we note first that the elements of a field F' obtained from F by a quadratic adjunction consist of numbers of the form $a + \sqrt{c}$, where a and c are in F. Then we need to give a construction for imaginary square roots. First, if c is a negative real number, then $\sqrt{c} = i\sqrt{-c}$ and using the Argand representation (see Section 3 of Chapter I) it can be represented by the point $(0, \sqrt{-c})$ in the plane. It remains to show how to construct $\sqrt{c}$ when $c = s + it$, for s and t real numbers and t not zero, assuming that s and t are themselves constructible. Recalling Section 3 of Chapter II, we write

$$s + it = r(\cos \theta + i \sin \theta),$$

where $r = \sqrt{s^2 + t^2}$ and θ between 0 and 2π is defined by $s = r \cos \theta$, $t = r \sin \theta$. If (see Figure VI.2) A has the coordinate $s + it$, then B with the coordinate $\sqrt{s + it}$ is on the bisector of the angle between OA and the x-axis, and the distance of B from the origin is $\sqrt{r}$. Thus we have justified the omission in Theorem 2.2 of any mention of real or imaginary numbers.

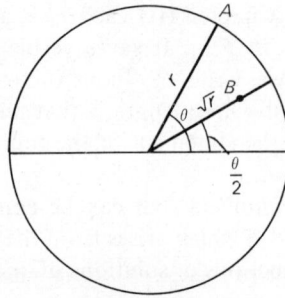

Figure VI-2

We have made progress, but we do not yet have quite enough information to deal with the problem. For instance, consider the classical problem of "duplicating the cube." This is to construct the side of a cube whose volume is twice that of a given one. This amounts to constructing a root of $x^3 = 2$. Now, $x^3 - 2$ is irreducible over the field of rational numbers. (Why?) For a root to be constructible, it would have to be also a root of some quadratic equation whose coefficients are constructible. This seems unlikely. One can show this by somewhat specialized methods for this equation, but it is simpler to prove the following theorem, which applies to a much wider class of construction problems.

Theorem 2.3. If a number is constructible starting with a given unit, it must be a zero of an irreducible polynomial over the field $\mathbf{Q}$ of rational numbers whose degree over $\mathbf{Q}$ is a power of 2.

To prove this theorem, we need to look a little more deeply into properties of algebraic numbers leading up to a theorem that has Theorem 2.3 as an immediate consequence. Recall (see Section 14 of Chapter III) that if F is a field and $f(x)$ an irreducible polynomial with coefficients in F, then, for θ a zero of $f(x)$, $F(\theta)$ consists of all linear combinations of powers of θ with coefficients in F. Remember also that if $f(x)$ is of degree n, we say that $F(\theta)$ is of degree n over F and $F(\theta)$ is algebraic over F; that is, $F(\theta)$ is a simple algebraic extension of F of degree n. But here we are concerned not just with simple extensions but finite sequences of simple extensions, that is, algebraic extensions that are not necessarily simple. To be sure, in Section 5 we shall prove that over every field of characteristic zero, every algebraic extension is simple. But that limits the field. Such limitation is not very serious, since our applications are to subfields of the field of complex numbers, but the proof in Section 5 is a little tricky. All in all, it is much more satisfactory to take the vector-space approach. Now we define a new term and will show shortly that it is equivalent to the term "algebraic extension."

Definition.   *Let K be a field containing a field F. Here K is called a* finite extension *of F if there are elements $\alpha_1, \alpha_2, \ldots, \alpha_n$ of K which are linearly independent over F such that every element of K is a linear combination of the α_i with coefficients in F. The number n is called the* degree *of K over F and is written $n = [K:F]$. Recall that the set of α_i is called a* basis *of K over F.*

Notice from Section 4 of Chapter IV that under the above conditions, K is a vector space of dimension n over F. Since we have already defined the degree of a simple algebraic extension, it is important to note that there is no conflict with the above, for if $K = F(\theta)$ and is of degree n over F, the elements

$$1, \theta, \theta^2, \ldots, \theta^{n-1}$$

form a basis of K over F. On our way is the following result.

Theorem 2.4.   If K is a finite extension of H and H a finite extension of F, then K is a finite extension of F and

$$[K:H][H:F] = [K:F].$$

Proof: Let $\alpha_1, \alpha_2, \ldots, \alpha_n$ be a basis of H over F and $\beta_1, \beta_2, \ldots, \beta_s$ a basis of K over H. We need to prove that

(2.1) $\{\alpha_i \beta_j\}$ for $1 \leq i \leq n$ and $1 \leq j \leq s$

is a basis of K over F.

First, to show that the elements (2.1) are linearly independent over F suppose that

(2.2) $\sum c_{ij} \alpha_i \beta_j = 0,$

where the c_{ij} are in F and the sum is over i and j in (2.1). Since the β_j are linearly independent over H and, for each j, the coefficient of β_j is in H, we see that

$$\sum_i c_{ij} \alpha_i = 0   \text{for each } j.$$

But the α_i are linearly independent over F. This implies that $c_{ij} = 0$ for each i and j. Hence (2.2) implies that c_{ij} are all zero, and we have shown that the set (2.1) is linearly independent.

Second, to show that the elements (2.1) span K over F, notice that since the β_j are a basis of K over H, every element of K can be represented in the form

$$\beta = \sum b_j \beta_j,$$

where the b_j are in H. But each b_j, being in H, is a linear combination of the α_i with coefficients in F. Thus, for d_{ji} in F:

$$b_j = \sum d_{ji} \alpha_i   \text{and}   \beta = \sum d_{ji} \alpha_i \beta_j,$$

where $1 \leq i \leq n$ and $1 \leq j \leq s$. The proof is complete.

Now we can tie together the concepts of algebraic extension and finite extension.

Theorem 2.5. A field K is a finite extension of a field F if and only if it is an algebraic extension.

Proof: Suppose that K is a finite extension with a basis $\alpha_1, \alpha_2, \ldots, \alpha_n$. Then $K \supseteq F(\alpha_1) = F_1$. If $K = F(\alpha_1)$, then K is a simple extension of F, and the theorem holds. If not, one of $\alpha_2, \ldots, \alpha_n$ is not in $F(\alpha_1)$. We change notation if necessary to choose α_2 not in F_1 and see that $F_1(\alpha_2) = F_2 \subseteq K$. If we continue this process until every remaining α_i is in the field, we see that K can be obtained from F by a finite sequence of simple adjunctions and hence is an algebraic adjunction.

Conversely, suppose that K is an algebraic extension of F, that is, obtainable from F by a sequence of r simple extensions. If $r = 1$, there is no problem. To apply an induction argument, assume that the theorem is true for a sequence of less than r simple extensions. Thus we may take $K = F_0(\alpha)$, where F_0 is obtained from F by a sequence of $r - 1$ simple extensions. By the induction hypothesis, F_0 is a finite extension of F. We call u its degree and let K be of degree v over F_0. By Theorem 2.4, K is of degree uv over F with a basis like (2.1), and hence a finite extension. The theorem is proved.

To return to Theorem 2.3, let K be any field obtained from another field F by a finite number of quadratic extensions. We have just shown that $[K:F] = 2^k$, for some positive integer k. Suppose that γ is any element of K and let its degree over F be m. Now, γ defines a field $F(\gamma)$ of degree m over F. Then, by Theorem 2.4,

$$2^k = [K:F] = [K:F(\gamma)][F(\gamma): F] = [K:F(\gamma)]m.$$

Thus we have proved the following theorem.

Theorem 2.6. If K is a field of degree 2^k over a field F, then the degree of every element of K over F is a power of 2.

Theorem 2.3 follows from this theorem and Theorem 2.1. It should be noted here that the converse of Theorem 2.3 is not true. To show this, it would be sufficient to show that the zeros of some irreducible quartic polynomials are not constructible. This should seem reasonable, but the proof would be rather difficult at this stage. We postpone such a proof until Section 13.

3. *Constructibility*

Theorem 2.3 shows that the roots of $x^3 = 2$ cannot be constructed, since any root is a zero of an irreducible polynomial of degree 3 instead of a power of 2. We can also apply the theorem to show the impossibility of trisection of certain angles. We recall the following formula from trigonometry:

$$\cos 3\theta = 4 \cos^3 \theta - 3 \cos \theta.$$

That is, $\cos \theta$ is a solution of the equation

(3.1) $$4x^3 - 3x - \cos 3\theta = 0.$$

Now if $3\theta = 90° = \pi/2$ radians, $\cos 3\theta = 0$ and Equation (3.1) is reducible. Its roots are 0, $\pm\sqrt{3}/2$, which jibes with our knowledge that $\cos(\pi/6) = \sqrt{3}/2$. Therefore, as we know, the angle 90° can be trisected with straightedge and compass. But if $3\theta = 120° = 2\pi/3$, Equation (3.1) becomes

(3.2) $$4x^3 - 3x + \tfrac{1}{2} = 0,$$

which has no rational roots and hence is irreducible over the field of rational numbers. (Why?) Thus the angle of 120° cannot be trisected with straightedge and compass. This also shows that the regular nine-sided polygon cannot be constructed because the central angle of a side is 40°.

What about the constructibility of regular polygons in general? We know that the equilateral triangle can be constructed, since, as we just showed, the angle of 30°; is constructible. Also, the regular pentagon can be constructed from high school geometry. To be more systematic, notice first that if a regular polygon of rs sides can be constructed, so can one of r sides by taking every sth vertex. Hence if a regular n-sided polygon can be constructed, so can a regular p-sided polygon for p any prime factor of n. We first state and prove a result for p-sided polygons.

Theorem 3.1.   If a regular p-sided polygon can be constructed, for p a prime number, then $p - 1$ is a power of 2.

Proof: Consider the equation $x^p - 1 = 0$. Its roots are the pth roots of unity, and (see Section 3 of Chapter II) these roots will be the vertices of a regular p-sided polygon, plotted as complex numbers. If a regular p-sided polygon can be constructed, then $\cos(2\pi/p)$ is constructible. This means that $\cos(2\pi/p)$, by Theorem 2.3, must be a zero of an irreducible polynomial over the rationals whose degree is a power of 2. Thus

(3.3) $$\cos \frac{2\pi}{p} + i \sin \frac{2\pi}{p}.$$

being a zero of $z^2 - 2(\cos[2\pi/p])z + 1$, is also a zero of an irreducible polynomial whose degree is a power of 2. Now, the number (3.3) is a zero of the polynomial $g(x) = (x^p - 1)/(x - 1)$. Since p is a prime number, $g(x)$ is irreducible, by Theorem 13.7 in Chapter III.

Thus we have shown that if a regular p-sided polygon can be constructed, then the degree of $g(x)$ must be a power of 2. This means that $p - 1$ is a power of 2 and completes the proof.

Suppose that $p - 1 = 2^r$, that is, $p = 1 + 2^r$. If p is to be a prime number, r must be a power of 2; for if r had an odd factor b greater than 1, then $1 + 2^{r/b}$ would be a factor of p, since $1 + x^{r/b}$ is a factor of $1 + x^r$ when b is odd. So we have the following corollary.

Corollary. If a regular p-sided polygon can be constructed, for p a prime number, then p is of the form

$$p = 2^{2^k} + 1 \qquad \text{for some nonnegative integer } k.$$

These numbers are called *Fermat numbers*. The following table gives the primes p for the first five values of k:

k	0	1	2	3	4
$2^{2^k} + 1$	3	5	17	257	65,537

No other such primes have been found. Notice that we have not yet proved that all such regular polygons can be constructed—we have only narrowed our consideration to these.

It can also be shown that a regular polygon of n sides cannot be constructed if n is of the form p^k, where p is an odd prime and k is greater than 1. This is accomplished by showing that $\cos(2\pi/p^k) + i\sin(2\pi/p^k)$ is a zero of an irreducible polynomial of degree $p^{k-1}(p - 1)$, which is not a power of 2 for an odd prime. But the proof would carry us too far afield. A proof may be found in van der Waerden's *Modern Algebra*, vol. I, pp. 113, 114, and 162 (see the Bibliography).

We shall show in the next three sections that if p is a prime for which $p - 1$ is a power of 2, then a regular p-sided polygon can be constructed. This is done by showing that for such equations the converse of Theorem 2.3 is valid. But much additional theory must be developed before this can be done.

Exercises

1. Justify the construction in Figure VI.1 of a segment of length ab, given segments of lengths a and b.

2. Give a construction of the number b/c when b and c are lengths of given segments.

3. Why is Equation (3.2) irreducible over the field of rational numbers?

4. Which of the following can be constructed? Why?
   (a) The segment whose length is the diagonal of a cube of unit side.
   (b) The length of the altitude of a triangle the lengths of whose sides are given.
   (c) The length of the side of a square whose area is equal to that of a triangle with given sides.
   (d) The length of a side of a cube whose volume is equal numerically to the area of a triangle whose sides are given.
   (e) The regular seven-sided polygon.
   (f) An angle of $15°$.

5. Given a triangle of sides a, b, and c. Show that the length of any median of the triangle is obtained from $Q(a, b, c)$ by a finite sequence of adjunctions of square roots.

6. Show that the result of Exercise 5 holds if "median" is replaced by "the line segment from any vertex to a trisection point of the opposite side."

7. Show that $x^3 - 2$ is irreducible over the field of rational numbers.

8. Can the zeros of $x^4 - 2$ be constructed?

9. Given a segment of length c, where c is a positive real number, show how one can construct a segment of length $\sqrt{c}$.

10. Let c in Exercise 9 be an imaginary number $a + bi$, where a and b are real numbers. Show how $\sqrt{c}$ can be constructed from a and b.

11. The roots of $x^3 + x + 1 = 0$ are imaginary. Show how these roots can be constructed.

12. Let $f(x)$ be a quadratic polynomial with real coefficients and imaginary zeros. Show how these roots can be constructed if the coefficients can be constructed.

13. Repeat Exercise 12 for complex coefficients of the quadratic polynomial.

14. In Step 2 of Section 2 suppose that some or all of a, b, c, the coefficients of the linear equation, and d, e, f, coefficients of the quadratic expression, are imaginary numbers. Show that Step 2 is still justified.

15. Prove that the polynomial $4x^3 - 3x + 1/p$ is irreducible over the field of rational numbers for every prime p.

16. Let K be a field that is algebraic over a field F and of prime degree over F. Show that there is no field H such that $K \supset H \supset F$.

17. Show that a regular polygon of 2^r sides can be constructed for all positive integers r.

18. Give a construction of a regular pentagon. [One method is to show that the roots of $(x^5 - 1)/(x - 1) = x^4 + x^3 + x^2 + x + 1 = 0$ can be obtained from the rational numbers by two successive adjunctions of zeros of quadratic equations. The substitution $x + 1/x = y$ is useful.]

19. Can the roots of the equation $x^4 + 3x^3 + 5x^3 + 3x + 1 = 0$ be constructed?

20. Does Theorem 2.6 hold also for a power of any prime number? Why or why not?

21. Just before Theorem 3.1 we showed that if a regular n-sided polygon can be constructed, so can a regular polygon for p any prime factor of n. Is it true that if a regular polygon of p sides, p being a prime, is constructible, then a regular polygon of p^k sides, for $k > 1$, can be constructed? Establish your result by proof or counterexample.

22. Here is a method of trisection of any angle that uses a mark on a ruler. Let a ruler have a mark M that is r units from one end and let angle $TOA = \beta$ be the angle to be trisected. Place the ruler so that its end E is on the line TO, with O between T and E; the mark M is on the circle of radius r; and the ruler passes through A as in Figure VI.3. Then the angle AEO is one third of angle β. Prove it.

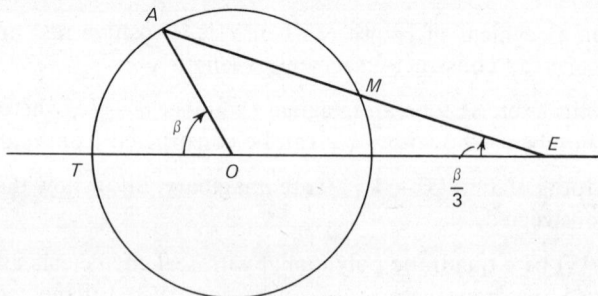

Figure VI-3

*23. Show without the use of Theorem 2.3 that a root of $x^3 = 2$ is not constructible.

4.   On the Way Toward Galois Theory

In Section 3 we showed that if a regular p-sided polygon, for p a prime number, is constructible, then $p - 1$ is a power of 2. To show that for all such primes, regular p-sided polygons are indeed constructible requires more basic theory. For although we have shown that constructibility leads to an extension field whose degree is a power of 2, it is not at all clear (and, in fact, is not always true, as noted at the end of Section 2) that if the degree of the extension is a power of 2, it can be accomplished by a sequence of *quadratic* extensions. One could perhaps show some of this by specialized methods, but it is much more fruitful to develop some basic theory not only because it is useful in solving the second problem of this chapter, but also because it adds to our understanding of the inner workings of important algebraic structures.

The basic idea of Galois in dealing with these problems was to set up a relationship between groups and fields, so that in the setting of groups one can solve the problems at hand more expeditiously. We now begin to develop such a connection. This section is largely exploratory to motivate and point toward the theory dealt with in the following section. The student who does not feel the need of such motivation or exploration may prefer to read this section lightly and go on to Section 5.

In Section 8 of Chapter II we dealt with the automorphisms of groups. The group that Galois associated with a field is the group of its automorphisms, that is, the isomorphisms of the field with itself. (See Section 2 of Chapter III.) We give a formal definition.

Definition.   *Let σ be a one-to-one correspondence between the elements of a field F and itself and denote by $a\sigma$ the element of F that corresponds to a in F under σ. Then σ is called an* automorphism *if, for all a and b in F,*

1. $(a + b)\sigma = a\sigma + b\sigma$.
2. $(ab)\sigma = (a\sigma)(b\sigma)$.

An automorphism may be described as a one-to-one correspondence of a field with itself which preserves sums and products. We leave as an exercise the proof of the following theorem.

Theorem 4.1.   The set of all automorphisms of a field form a group.

We can now be more explicit about Galois's fundamental idea. Suppose that a field K is obtained from a field F by a finite number of algebraic adjunctions, that is, K is a finite extension of F, and let G be the set of automorphisms of K that leave fixed each element of F. Then, at least for certain types of fields K, there is a one-to-one correspondence between the subfields of K that contain F and the subgroups of G.

Now we continue our exploration of automorphisms of fields. First, we do this for the rational field **Q**. Let σ be an automorphism and see that from properties 1 and 2, $(0)\sigma = 0$ and $(1)\sigma = 1$. (See Exercise 12 at the close of this section.) Thus, from property 1, $n\sigma = n$ for all positive integers n. Also

$$0 = (0)\sigma = [a + (-a)]\sigma + a\sigma + (-a)\sigma = a + (-a)\sigma$$

shows that $(-a)\sigma = -a$; hence $n\sigma = n$ for all integers. Similarly,

$$1 = 1\sigma = \left[a\left(\frac{1}{a}\right)\right]\sigma = a\sigma\left[\left(\frac{1}{a}\right)\sigma\right] = a\left[\left(\frac{1}{a}\right)\sigma\right]$$

implies $(1/a)\sigma = 1/a$. Hence the following theorem holds:

Theorem 4.2. The group of automorphisms of the field of rational numbers is the group (ϵ) consisting of the identity element ϵ alone.

At this point we have nothing very productive. Next consider the quadratic extension of **Q** obtained by adjoining a root of the equation $x^2 + ax + c = 0$, where the coefficients are in **Q** and the polynomial is irreducible over **Q**. The use of the quadratic formula indicates that the roots are

$$\frac{-a \pm \sqrt{a^2 - 4c}}{2}.$$

But the field obtained by adjoining one of these numbers to **Q** is the same as that obtained by the adjunction of $\sqrt{a^2 - 4c}$ to **Q**. Thus we can, without loss, confine ourselves to roots of the simpler equation

(4.1) $$x^2 - b = 0,$$

where b is in **Q** but is not the square of an element of **Q**. Let x_1 be one of the roots of (4.1) and note that the other root is $-x_1$. Then $\mathbf{Q}(x_1) = \mathbf{Q}(-x_1)$, and it makes no difference which root we use for the extension. Let σ be an automorphism of $\mathbf{Q}(x_1)$. We know by the proof of Theorem 4.2 that it leaves each element of **Q** unchanged. (See Exercise 14.) Using properties 1 and 2 of an automorphism, we have

$$0 = (x_1^2 - b)\sigma = x_1^2\sigma - b = (x_1\sigma)^2 - b.$$

So, $(x_1)\sigma$ is a root of (4.1). Thus there are just two possibilities for σ:

$$(x_1)\sigma_1 = x_1, \qquad (-x_1)\sigma_1 = -x_1$$

and

$$(x_1)\sigma_2 = -x_1, \qquad (-x_1)\sigma_2 = x_1.$$

Also, σ_1 is the identity automorphism, $\sigma_2^2 = \sigma_1$, and the group of automorphisms of $\mathbf{Q}(x_1)$ has order 2.

We can begin to see how things are going if we consider the field obtained by adjoining to $\mathbf{Q}(x_1)$ a root of the equation

(4.2) $$x^2 - c = 0,$$

where c is in $\mathbf{Q}(x_1)$ but not in **Q** and $\sqrt{c}$ is not in $\mathbf{Q}(x_1)$. Let x_2 be a root of (4.2) and define the field F to be $\mathbf{Q}(x_1, x_2)$.

Now we shall show that $Q(x_2) = Q(x_1, x_2)$ and that F is of degree 4 over Q. First, notice that c, being in $Q(x_1)$, is of the form $ax_1 + d$, where a and d are in Q with $a \neq 0$. Let $c' = -ax_1 + d$; that is, c' is obtained from c by replacing x_1 by its conjugate $-x_1$. Then define

$$(4.3) \qquad\qquad g(x) = (x^2 - c)(x^2 - c').$$

Since the coefficients of $g(x)$ are symmetric in x_1 and its conjugate $-x_1$, they are in Q. In fact, $c + c' = 2d$ and $cc' = d^2 - a^2 \cdot x_1^2 = d^2 - a^2 b$. To show that g is irreducible over Q, we need merely show, as in the proof of Theorem 2.4, that

$$1, x_2, x_2^2 = c, x_2^3 = cx_2$$

are linearly independent over Q. Now,

$$e_0 + e_1 x_2 + e_2 c + e_3 x_2 c = (e_0 + e_2 c) + (e_1 + e_3 c)x_2 = 0,$$

for e_i in Q implies that $e_0 + e_2 c = 0 = e_1 + e_3 c$, since x_2 is not in $Q(x_1)$. But since c is not rational, the previous equations show that the e_i are all zero. Also, $x_2^2 = c = ax_1 + d$ with $a \neq 0$ implies that x_1 is in $Q(x_2)$. Thus we have shown that $Q(x_1, x_2) = Q(x_2) = F$, which is of degree 4 over Q.

So far so good, but at this point we have something different from the previous cases. Not all the roots of (4.3) need be in F. (An example will be asked for in Exercise 9.) In this case it is true, as is shown in the next theorem, that there are two automorphisms of F: the identity and $(x_2)\sigma = -x_2$, since x_2 and $-x_2$ are the only roots of (4.3) in F. The difficulty, then, is that the order of G is 2, whereas the degree of K over Q is 4. Now we first prove a theorem which shows that in this case the group of automorphisms G is of order 2. Thus we have a sequence of three fields: $Q(x_2) \supset Q(x_1) \supset Q$. But $G = (\epsilon, \sigma)$ has only one proper subgroup. Thus there cannot be a one-to-one correspondence between the subgroups of G and the subfields of $Q(x_2)$. After this proof we show how to get around the difficulty just mentioned.

Theorem 4.3. Let F be a field, $f(x)$ an irreducible polynomial over F, α a zero of $f(x)$, and β an element of an algebraic extension of F. Let σ be a linear transformation of the elements of $F(\alpha)$ which leaves fixed all elements of F. First, if $\alpha\sigma = \beta$ defines an isomorphism of $F(\alpha)$ onto $F(\beta)$, then β is a conjugate of α over F. Second, if $f(\beta) = 0$ and $\alpha\sigma = \beta$, then σ is an isomorphism of $F(\alpha)$ onto $F(\beta)$.

Proof: Let σ be an isomorphism of $F(\alpha)$ onto a field F' that leaves fixed all elements of F. Then $[f(\alpha)]\sigma = 0$ implies that $f(\alpha\sigma) = 0$; thus $\beta = \alpha\sigma$ is a zero of $f(x)$, and hence β is a conjugate of α. Conversely, if $\alpha\sigma = \beta$ and n is the degree of $f(x)$, then

$$(a_0 + a_1\alpha + \cdots + a_{n-1}\alpha^{n-1})\sigma = a_0 + a_1\beta + \cdots + a_{n-1}\beta^{n-1},$$

which shows that the correspondence is preserved under the transformation. This completes the proof.

Corollary. Under the conditions of the theorem, every automorphism of $F(\alpha)$ takes α into a conjugate of α over F.

In order to get more automorphisms, the thing to do is to enlarge the field $F = \mathbf{Q}(x_2)$ so that it contains *all* the roots of (4.3). Let x_3 be a root of $x^2 - c' = 0$ and $K = \mathbf{Q}(x_2, x_3) = F(x_3)$; note that now K contains all the roots of (4.3). By the same argument as before, it can be shown (Exercise 10) that K is spanned by

$$(4.4) \qquad 1, x_2, x_2^2, x_2^3, x_3, x_3x_2, x_3x_2^2, x_3x_2^3.$$

Thus an automorphism of K will be determined by the images of x_2 and x_3. Now, by the corollary of Theorem 4.3, the only possible images of x_2 are the roots of (4.3), namely, $x_2, -x_2, x_3$, and $-x_3$. The same can be said for the images of x_3, but, once the image of x_2 is determined, we do not have complete liberty for the image of x_3, since $x_2^2 + x_3^2 = c + c' = 2d$, a rational number, implying that $(x_2^2 + x_3^2)\sigma = x_2^2 + x_3^2$. Hence

$$(x_3^2)\sigma = x_2^2 + x_3^2 - (x_2^2)\sigma.$$

Thus if $(x_2^2)\sigma = x_2^2$, we have $(x_3^2)\sigma = x_3^2$, and hence $x_3\sigma = x_3$ or $-x_3$. If $(x_2^2)\sigma = x_3^2$, we have, similarly, $x_3\sigma = x_2$ or $-x_2$. Then we prepare the following table of possible automorphisms:

i	1	2	3	4	5	6	7	8
$\sigma_i(x_2)$	x_2	$-x_2$	x_2	$-x_2$	x_3	$-x_3$	x_3	$-x_3$
$\sigma_i(x_3)$	x_3	x_3	$-x_3$	$-x_3$	x_2	x_2	$-x_2$	$-x_2$

Now we need to show that each σ_i is an automorphism. In order to deal with all eight possibilities at once, we suppress the subscripts of σ and obtain

$$x_2\sigma = e_1x_j, \qquad x_3\sigma = e_2x_k,$$

where each e_i is 1 or -1 and j and k are 2 and 3, in some order. The images of the basis elements (4.4) are defined as follows:

$$(x_2^r x_3^s)\sigma = (x_2^r)\sigma(x_3^s)\sigma = e_1^r x_j^r e_2^s x_k^s.$$

Then, to achieve linearity, we define

$$[a_1 + a_2x_2 + a_3x_2^2 + a_4x_2^3 + x_3(b_1 + b_2x_2 + b_3x_2^2 + b_4x_2^3)]\sigma$$
$$= [a_1 + a_2e_1x_j + a_3x_j^2 + a_4e_1x_j^3 + e_2x_k(b_1 + b_2e_1x_j + b_3x_j^2 + b_4e_1x_j^3)].$$

Thus σ is a one-to-one linear transformation of K onto itself. It immediately follows that property 1 of an automorphism holds. We made our definition above so that property 2 holds for the product of any two basis elements. Since a product of any two elements of K is a sum of products of basis elements, linearity and property 1 imply that property 2 holds also. We have shown that such a σ is an automorphism and hence that every σ_i in the preceding table is an automorphism.

By adjoining x_3 to F, we have obtained a group G of *eight* automorphisms, which is the degree of K over $\mathbf{Q}$. We shall find that there is a one-to-one correspondence between the subgroups of G and the subfields of K. Note that G is the octic group dealt with in Section 9 of Chapter I.

Now we list four of the subfields F_i of K and associate with each F_i the subgroup G_i of G, which leaves fixed each element of F_i:

Subfield	$\mathbf{Q}(x_2) = F_1$	$\mathbf{Q}(x_3) = F_2$	$\mathbf{Q}(x_1) = F_3$	$\mathbf{Q}(x_2 x_3) = F_4$
Subgroup	$\{\sigma_1, \sigma_3\} = G_1$	$\{\sigma_1, \sigma_2\} = G_2$	$\{\sigma_1, \sigma_2, \sigma_3, \sigma_4\} = G_3$	$\{\sigma_1, \sigma_4, \sigma_5, \sigma_8\} = G_4$

The last is a field, since $(x_2 x_3)^2 = cc'$, which is in $\mathbf{Q}$. Note that in each case the degree of the field over $\mathbf{Q}$ is equal to the index of its group in G. To justify G_3, note from our definitions that $x_2^2 = ax_1 + d$. Thus G_3 is the set of automorphisms σ_i such that $x_2^2 \sigma_i = x_2^2$. This is true for $i = 1, 2, 3, 4$.

Let us try the correspondence in the other direction. Another subgroup of G is $G_5 = \{\sigma_1, \sigma_5\}$. This should be associated with a subfield of degree 4. It does not take much looking to see that the elements of G_5 leave fixed the following elements of K:

$$(4.5) \qquad 1, \, x_2 + x_3, \, x_2 x_3, \, x_2 x_3 (x_2 + x_3).$$

We now show that the numbers of (4.5) span a field of degree 4 over $\mathbf{Q}$. We saw above that 1 and $x_2 x_3$ span a field F. Now $x_2 + x_3$ satisfies the following equation over F_4:

$$(x_2 + x_3)^2 = c + c' + 2 x_2 x_3.$$

But $x_2 + x_3$ is not in F_4, since $x_2 + x_3 = r(x_2 x_3) + s$ with r and s in $\mathbf{Q}$ would imply that x_3 is in $\mathbf{Q}(x_2)$, which is false. Hence the elements of (4.5) span a space obtained from $\mathbf{Q}$ by two successive quadratic adjunctions. Thus we have found a subfield associated with the subgroup G_5.

Now, G is called the *Galois group* of K over $\mathbf{Q}$. (The general definition will be given in Section 5.) From the preceding discussion it appears that there is a one-to-one correspondence between the subgroups of G and the subfields of K such that to each subgroup G_i of G corresponds a field F_i of K with the property that every element of G_i leaves fixed every element of F_i. Furthermore, the degree of F_i over $\mathbf{Q}$ is equal to the index of G_i in G. This we shall establish in general in Section 5.

Before leaving the example, we should notice two things. First, the set

(4.5) corresponds to another factorization of $g(x)$, namely, $g(x) = h(x)h'(x)$, where

$$h(x) = (x - x_2)(x - x_3) \qquad \text{and} \qquad h'(x) = (x + x_2)(x + x_3).$$

The coefficients of $h(x)$ and $h'(x)$ are left fixed by the elements of G_5, $g(x)$ is reducible over F_4, and x_2 and x_3 are zeros of the quadratic polynomial $h(x)$ over F_4.

It is also true, as we shall see in general, that $K = \mathbf{Q}(x_4)$ for a proper choice of x_4. In other words, we could express the basis of K as $\{x_4^i\}$ for $0 \le i \le 7$. This would have some advantages in identifying the automorphisms.

Exercises

1. Show that the group of eight automorphisms described in the table after Theorem 4.3 is the octic group.

2. In the example of Section 4, where $K = \mathbf{Q}(x_2, x_3)$, find the quadratic equation over $\mathbf{Q}(x_2)$ that is satisfied by x_3.

3. In Exercise 9 of Section 9 in Chapter I you were asked to find all the subgroups of the octic group. For each subgroup not dealt with in the present section, find the corresponding subfield of K.

4. Which of the subgroups of the octic group are normal subgroups? (See Section 10 of Chapter II.) What distinguishes the subfields associated with normal subgroups from those associated with subgroups that are not normal?

5. In the example of Section 4 show that $\{\sigma_1, \sigma_4\}$ is the center of the octic group.

6. Find a number x_4 referred to at the close of this section such that $\mathbf{Q}(x_4) = \mathbf{Q}(\sqrt{2}, \sqrt{5})$.

7. Prove Theorem 4.1.

8. In the example of Section 4 find the Galois group of K over $\mathbf{Q}(x_1)$. What are the subgroups and the corresponding fields?

9. Find a numerical value of c in Equation (4.3) such that not all the zeros of $g(x)$ are in $F_1 = \mathbf{Q}(x_2)$.

10. Show that the elements of (4.4) span K as defined in this section.

11. In the example of Section 4, is K normal over F_4? Why or why not?

12. Show that if σ is an automorphism of the rational field $\mathbf{Q}$, then $(0)\sigma = 0$ and $(1)\sigma = 1$.

13. Why is it that Theorem 4.2 does not directly imply that if σ is an automorphism of $\mathbf{Q}(x_1)$ with $x_1^2 = b$ in $\mathbf{Q}$, then σ leaves fixed each element of $\mathbf{Q}$? Nevertheless, show that the proof of Theorem 4.2 does imply this result.

14. Let K be a simple algebraic extension of $\mathbf{Q}$. Show that every automorphism of K takes each element of $\mathbf{Q}$ into itself.

*15. Let $f(x)$ and $g(x)$ be two distinct irreducible polynomials over $\mathbf{Q}$ with leading coefficient 1. Let $f(\alpha) = 0 = g(\beta)$ and let r and s be the respective degrees of $f(x)$ and $g(x)$ with r and s relatively prime. Prove that $\mathbf{Q}(\alpha, \beta)$ is of degree rs over $\mathbf{Q}$.

5.   The Fundamental Theorem of Galois Theory

Now, on the basis of the above exploration we can proceed to develop what we need of Galois theory. In line with what we noted in the previous section, we first prove the following theorem.

Theorem 5.1.   Let F be a field of characteristic zero, that is, which contains $\mathbf{Q}$, the field of rational numbers; $F(\theta)$, a simple algebraic extension of F of degree r; and $F(\theta, \eta)$, a simple algebraic extension of $F(\theta)$ of degree s. Then there is an element α of $F(\theta, \eta)$ such that $F(\theta, \eta) = F(\alpha)$.

Proof: Let $\theta_1 = \theta$ and $\theta_2, \theta_3, \ldots, \theta_r$ be the conjugates of θ over F; let $\eta_1 = \eta$ and $\eta_2, \eta_3, \ldots, \eta_s$ be the conjugates of η over $F(\theta)$. Then let $\alpha = \theta + b\eta$, where b is an element of F later to be determined. Define

(5.1) $$\alpha_{ij} = \theta_i + b\eta_j, \qquad 1 \le i \le r, \quad 1 \le j \le s,$$

where $\alpha_{11} = \alpha$. Now we choose b so that no two α_{ij} are equal. Why can this be done? Suppose that $\theta_i + b\eta_j = \theta_k + b\eta_t$. Then $(\theta_i - \theta_k) = b(\eta_t - \eta_j)$. Thus for any given i, j, k, and t there is at most one element b of F for which the equality holds. Since F contains an infinite number of elements, we can avoid the finite number of values of b which make two elements of (5.1) equal. Assume that b has been chosen so that the elements (5.1) are distinct. Now since $\alpha = \theta + b\eta$, it is true that $F(\alpha) \subseteq F(\theta, \eta)$. We want to show that $F(\theta, \eta) \subseteq F(\alpha)$. To this end let

(5.2) $$g(x) = \prod_{i,j} (x - \alpha_{ij}).$$

Suppose that β is some element of $F(\theta, \eta)$. Then $\beta = f(\theta, \eta)$ for some polynomial over F. Define $\beta_{ij} = f(\theta_i, \eta_j)$ with $\beta_{11} = \beta$. Then set

(5.3) $$h(x) = g(x)\left(\frac{\beta}{x - \alpha} + \frac{\beta_{12}}{x - \alpha_{12}} + \cdots + \frac{\beta_{ij}}{x - \alpha_{ij}} + \cdots + \frac{\beta_{rs}}{x - \alpha_{rs}} \right)$$

and note from (5.2) that $h(x)$ is a polynomial with coefficients in F, since its coefficients are symmetric in θ, η, and their conjugates. Now, $g(x) = (x - \alpha)k(x)$ for some polynomial $k(x)$ with coefficients in an algebraic extension of F. A glance at (5.3) shows that unless $i = j = 1$, $g(x)\beta_{ij}/(x - \alpha_{ij})$ has a factor $x - \alpha$ and hence, if $\alpha \neq \alpha_{ij}$, $g(\alpha)\beta_{ij}/(\alpha - \alpha_{ij}) = 0$. Thus (5.3) shows that $h(\alpha) = \beta k(\alpha)$. Using a prime to designate the derivative (see Section 16 of Chapter III), we have

$$g'(x) = k(x) + (x - \alpha)k'(x),$$

which implies that $g'(\alpha) = k(\alpha)$. Thus $h(\alpha) = \beta g'(\alpha)$.

But $g'(\alpha) \neq 0$, since (see Theorem 16.2 of Chapter III) α is not a zero of $g(x)/(x - \alpha)$. Hence

$$\beta = \frac{h(\alpha)}{g'(\alpha)},$$

which shows that β is in $F(\alpha)$ and completes the proof.

By applying the theorem a finite number of times, one can establish the following corollary.

Corollary. Let θ_1 be algebraic over a field F containing $\mathbf{Q}$, θ_2 algebraic over $F(\theta_1)$, θ_3 algebraic over $F(\theta_1, \theta_2)$, ..., θ_k algebraic over $F(\theta_1, \theta_2, \ldots, \theta_{k-1})$. Then there is a number α such that

$$F(\alpha) = F(\theta_1, \theta_2, \ldots, \theta_k).$$

That is, if F is a field of characteristic zero, then every finite extension of F is a simple algebraic extension of F.

Note that $g(x)$ defined by (5.2) is irreducible over F, since it has α as a zero, its coefficients are in F, and the degree of $F(\alpha)$ over F is rs from Theorem 2.4.

We should digress a moment to point out the reasons for the restriction that the field F have characteristic zero. One reason is that we needed in the proof to have a field with "enough elements." That is, to choose b so that the α_{ij} are distinct, we needed to know that the field contained more than rs elements. Also, the conjugates of θ and η had to be distinct for the proof to work. As was pointed out in Theorem 16.2 of Chapter III, if a field has characteristic zero, then if a polynomial $f(x)$ is irreducible over F, the conjugates of every zero of $f(x)$ are distinct. To be sure, we could provide for both these contingencies by a lesser restriction on the field, but since our applications are to fields of characteristic zero, it seems simpler to impose the restriction in the theorem.

We noted in Section 4 that in an approach to Galois theory, one must consider not just $F(\theta)$, where θ is an irreducible polynomial $f(x)$ over F, but one must also consider the field containing all the zeros of $f(x)$, that is, the root field as defined in Section 14 of Chapter III. For convenience we now repeat the definition and that of an allied term.

Definition.   *Given an irreducible polynomial $f(x)$ over a field F, the root field of $f(x)$ over F is the field obtained by adjoining to F all the zeros of $f(x)$.*

In the root field of $f(x)$, the polynomial can be expressed as a product of linear factors. For this reason, as we noted earlier, the root field is also called the *splitting field* of the polynomial.

As we noted in Section 14 of Chapter III, one needs to be a little careful about using the article "the" to refer to *the* root field, since for arbitrary fields one needs a proof that all root fields of a given polynomial are isomorphic. We assume this result without proof, secure in the knowledge that the applications we shall make are to algebraic extensions of **Q**. Here the fundamental theorem of algebra and Section 14 of Chapter III assure us that in the field of complex numbers, root fields are indeed unique up to an isomorphism.

To apply Theorem 5.1 to root fields, let $f(x)$ be an irreducible polynomial over a field F containing **Q** and let θ be a zero of $f(x)$. Let K be the root field of $f(x)$ and, using Theorem 5.1, let $K = F(\alpha)$ for α some element of K. We know that K contains all the conjugates of θ by its construction. But does K contain all the conjugates of α? Since it is useful to define the automorphisms of K by means of the images of α, we need to prove that the answer to the question is "yes."

Theorem 5.2.   Let F be a field of characteristic zero, $f(x)$ an irreducible polynomial of degree n over F, and K the root field of $f(x)$. We know that $K = F(\alpha)$ for some α. Then K contains all the conjugates of α over F.

Proof: Let α be a zero of the irreducible polynomial $g(x)$ over F and suppose that β is another zero of $g(x)$. We want to show that β is in K. Let τ be a linear transformation such that $\alpha\tau = \beta$ and, for all c in F, $c\tau = c$. Then τ is an isomorphism of K onto $K' = F(\beta)$, by Theorem 4.3. We want to show that $K = K'$. Now, $f(\theta) = 0$ implies that $[f(\theta)]\tau = 0 = f(\theta\tau) = 0$. This shows that if θ is a zero of $f(x)$, so is $\theta\tau$. Hence τ permutes the zeros of $f(x)$. Since α is in K, the root field,

$$\alpha = h(\theta_1, \theta_2, \ldots, \theta_n),$$

where h is a polynomial in the zeros of $f(x)$ with coefficients in F. Thus $\beta = \alpha\tau$ is the same function of a permutation of the θ_i. Since every θ_i is in $F(\alpha)$, the proof is complete.

In fact, the field K of Theorem 5.2 contains the conjugates of *every* element in K over F as stated in the following theorem, which is an extension of Theorem 5.2.

Theorem 5.2a. Let F be a field of characteristic zero, $f(x)$ an irreducible polynomial of degree n over F, and K the root field of $f(x)$. Then if β is any element of K, its conjugates over F are also in K.

Proof: Since $K = F(\alpha)$, we can write β in the form

$$\beta = \sum c_i \alpha^i,$$

where the c_i are in F and the sum is from $i = 0$ to $i = n - 1$. Let $\alpha = \alpha_1$ and α_j for $2 \leq j \leq n$ be the conjugates of α over F and define

$$\beta_j = \sum c_i \alpha_j^i, \quad 2 \leq j \leq n, \quad \beta = \beta_1.$$

Let

$$g(x) = \prod_{j=1}^{n} (x - \beta_j).$$

Then $g(x)$ has coefficients that are symmetric functions of α and its conjugates and hence are in F. Thus $g(x)$ is a polynomial with coefficients in F that has β as a zero. Let $h(x)$ be the polynomial of least degree over F which has β as a zero. It must be irreducible and also a factor of $g(x)$. But the conjugates of β are all zeros of $h(x)$. Hence the conjugates of β must be among the β_j. But the conjugates of α are all in K. Thus every β_j is in K and hence the conjugates of β are in K. This completes the proof.

Definition. *Let F be a field and $K = F(\alpha)$ an algebraic extension of F. The field K is called a* normal extension *of F if K contains all the conjugates of α. An equation over F is called* normal *if it has a root θ such that all its roots are in $F(\theta)$.*

Corollary. A normal extension K of a field F contains the conjugates over F of every element of K.

We now define the Galois group.

Definition. *Let F be a field, $f(x)$ a polynomial over F, and K the root field of $f(x)$. The group of automorphisms of K that leave fixed every element of F is called the* Galois group *of K over F and is written $G(K:F)$. The group G is also called the* Galois group *of the equation $f(x) = 0$.*

Notice that from Theorem 4.3 every element of the Galois group G permutes the zeros of the polynomial $f(x)$ in the definition above. Hence $G(K:F) \subseteq S_n$, where S_n is the symmetric group on n symbols and n is the degree of $f(x)$. Now we prove the fundamental theorem.

Theorem 5.3. (Fundamental theorem of Galois theory.) Let F be a field of characteristic zero, $f(x)$ an irreducible polynomial over F, and K the root field of $f(x)$. Let G be the group of automorphisms of K that leave fixed each element of F, that is, the Galois group of K over F. Then

1. For every field F_1 such that $F \subseteq F_1 \subseteq K$, there is a subgroup G_1 of G consisting of all those elements of G that leave fixed each element of F_1.

2. For G_1 as defined in part 1, the order of $G_1 = G(K:F_1)$ is the degree of K over F_1, that is, the index of G_1 in G is the degree of F_1 over F. In notation,

$$o[G(K:F_1)] = [K:F_1] \qquad \text{and} \qquad [G:G_1] = [F_1:F];$$

in particular,

$$o(G) = o[G(K:F)] = [K:F].$$

3. For every subgroup G_1 of G there is a subfield F_1 of K consisting of all those elements of K left fixed by every element of G_1.

4. There is a one-to-one correspondence between the subgroups of G and the subfields containing F, defined as above.

Proof: To prove part 1, let S be the set of automorphisms of K that leave fixed each element of F_1. If σ_1 and σ_2 are in S, then $\sigma_1\sigma_2$ is in S and the inverse of σ_1 is in S. Hence S is a group; call it G_1, which is a subgroup of G since if G_1 leaves fixed each element of F_1, it certainly leaves fixed each element of F.

To prove part 2, let F_1 be a field such that $F \subseteq F_1 \subseteq K$ and define G_1 as in part 1; that is, G_1 is the subgroup of G that leaves fixed every element of F_1. [We do not know that the elements of F_1 are the *only* elements of K left fixed by every element (automorph) of G_1. That proof comes later.] From Theorem 5.1 there is an α in K such that $K = F(\alpha)$. Then if K is of degree t over F_1, denote the conjugates of α over F_1 by $\alpha_1, \alpha_2, \ldots, \alpha_t$, with $\alpha = \alpha_1$. On the one hand, since the powers of α span K over F_1, every automorph of K over F_1 is determined by the image of α. On the other hand, Theorem 4.3 tells us that the image of α must be one of its conjugates. Hence there are exactly t automorphisms in G_1. They may be defined by

$$\alpha\sigma_i = \alpha_i, \qquad 1 \leq i \leq t.$$

Thus we have shown that $[K:F_1] = o(G_1) = o[G(K:F_1)]$. In particular, for $F = F_1$ we have $[K:F] = o(G) = o[G(K:F)]$. Now

$$[K:F] = [K:F_1][F_1:F] \qquad \text{and} \qquad o(G) = o(G_1)[G:G_1],$$

which completes the proof of part 2.

To prove part 3, let G_1 be a subgroup of G and H the set of elements of K left fixed by every element of G. We want to prove that H is a field. Then if r_1 and r_2 are in H, it follows that for each σ in G_1, $r_1\sigma = r_1$ and $r_2\sigma = r_2$. This implies that

$$(r_1 + r_2)\sigma = r_1 + r_2 \qquad \text{and} \qquad (r_1 r_2)\sigma = r_1 r_2.$$

Hence H is closed under addition and multiplication. As for division, if $r_1 \neq 0$,

$$(r_1 r_1^{-1})\sigma = (r_1 \sigma)(r_1^{-1}\sigma) = (1)\sigma = 1$$

shows that $(r_1^{-1})\sigma = (r_1\sigma)^{-1} = r_1^{-1}$. Hence the set of nonzero elements of H is closed under division. Since H is a subset of the field K, the elements of K left fixed by all elements of G form a subfield of K.

To prove part 4, let G_1 be defined as in part 1. To show the one-to-one correspondence desired, we must prove that the elements of F_1 are the *only* elements left fixed by G_1. Now suppose that β is an element of K left fixed by all automorphisms of G_1; that is, $\beta\sigma_j = \beta$ for all σ_j in G_1. Since β is in $K = F(\alpha)$,

$$\beta = \sum_{i=0}^{n-1} c_i \alpha^i$$

for c_i in F. From the proof of part 2, each σ_j of G_1 is defined by $\alpha\sigma_j = \alpha_j$ for α_j the conjugates of α over F. Hence

$$\beta = \beta\sigma_j = \sum_{i=0}^{n-1} c_i \alpha_j^i, \qquad 1 \le j \le t.$$

Summing over all j, we have

$$t\beta = \sum_j \sum_i c_i \alpha_j^i = \sum_i c_i \left[\sum_j \alpha_j^i \right], \qquad 0 \le i \le n-1, \quad 1 \le j \le t.$$

Now the coefficient of each c_i is a symmetric polynomial in α and its conjugates over K_1 and hence is in F_1. Hence $t\beta$ and β are in F_1. This completes the proof of the theorem.

This fundamental theorem holds for some fields with finite characteristic. You will find a proof for separable fields in the book by Birkhoff and Mac Lane listed in the Bibliography.

6. Constructibility of Regular Polygons

Theorem 3.1 stated that if a regular p-sided polygon can be constructed for p a prime number, then $p - 1$ is a power of 2. In this section we use Galois theory to prove the converse.

Theorem 6.1. If p is a prime number such that $p - 1$ is a power of 2, then a regular p-sided polygon can be constructed.

To prove this, we need to investigate the Galois group of the equation

$$g(x) = \frac{x^p - 1}{x - 1},$$ (6.1)

since, as we noted in Section 3 of Chapter II, the zeros of $g(x)$ are complex coordinates of the vertices of a regular polygon with p sides inscribed in the unit circle. Let r be a zero of $g(x)$. Then the elements of $\mathbf{Q}(r)$ are the linear combinations with rational coefficients of

$$1, r, r^2, \ldots, r^{p-1}.$$ (6.2)

Furthermore, from Theorem 13.7 in Chapter III, $g(x)$ is irreducible over $\mathbf{Q}$ and the powers of r are its conjugates. Since the zeros of $g(x)$ are the elements of (6.2) with 1 omitted, $\mathbf{Q}(r)$ is a normal extension of $\mathbf{Q}$ and is the splitting field of $g(x)$. Since by Theorem 4.3 every automorphism σ of $\mathbf{Q}(r)$ takes r into a conjugate, we know that $r\sigma = r^k$ for some k between 1 and $p - 1$ inclusive. Furthermore, since (6.2) span $\mathbf{Q}(r)$, we see that the image of r determines the automorphism. Hence we have shown the first sentence of the following theorem.

Theorem 6.2.   The Galois group G of $g(x)$ defined in (6.1) consists of the automorphisms σ_i defined by

$$r\sigma_i = r^i, \quad 1 \le i \le p - 1.$$ (6.3)

This Galois group is cyclic of order $p - 1$.

Proof: To show that G is cyclic, we first prove by mathematical induction that

$$r\sigma_i^k = r^{ik}, \quad 1 \le i \le p - 1.$$ (6.4)

If $k = 1$, this is (6.3). Now we assume (6.4) for k and show that it holds for $k + 1$. The following sequence accomplishes this:

$$r\sigma_i^{k+1} = (r\sigma_i^k)\sigma_i = r^{ik}\sigma_i = (r\sigma_i)^{ik} = (r^i)^{ik} = r^{ik+1}.$$

Now from Theorem 12.4 of Chapter III we know that for each prime p, the field $\mathbf{Z}_p$ of integers (mod p) is cyclic. That is, there is a number a whose powers (mod p) are the nonzero elements of $\mathbf{Z}_p$. To prove G cyclic, we now show that $\tau = \sigma_a$ is a generator of G. Let $\tau^0 = \epsilon$, the identity automorphism, and calculate, using (6.4), the following table of images of r under powers of τ:

$$r\tau^0 = r, \ r\tau = r^a, \ r\tau^2 = r^{a^2}, \ldots, r\tau^s = r^{a^s}, \ldots, r\tau^{p-2} = r^{a^{p-2}}.$$ (6.5)

By way of checking this, note that by Fermat's theorem $a^{p-1} \equiv 1 \pmod{p}$ and hence $r\tau^{p-1} = r$, which shows that $\tau^{p-1} = \tau^0 = \epsilon$. The powers of r in (6.5) are distinct, since the powers of a are distinct (mod p). This shows that the powers of τ are distinct and hence that τ generates G. This completes the proof of the theorem.

We have just shown that if $p - 1 = 2^u$, then G is a cyclic group of order 2^u (p being a prime). By Theorem 5.2 of Chapter II we know that for every w such that $1 \leq w \leq u$, G has a cyclic subgroup of order 2^w. In particular, if $w = u - 1$, G has a subgroup G_1 of order 2^{u-1}. Similarly, G_1 has a subgroup G_2 of order 2^{u-2}. Continuing, we see that there is a sequence of subgroups of G:

$$(6.6) \qquad G \supset G_1 \supset G_2 \supset \cdots \supset G_{u-1} \supset G_u = (\epsilon),$$

each of which is of index 2 in its predecessor. Hence if $K = \mathbf{Q}(r)$, the fundamental theorem of Galois theory affirms that there is a corresponding sequence of fields:

$$\mathbf{Q} \subset F_1 = \mathbf{Q}(x_1) \subset F_2 = F_1(x_2) \subset \cdots \subset K = F_{u-1}(x_u),$$

where each field is of degree 2 over its predecessor and every element of G_i leaves fixed every element of F_i. Thus K is obtained starting with $\mathbf{Q}$ by a finite number of quadratic extensions. We have proved Theorem 6.1.

Theorems 6.1 and 3.1 imply the following theorem:

Theorem 6.3. If p is a prime number, a regular polygon of p sides can be constructed if and only if

$$p = 2^{2^t} + 1$$

for some nonnegative integer t. (See the end of Section 3.)

To complete the listing of constructible polygons, we state and prove the following theorem.

Theorem 6.4. If regular polygons of r and s sides can be constructed, where r and s are relatively prime integers, then a regular polygon having rs sides can also be constructed.

Proof: The hypothesis shows that the angles $2\pi/r$ and $2\pi/s$ are constructible. From Corollary 2 of Theorem 10.2 in Chapter I, we know that there are integers x and y such that $rx + sy = 1$. Then

$$\left(\frac{2\pi}{r}\right)y + \left(\frac{2\pi}{s}\right)x = \frac{2\pi(sy + rx)}{rs} = \frac{2\pi}{rs}.$$

This shows that the conclusion stated is correct.

Recall that in Section 3 we showed a little more than the converse of Theorem 6.4, namely, that if a regular polygon of rs sides can be constructed, so can one of r sides and one of s sides; here r and s need not be relatively prime. You were asked to show in Exercise 17 after Section 3 that a regular polygon of 2^u sides can be constructed. Assuming, as stated but not proved in Section 3, that regular polygons of p^k sides cannot be constructed if p is an odd prime and $k > 1$, we have the following theorem.

Theorem 6.5. A regular polygon of n sides can be constructed if and only if

$$n = 2^u p_1 p_2 p_3 \cdots p_v,$$

where the p_i are distinct primes of the form $2^{2^t} + 1$, and u is a nonnegative integer.

As noted earlier, Theorem 6.1 can be established without the use of Galois theory by finding step-by-step subfields of index 2 in their predecessor, but the process is somewhat involved. Even for the 17-sided polygon dealt with in the next section, the construction of the intermediate quadratic fields would be difficult to motivate without Galois theory. The fact that Gauss did this is a tribute to his genius.

7. Construction of the Regular 17-sided Polygon

To illustrate these ideas, we now show how to calculate a sequence of quadratic extensions leading to the construction of the regular 17-sided polygon, somewhat as Gauss did before Galois was born, but using the above results to guide us on our way and bypass some of the calculations. It is rather fascinating to see how neatly the theory works out in this case. Since a little experimentation shows that 3 (but not 2) is a generator of $\mathbf{Z}_{17}$, we see from (6.5) that $\tau = \sigma_3$ is a generator of G and from (6.3) we have that

(7.1) $$r\tau = r^3,$$

where r is an imaginary 17th root of unity.

The first step is to construct a subgroup G_1 of index 2 in G and then a corresponding quadratic extension of $\mathbf{Q}$. Now the elements of G are the powers of τ from 0 to and including 16, since $\tau^{16} = \tau^0 = \epsilon$, and we have

$$G = \{\tau^i\}, \qquad 0 \le i \le 15.$$

So we choose G_1 to be the even powers of τ; that is,

$$G_1 = \{\epsilon, \tau^2, \tau^4, \tau^6, \tau^8, \tau^{10}, \tau^{12}, \tau^{14}\}.$$

We want to find an element s_1 of $K = \mathbf{Q}(r)$ left fixed by all elements of G_1. A way to accomplish this is to take s_1 to be the sum of the images of r under the elements (automorphisms) of G_1. Now by (6.4), $r\tau^2 = r\sigma_3^2 = r^9$ and

$$r(\tau^2)^k = r^{9^k}.$$

Hence we choose

$$s_1 = \sum_{k=0}^{7} r^{9k}.$$

A more satisfactory form of s_1 can be obtained by use of the following table of the powers of 9 (mod 17):

(7.2)

k	0	1	2	3	4	5	6	7
9^k (mod 17)	1	9	13	15	16	8	4	2

Thus we have

$$s_1 = r + r^9 + r^{13} + r^{15} + r^{16} + r^8 + r^4 + r^2.$$

Since $s_1\tau^2 = s_1$ it follows that s_1 is left fixed by all automorphisms of G_1. Now

$$s_1\tau = r^3 + r^{10} + r^5 + r^{11} + r^{14} + r^7 + r^{12} + r^6.$$

Define $s_1' = s_1\tau$ and see that $s_1 + s_1'$ is the sum of the powers of r between 1 and 16 inclusive. That is, $s_1 + s_1'$ is the sum of the zeros of $g(x) = (x^{17} - 1)/(x - 1)$. Since 1 is the coefficient of x^{15} in $g(x)$, we have

$$s_1 + s_1' = -1.$$

Now $(s_1 s_1')\tau = (s_1\tau)(s_1'\tau) = (s_1\tau)(s_1\tau^2) = s_1's_1$ shows that $s_1 s_1'$ is fixed by τ, hence by every element of G and thus is in **Q**. Therefore, s_1 and s_1' are the roots of the equation

(7.3) $$x^2 - (s_1 + s_1')x + s_1 s_1' = x^2 + x + s_1 s_1' = 0,$$

where $s_1 s_1'$ is in **Q**. (You will be asked in Exercise 8 to calculate $s_1 s_1'$, but such calculation is not necessary to show the existence of a quadratic extension.) Thus the field $F_1 = \mathbf{Q}(s_1)$ is of degree 2 over **Q** and consists of the elements of $K = \mathbf{Q}(r)$ left fixed by all elements of G_1. Also, we have the following correspondence between groups and fields:

$$G \supset G_1,$$
$$\mathbf{Q} \subset F_1 = \mathbf{Q}(s_1).$$

The second step is to define

(7.4) $$G_2 = \{\epsilon, \tau^4, \tau^8, \tau^{12}\}$$

and, looking at the table (7.2), for k even

$$s_2 = r + r^{13} + r^{16} + r^4,$$
$$s_2' = s_2\tau^2 = r^9 + r^{15} + r^8 + r^2.$$

It follows that $s_2\tau^4 = s_2$ and $s_2 + s_2' = s_1$, which is left fixed by every element of G_1. As in the first step,

$$(s_2 s_2')\tau^2 = (s_2\tau^2)(s_2'\tau^2) = (s_2\tau^2)(s_2\tau^4) = (s_2\tau^2)s_2 = s_2's_2$$

shows that $s_2 s_2'$ is left fixed by every element of G_1. Hence s_2 and s_2' are roots of

(7.5) $$x^2 - (s_2 + s_2')x + s_2 s_2' = 0,$$

whose coefficients, being left fixed by G_1, are in $F_1 = \mathbf{Q}(s_1)$. Furthermore, $F_2 = F_1(s_2)$ is a quadratic extension of F_1 and every element of F_2 is left fixed by all elements of G_2.

The third step is to define G_3 to be the following subgroup of G_2:

$$G_3 = \{\epsilon, \tau^8\}$$

and $s_3 = r + r^{16}$, $s_3' = s_3 \tau^4 = r^{13} + r^4$. Then, as above, s_3 and s_3' are roots of

(7.6) $$x^2 - (s_3 + s_3')x + s_3 s_3' = 0,$$

where $s_3 + s_3'$ and $s_3 s_3'$ are in F_2, being left fixed by elements of G_2. Furthermore, $F_3 = F_2(s_3)$ is a quadratic extension of F_2, and each element of F_3 is left fixed by all elements of G_3.

Finally, r is a root of

(7.7) $$x^2 - (r + r^{16})x + r \cdot r^{16} = x^2 - s_3 x + 1$$

with coefficients in F_3.

Thus we have shown that K, the root field of $g(x)$, can be obtained from $\mathbf{Q}$ by a sequence of four quadratic adjunctions, and hence the elements of K are constructible. In fact, by calculating the coefficients of the various quadratic equations, one could give an explicit construction as Gauss did of the regular 17-sided polygon.

Exercises

1. List all the values of n less than 50 for which a regular n-sided polygon can be constructed.

2. Suppose that $f(x)$ is an irreducible polynomial over $\mathbf{Q}$ whose Galois field G is cyclic and of order p^k, for p a prime number and k a positive integer. Prove that K, the root field of $f(x)$, can be obtained from $\mathbf{Q}$ by k successive adjunctions of roots of equations of degree p.

3. Suppose that there is an instrument I that can be used to construct the zeros of any cubic equation all of whose roots are real and whose coefficients are constructible by use of I a finite number of times. (See Exercise 22 after Section 3.) Prove that the roots of $f(x) = 0$ in Exercise 2 for $p = 3$ can be constructed with I provided that all the roots of $f(x) = 0$ are real.

4. Suppose that the root field K of a cubic equation over $\mathbf{Q}$ is of degree 6. Then the Galois group G of the cubic is of order 6. Can G be a cyclic group? Why or why not?

5. Show that $\lambda = \sigma_6$ is another generator of the Galois group G of Section 7. Define G_1 to be the subgroup of G that consists of all even powers of λ. Show that

$$r(\lambda^2)^k = r^{2k}$$

and, just as for τ, find a t_1 that is left fixed by every element of G_1. Show that $t_1 = s_1$.

6. Suppose that μ is any generator of the Galois group G in Section 7. Show that $s_1\mu = s_1$, where s_1 is as defined in Section 7. Thus show that the same sequence of quadratic equations is obtained independently of what generator of G is chosen.

7. Find a number x_4 such that $Q(x_4) = Q(1 + \sqrt{2}, 1 - \sqrt{2})$.

8. Calculate each of $s_1's_1$, $s_2's_2$, and $s_3's_3$ in Section 7 and hence find explicitly the quadratic equations that define the successive quadratic adjunctions.

9. Show how the proof of Theorem 5.2a can be shortened by use of Lemma 1 in Chapter V.

10. Show that each of the groups G_i in (6.6) is a normal subgroup of its predecessor. Would the same result hold if the Galois group were that described in Exercise 2?

8.   The Second Problem

Now that we have solved the problem on constructibility, we begin with the second problem of the chapter: solution of equations by radicals. By way of leading up to a formulation of the problem, note first that there is a formula for the solution of any quadratic equation once the coefficients are known. It is a linear combination with rational coefficients of the coefficients of the equation and the square root of such a combination. Phrasing this in terms of algebraic fields, this means, as we saw earlier in this chapter, that the roots of any quadratic equation with coefficients in a field lie in a quadratic extension of that field, in fact by adjoining a square root to the field. Analogously, one might expect that the roots of any cubic equation lie in a field obtained by adjoining to the field of coefficients a finite number of square roots and cube roots. For quartic equations (of the fourth degree) one would expect to need only extensions of square roots, cube roots, and fourth roots. Thus we would hope to have a formula for solutions that involve roots of various orders, of expressions in the coefficients of the original equation. This is what is meant by having solutions by radicals.

There is, however, a little difficulty in the above terminology. The symbol $\sqrt[3]{5}$, for instance, customarily does not mean just any solution of the equation

$x^3 = 5$ but the real number satisfying the equation. If we want to include other roots, we must talk in terms of the equation. Furthermore, an equation such as $x^6 = -3$ has, as we know from Section 13 of Chapter III, six different roots, none of which is real. We could specify the roots by trigonometric expressions, but even then there is some ambiguity, since one cannot distinguish one square root of -1 from the other except by saying that one is the negative of the other. Hence it is more precise to describe the problem in terms of solutions of equations of certain types, as in the following definition.

Definition.   *An equation with coefficients in a field F is said to be* solvable by radicals *over F if every solution is in a field K obtainable by a finite sequence of adjunctions of roots of equations of the type $x^n = a$. In notation this means that*

$$(8.1) \qquad F_1 = F(b_1), \qquad F_2 = F_1(b_2), \ldots, K = F_r(b_{r+1}),$$

where each b_{i+1} is a root of an equation $x^{n_i} = a_i$, a_i is in F_i, and $F_0 = F$. We say that F_i is obtainable from F_{i-1} by an adjunction of a radical.

Notice that this restriction is a much more natural one than that for constructibility by ruler and compass. To be sure, we are restricting the form that solutions may take, but without such a restriction we might just as well adjoin to F the solutions of the equation itself. This would beg the whole question.

To phrase the problem in terms of the definition, it is to find what equations are solvable by radicals. We shall see that every cubic and quartic equation is solvable by radicals. It was one of the surprises of nineteenth-century mathematics, as Abel and Galois found, that there are quintic equations that are not solvable by radicals. This we show in a later section after dealing with the cubic and quartic equations.

We need to keep track of what kinds of fields we are dealing with. Instead of making any basic restriction throughout the chapter, we shall be careful to specify any condition on the field under consideration. Wherever we use the fundamental theorem of Galois theory we must restrict the field to have characteristic zero. This is no great disadvantage since our applications are to such fields. At times we shall want to restrict the field further.

9.   The Galois Group of the Cubic Equation

Before proceeding to the solution of the general cubic equation, we first apply part of what we have learned about Galois theory to get some information about roots of cubic equations with coefficients in a field F of characteristic zero. It is convenient to begin by simplifying a little the form of the cubic equation $f(x) = ax^3 + bx^2 + cx + d = 0$, where a, b, c, and d are in a field F. Let $x = y + h$ and see that in $f(y + h)$ the coefficient of y^2 is $3ah + b$. If we choose $h = -b/3a$ and then divide by a, $f(x)$ becomes

(9.1) $$y^3 + py + q = 0 \quad \text{for some } p \text{ and } q \text{ in } F.$$

We assume that the polynomial of (9.1) is irreducible over F, since otherwise it would have a zero in F and the other two roots would be roots of a quadratic equation with coefficients in F. Let G be the Galois group of (9.1) and θ_1, θ_2, and θ_3 be the roots with $\theta_1 = \theta$. From the statement before Theorem 5.3, the group G is a subgroup of S_3 and hence its order is a factor of 3!. Hence the order of G is 2, 3, or 6. Now the fundamental theorem of Galois theory informs us that the order of the Galois group G is equal to the degree of the root field K over F. We give two proofs of the following theorem, the first of which uses the fundamental theorem and the second of which shows more of the mechanism of what is happening.

Theorem 9.1. The Galois group G of (9.1) cannot be of order 2.

Proof I: As noted above, if G is of order 2, the root field is of degree 2 over F. This means that θ satisfies a quadratic equation over F, which contradicts our assumption that it is the zero of an irreducible cubic polynomial.

Proof II: Suppose that G is of order 2 and $G = \{\epsilon, \sigma\}$. Then by Theorem 4.3, σ takes θ into another root of (9.1). So we may assume that $(\theta)\sigma = \theta_2$. If G were of order 2, σ^2 would be ϵ and $\theta_2\sigma$ would be θ. But then $(\theta_2 + \theta)\sigma = \theta + \theta_2$, which implies that $\theta + \theta_2$ is in F. In this case $\theta_3 = -\theta_2 - \theta$ is in F, contrary to our assumption that the cubic is irreducible over F. The proof is complete.

Now let us find necessary and sufficient conditions that the Galois group be of order 3. This will be true if and only if the root field K is of degree 3 over F, that is, if and only if $K = F(\theta)$. In this case, from Section 6 of Chapter II and Exercise 2 after Section 9 of Chapter I, $G = A_3$, the alternating group on three symbols. It would be enlightening to find in terms of the coefficients a condition that $G = A_3$, that is, that $F(\theta)$ be the root field. This could be done by brute force, but a flanking attack is much easier, as well as being more enlightening.

So suppose that A_3 is the Galois group of the cubic equation. We saw in the proof of Theorem 6.2 in Chapter II that A_3 leaves unchanged the expression

(9.2) $$(\theta_1 - \theta_2)(\theta_2 - \theta_3)(\theta_3 - \theta_1) = d.$$

Since only the elements of F are left unchanged by all elements of the Galois group, it follows that if A_3 is the Galois group, then d must be in F. We have proved half of the following theorem.

Theorem 9.2.   The Galois group of the cubic (9.1) is of order 3, that is, is the alternating group A_3, if and only if d as defined in (9.2) is in F.

Proof: It remains to show that if d is in F, then A_3 is the Galois group of (9.1). Now A_3 leaves all elements of F fixed, since it is a subgroup of S_3. We need to show that only the elements of A_3 leave fixed the elements of F. The transposition (12) replaces d by $-d$ and hence, since d is not zero, it is an element of F not fixed by (12). The same holds for the other odd permutations (13) and (23). This completes the proof.

Corollary.   The cubic polynomial (9.1) is normal (see Section 5) if and only if d in (9.2) is in F.

We now show the corollary directly without use of Galois theory on the way toward calculating d. Notice that $\theta_1 + \theta_2 + \theta_3 = 0$. Taking $\theta = \theta_1$, we have

$$d = (\theta_2 - \theta_3)(\theta\theta_2 + \theta\theta_3 - \theta^2 - \theta_2\theta_3) = (\theta_2 - \theta_3)(-2\theta^2 - \theta_2\theta_3).$$

Since $\theta_2\theta_3 = -q/\theta$, the above implies that

$$(9.3) \qquad\qquad \theta_2 - \theta_3 = \frac{d}{-2\theta^2 + q/\theta}.$$

Then $\theta_2 - \theta_3$ is in $F(d, \theta)$, and since $\theta_2 + \theta_3 = -\theta$, it follows that θ_2 and θ_3 are both in $F(d, \theta)$. This shows that if d is in F, then θ_2 and θ_3 are in $F(\theta)$. This proves directly the corollary above.

Now, d^2 is a symmetric function of the roots of (9.1) and hence by the fundamental theorem on symmetric functions, we should be able to express it in terms of p and q, the coefficients of (9.1); in fact, d^2 must be a polynomial in p and q with integer coefficients. To find this polynomial, note first that since θ is a root of (9.1), we can write the right side of (9.3) as $d\theta/(2p\theta + 3q)$, and thus we have

$$d - (\theta - \theta_2)(\theta_2 - \theta_0)(\theta_0 - \theta) = \frac{-d^3q}{(2p\theta + 3q)(2p\theta_2 + 3q)(2p\theta_3 + 3q)}$$

and

$$D = d^2 = -\frac{(2p\theta + 3q)(2p\theta_2 + 3q)(2p\theta_3 + 3q)}{q}$$

$$(9.4) \qquad\qquad = -\frac{8p^3(-q) + 27q^3 + 4p^2(3q)p + 18pq^2 \cdot 0}{q}$$

$$= -(4p^3 + 27q^2).$$

Note that the above computations do not depend on whether the polynomial $f(x)$ in (9.1) is irreducible or not.

The number D in (9.4) is called the *discriminant* of the cubic $y^3 + py + q$. It has several properties in common with the discriminant of the quadratic equation. They are given in the following theorem, in which the field of coefficients is restricted to be a subfield of the field of real numbers.

Theorem 9.3. Let θ_1, θ_2, and θ_3 be the zeros of $y^3 + py + q$, where p and q are real numbers. Then the discriminant D in (9.4) has the following properties:

1. If two of the θ_i are equal, then $D = 0$.
2. If the roots are all real and distinct, then $D > 0$.
3. If two of the roots are imaginary, then $D < 0$.

Proof: The first two properties follow from $D = d^2$, where d is given by (9.2). Suppose that two of the roots are imaginary. Let $\theta_1 = \theta$ be real and $\theta_2 = a + bi$, where a and b are real with b not zero. Then $\theta_3 = a - bi$ and $\theta_2 - \theta_3 = 2bi$. Thus (9.3) shows that $d = ir$ for r a nonzero real number. Hence $d^2 = D < 0$. This completes the proof.

Since the roots of any cubic equation with real coefficients must be in one of the three categories of the theorem, the corollaries below follow.

Corollary 1. The converse of each of the three statements in the above theorem is true.

Corollary 2. If F is a subfield of $\mathbf{R}$, an irreducible cubic equation over F has a Galois group of order 6 unless all the zeros are real and $\sqrt{D}$ is in F.

Formula (9.4) for the discriminant together with Theorem 9.2 gives us a simple means of determining when the Galois group of (9.1) is A_3. The following is just a restatement of Theorem 9.2.

Theorem 9.2a. The Galois group of (9.1) is A_3 if and only if the discriminant given by (9.4) is the square of an element of F.

For instance, if $f(x) = x^3 - 3x - 1$, then $D = 81$ and hence A_3 is the Galois group of $f(x)$ and $f(x)$ is normal. But for $g(x) = x^3 - 6x - 1$, $D = 837$, which shows that all the roots are real but that S_3 is the Galois group and $g(x)$ is not normal.

To get a better understanding of what the Galois group of a cubic equation is, let us look at the above from a somewhat different point of view. Let σ be an automorphism of the root field K of $f(x)$. Thus $K = F(\theta_1, \theta_2, \theta_3) = F(\theta_1, \theta_2)$. Now σ must take θ_1 into one of its conjugates; that is, $\theta_1\sigma = \theta_1$, θ_2, or θ_3. If θ_2 and θ_3 are in $F(\theta_1)$, the image of θ_1 determines the automorphism, and hence there are just three automorphisms of K that leave fixed each element of F. However, if θ_2 is not in $F(\theta_1)$, since $K = F(\theta_1, \theta_2)$ is of degree 6 over F, θ_2 must be of degree $\frac{6}{3} = 2$ over $F(\theta_1)$. For each of the three possibilities for $(\theta_1)\sigma$, there must be two possibilities for $(\theta_2)\sigma$, since 6 is the order of the Galois group.

10.   The Solution of the Cubic Equation

One way to solve a cubic equation is to make a substitution used by La Vieta in 1591, namely, $y = z + k/z$, where k is chosen to effect a simplication. Then Equation (9.1) becomes

(10.1) $$z^3 + 3zk + \frac{3k^2}{z} + \frac{k^3}{z^3} + pz + \frac{pk}{z} + q = 0.$$

As luck would have it, by choosing k so that $3k + p = 0$, we not only eliminate the term in z but that in $1/z$ as well. Then by the substitution $y = z - p/3z$, the cubic $y^3 + py + q$ becomes

(10.2) $$z^3 - \frac{p^3}{27z^3} + q = 0,$$

which is a quadratic equation in z^3. Thus

(10.3) $$z^3 = \frac{-q \pm \sqrt{-D/27}}{2},$$

where D is determined by (9.4).

Since $z^3 = t$ has three different solutions for each t not zero, Equation (10.3) determines six different z's, three for each of the signs $\pm$. But for each z satisfying (10.2),

(10.4) $$y = z - \frac{p}{3z}$$

gives a solution of (9.1). Since there are three roots of (9.1), it must be possible to pair the solutions of (10.3) so that the two z's of each pair give the same y. In fact, if

(10.5) $$z_1^3 = \frac{-q + \sqrt{-D/27}}{2} \quad \text{and} \quad z_2^3 = \frac{-q - \sqrt{-D/27}}{2},$$

it is true, as you are asked to show in Exercise 14 after Section 12, that

$$(z_1 z_2)^3 = -\frac{p^3}{27},$$

and hence $z_1 z_2 = (-p/3)\lambda$, where $\lambda^3 = 1$. Replacing z_2 by one of z_2, ωz_2, or $\omega^2 z_2$, where ω is an imaginary cube root of 1, we may assume $z_1 z_2 = -p/3$

without changing (10.5). Then it follows, as you are asked to show, that the roots of the cubic are given by

$$(10.6) \qquad y_1 = z_1 + z_2, \qquad y_2 = \omega z_1 + \omega^2 z_2, \qquad y_3 = \omega^2 z_1 + \omega z_2,$$

where ω is an imaginary root of $x^3 = 1$. The formulas (10.6) are called Cardan's formulas, since he published them in 1545, although he had obtained them previously under a promise of secrecy from Tartaglia.

Now, (10.3) and (10.4) show that the zeros of the cubic are expressible by radicals. That is, each zero of the cubic is in a field obtained from F by first adjoining $\sqrt{-D/27}$ and second, adjoining a cube root of the right side of (10.3). If $F \subseteq \mathbf{R}$ and $D < 0$, that is, if two roots are imaginary, one pair of roots z of (10.3) is real and the other two pairs are imaginary. If $D > 0$, that is, if all the roots are real, we still have a solution by radicals, but here, strangely enough, the *real* roots are expressed in terms of *imaginary* radicals. This strange phenomenon bothered the early mathematicians and hence was dubbed "the irreducible case." Although this is somewhat aside from the path that we have chosen to follow, we deal with this case in the next section.

11.   Cubic Equations with Three Real Roots

Let us formulate more carefully the situation in the irreducible case. Let F be any subfield of $\mathbf{R}$, the field of real numbers, and let $f(y) = y^3 + py + q$ be irreducible over F, where p and q are in F. Assume that all the zeros of $f(y)$ are real. Let K be some field obtained from F by adjoining successively a finite number of *real* radicals. More precisely, referring back to the notation in the definition of solvability by radicals, we require each b_i in (8.1) to be real. We show in the proof of the following theorem that the zeros of $f(y)$ cannot be in such a field K.

Theorem 11.1.   Let $f(y)$, F, and K be defined as in the previous paragraph. Then no zero of $f(y)$ lies in such a field K.

Proof: Let the sequence of fields leading to K be represented as follows:

$$(11.1) \qquad\qquad F \subset F_1 \subset F_2 \subset \cdots \subset F_{i+1} - K,$$

where each field F_i is obtained from its predecessor by adjoining a real radical, that is, a real root of an equation of the form $x^n = a_i$ with a_i in F_{i-1}. We begin by readjusting the sequence to more manageable form.

First, it turns out to be convenient to consider all the roots of the cubic instead of just one. We know from the corollary of Theorem 9.2 that if D is the discriminant of $f(y)$ and θ is one zero, all the zeros of $f(y)$ are in $F(\theta, \sqrt{D})$. If K in (11.1) does not contain $\sqrt{D}$, we can let $K' = K(\sqrt{D})$ and have a sequence (11.1) in which the last field contains $\sqrt{D}$. Then we can rearrange the sequence so that $F_1 = F(\sqrt{D})$, making whatever changes are needed in the rest of the sequence but still keeping the adjunctions to be by real radicals.

Second, we adjust the sequence (11.1) so that each field is of prime degree over its predecessor. Suppose we are adjoining a real root ρ of $x^{rs} = a$. Then $(\rho^r)^s = a$ shows that we can accomplish the adjunction in two steps by first adjoining a real root σ of $x^s = a$ and then a real root of $x^r = \sigma$.

Now, after these rearrangements, we may assume (11.1) with $F_1 = F(\sqrt{D})$ and that every field in the sequence (11.1) is of prime degree in its predecessor. Now suppose that θ, a zero of $f(y)$, is in K. We can without loss assume that θ is not in F_r, since otherwise we could stop the sequence at F_r or earlier. Now $f(y)$ is irreducible over F_r, since otherwise, being a cubic, it would have a zero in F_r. But since F_r contains $\sqrt{D}$, it contains all the zeros of $f(y)$, including θ. That is, K is normal over F_r. On the other hand, K is obtained from F_r by adjoining the real root of an equation $x^3 = a$. Since K is normal over F_r, it follows from Theorem 5.2a that it must contain all the roots of $x^3 = a$. But this is impossible, since all elements of K are real and two roots of $x^3 = a$ are imaginary. This is the contradiction that proves the theorem.

Although cubic equations with three real roots seem untractable from the above point of view, there is a well-known method of solution of such cubics using trigonometric functions. It makes use of the following formula, which we used in Section 3:

$$\cos(3\theta) = 4\cos^3 \theta - 3\cos \theta.$$

This tells us that the roots of

$$(11.2) \qquad\qquad 4x^3 - 3x - \cos(3\theta) = 0$$

are $\cos \theta$, $\cos(\theta + 2\pi/3)$, and $\cos(\theta + 4\pi/3)$. We want to fit Equation (11.2) to

$$(11.3) \qquad\qquad y^3 + py + q = 0.$$

This cannot be done directly, since p may not be $-\tfrac{3}{4}$. Therefore, we replace y by x/r and choose r and θ so that the equation does fit. Now the substitution $y = x/r$ takes (11.3) into $x^3 + pr^2x + qr^3 = 0$. Hence we must choose r and θ so that $pr^2 = -\tfrac{3}{4}$ and $qr^3 = -\cos 3\theta/4$, that is,

$$(11.4) \qquad\qquad p = -\frac{3}{4r^2} \quad \text{and} \quad q = -\frac{\cos 3\theta}{4r^3}.$$

The first equation of (11.4) is equivalent to $r = \pm\sqrt{-\tfrac{3}{4}p}$. Since the discriminant of (11.3) is positive, (9.4) shows that p must be negative. Thus r, determined in accordance with (11.4), is real. The second equation of (11.4) is equivalent to $\cos(3\theta) = -4r^3q$, where $-4r^3q$ is real. To prove that the angle θ is real, we need to show that $4r^3q$ is in absolute value not greater than 1. Thus we have

$$16r^6q^2 = \frac{16(27q^2)}{64(-p)^3} = \frac{27q^2}{-4p^3} < 1,$$

since $-D = 4p^3 + 27q^2 < 0$. Thus we have the following theorem.

Theorem 11.2. If the roots of the cubic equation $y^3 + py + q = 0$ are all real, they are $r^{-1}\cos(\theta)$, $r^{-1}\cos(\theta + 2\pi/3)$, and $r^{-1}\cos(\theta + 4\pi/3)$, where

$$r = \sqrt{\frac{-3}{4p}} \quad\text{and}\quad \cos(3\theta) = -4r^3q.$$

(Note that if the roots are all real, $p < 0$.)

12.   The Quartic Equation

In this section we show that every quartic equation over a field F is solvable by radicals, leaving the application of Galois theory to Section 13. To begin with, by a substitution similar to that for the cubic, we can reduce the quartic equation to

(12.1) $$y^4 + py^2 + qy + r = 0.$$

One way to solve this equation is, following Descartes,[†] to choose u, v, and w so that (12.1) takes the form

(12.2) $$\left(y^2 + \frac{u}{2}\right)^2 - (vy + w)^2 = 0,$$

since then the expression is factorable. If we equate coefficients of (12.2) to those of (12.1), we have

(12.3) $$p = u - v^2, \qquad q = -2vw, \qquad r = \frac{u^2}{4} - w^2.$$

We use the first two equations of (12.3) to get $u = p + v^2$ and $w = -q/2v$ and, when we substitute these into the third equation of (12.3), we have

(12.4) $$v^6 + 2pv^4 + (p^2 - 4r)v^2 - q^2 = 0.$$

This is a cubic in v^2 which is thus a root of

(12.5) $$x^3 + 2px^2 + (p^2 - 4r)x - q^2 = 0.$$

† René Descartes (1596–1650) is chiefly known among mathematicians as the founder of analytic geometry. Cartesian coordinates are named after him.

If we adjoin to the field F of coefficients of (12.1) a root of the cubic equation (12.5) and then the square root of this root, we can find v, w, and u in a field K of degree 6 over F. Then (12.2) leads to two quadratic equations over K. If the discriminants of these equations are squares of elements of K, we know that every root of (12.1) is in the field K. If the discriminant of one of the quadratics is not a square in K, a further quadratic extension is necessary. In view of the solution of the cubic equation, we have proved the following theorem.

Theorem 12.1. Every quartic equation with coefficients in a field F is solvable by radicals over F. In fact, each root of (12.1) is an element of a field K' obtained from F by at most an adjunction of a square root, then a cube root, and then two square roots.

Although the above theorem is the principal goal of this section, it is informative to find an explicit expression for the zeros of the quartic in terms of the zeros of the cubic (12.5), which is called the *resolvent cubic* of (12.1). We shall see that the resolvent cubic and the quartic have, conveniently, the same discriminant. Now the solutions of (12.1) are the solutions of (12.2) for v a root of (12.4), when u and w are chosen in accordance with (12.3). Let y_1, y_2, y_3, and y_4 be the roots of the quartic. From the form of the equation (12.1) it is clear that the sum of the y's is zero. Furthermore, (12.2) is equivalent to the following pair of equations:

$$(12.6) \qquad y^2 + \frac{u}{2} - (vy + w) = 0, \qquad y^2 + \frac{u}{2} + (vy + w) = 0.$$

First assume that the resolvent cubic has three distinct roots and let v_1^2, v_2^2. and v_3^2 be the roots of (12.5). Then (12.6) shows that each v_i is the sum of two y_j. So, reordering the y's and changing the sign of v_1 if necessary, we take

$$v_1 = y_1 + y_2 = -(y_3 + y_4).$$

Then since v_2 is neither v_1 nor $-v_1$, we may by changing the sign of v_2 and interchanging y_3 and y_4 if necessary take

$$v_2 = y_1 + y_3 = -(y_2 + y_4).$$

Finally, by changing the sign of v_3 if necessary we take

$$v_3 = y_1 + y_4 = -(y_2 + y_3).$$

Hence

$$\begin{aligned}
v_1 - v_2 &= y_2 - y_3, & v_1 + v_2 &= y_1 - y_4, \\
v_2 - v_3 &= y_3 - y_4, & v_2 + v_3 &= y_1 - y_2, \\
v_3 - v_1 &= y_4 - y_2, & v_3 + v_1 &= y_1 - y_3.
\end{aligned}$$

This implies that

$$(v_1^2 - v_2^2)(v_2^2 - v_3^2)(v_3^2 - v_1^2) = d$$
$$= (y_2 - y_3)(y_1 - y_4)(y_3 - y_4)(y_1 - y_2)$$
$$\cdot (y_4 - y_2)(y_1 - y_3)$$

and d^2 is the discriminant of (12.5) and also of (12.1).

The above equations can be used to get the following expressions for y_i in terms of the v_j:

(12.7)
$$\begin{aligned}
v_1 + v_2 + v_3 &= 3y_1 + y_2 + y_3 + y_4 = 2y_1, \\
v_1 - v_2 - v_3 &= 2y_2, \\
-v_1 + v_2 - v_3 &= 2y_3, \\
-v_1 - v_2 + v_3 &= 2y_4.
\end{aligned}$$

(You are asked to show this in Exercise 15.)

Now we must consider the case when the resolvent cubic has a repeated root. To show that the discriminants of the quartic and its resolvent cubic are equal in this case, we must show that if one is zero, the other must be also. Suppose that there are three distinct roots of the resolvent cubic; then its discriminant is not zero and hence the discriminant of the quartic, being the same, is not zero, which shows that the quartic has four distinct roots. Suppose that the quartic has four distinct roots. From equations (12.6) and the previous discussion, the sum of every pair of roots of the quartic is a root of (12.4). Now if no two of y_1, y_2, y_3, and y_4 are equal, no two of $y_1 + y_2$, $y_1 + y_3$, and $y_1 + y_4$ are equal. Furthermore, no one of the three can be the negative of any of the others, for, for instance, $y_1 + y_2 = -y_1 - y_3$ would imply that $y_1 + y_2 = y_2 + y_4$, which is impossible. Hence there are three distinct roots v_i^2 of the resolvent cubic. So we have shown that the discriminant of the quartic is not zero if and only if the discriminant of the resolvent cubic is not zero; this means that they are both zero or both not.

The equations (12.7) can be verified directly, independently of whether the discriminants are zero. This you will be asked to do in Exercise 15. [Incidentally, the equations giving each v_i as a sum of two y_j can be recovered from (12.7).] Thus we have shown the following theorem.

Theorem 12.2. The discriminant of a quartic equation and its resolvent cubic are equal. If v_i^2 are the zeros of the resolvent cubic, for proper choice of sign of the v_i, the equations (12.7) give the roots of the quartic.

Combining the results of previous sections and Theorem 12.1, we have the following theorem.

Theorem 12.3. Every equation of degree less than 5 is solvable by radicals.

We shall be devoting the rest of the chapter chiefly to showing that there are polynomial equations of degree 5 that are not solvable by radicals over the field of coefficients.

Exercises

1. Let θ_1 and θ_2 be the zeros of $ax^2 + bx + c$. Find $(\theta_1 - \theta_2)^2$ in terms of a, b, and c.

2. How are the solutions of (10.3) paired to give solutions of (9.1)?

3. Show that if $D < 0$, exactly one root of (10.3) is real. If $D < 0$ and z as defined by (10.3) is imaginary, can y in (10.4) be real? Why or why not?

4. Find by radicals the roots of $x^3 - 3x + 1 = 0$ and $x^3 - 7x + 7 = 0$. If θ is one root, express the others as rational functions of θ.

5. Show that if in a cubic equation in x, x is replaced by $x - k$ for some k, then the discriminant is not changed.

6. Is Theorem 9.3 valid if the cubic equation is $x^3 + ax^2 + bx + c = 0$, where $a \neq 0$, $D = d^2$, and d is defined as in (9.2)? Why?

7. Solve the cubics $x^3 + 6x + 2 = 0$ and $x^3 + 9x + 6 = 0$.

8. Find the trigonometric solutions of each of the following.
   (a) $x^3 - 3x + 1 = 0$.   (b) $x^3 - 7x + \frac{7}{3} = 0$.
   (c) $x^3 - 7x + \frac{19}{3} = 0$.

9. Suppose that $\cos(3\theta)$ is in **Q**. Find a necessary and sufficient condition on $\cos(3\theta)$ that $\cos(\theta + 2\pi/3)$ be expressible as a linear combination of 1, $\cos \theta$ and $\cos^2 \theta$, with rational coefficients.

10. Find the substitution that reduces a quartic equation to the form (12.1).

11. Find the resolvent cubic of the quartic: $y^4 + \frac{3}{4}y + 1$.

12. Let $f(x) = x^4 + 2x^2 + 2x + \frac{5}{4}$. Find its resolvent cubic $g(t)$ and show that it has three rational zeros. Does $f(x)$ have any rational zeros?

13. If $p = 0$, Equation (10.3) becomes $z^3 = (-q \pm q)/2$, which seems to indicate that $z = 0$ is a possibility. But this seems to be impossible if $q \neq 0$. What is wrong?

14. Show that if z_1 and z_2 are defined as in (10.5), then

$$(z_1 z_2)^3 = \frac{-p^3}{27}.$$

15. Derive the equations (12.7).

16. Derive Cardan's formulas (10.6).

17. Let $f(x)$ be a polynomial of degree n whose zeros are θ_i for $1 \le i \le n$. The *discriminant* of $f(x)$ is defined to be

$$D = \prod (\theta_i - \theta_j)^2,$$

where the product is over $1 \le i < j \le n$. Show that D is a polynomial in the coefficients of $f(x)$ with rational coefficients. Show that $D = 0$ if and only if two θ_i are equal. Show that if all the roots of $f(x) = 0$ are real, then D is positive.

18. Show that if in Exercise 17 each θ_i is replaced by $\theta_i - k$ for some k, the discriminant is not changed.

19. Prove that if θ, $1/(1 - \theta)$, and $(\theta - 1)/\theta$ are three roots of a cubic equation, it must be of the form $x^3 + cx^2 - (c + 3)x + 1 = 0$, and that, conversely, every cubic equation of this form has three roots related as given. (See Section 5 of Chapter I.) Why is such an equation normal?

20. Suppose that the quartic (12.1) has real coefficients and no zero root. If its discriminant D is positive, show that the v_i^2 are all real and either (a) all v_i^2 are positive or (b) one is positive and the other two are negative. In case (a) show that all the roots of the quartic are real. In case (b) show that all the roots are imaginary. Show that if the discriminant of the quartic is negative, two roots are imaginary and two real.

21. Find a condition on the coefficients of the quartic that distinguishes between cases (a) and (b) in Exercise 20.

22. Show that if the zeros of a quartic real polynomial are all real, then its discriminant is positive.

23. Use the construction described in Exercise 22 after Section 3 to show the roots of any cubic equation whose coefficients are rational numbers and whose roots are all real can be constructed with a marked straightedge and compass.

13.   Normal Extensions and Groups for Cubic and Quartic Equations

In the light of Galois theory let us look at the solutions of cubic and quartic equations over a field F_0 of characteristic zero. First, we showed in Theorem 9.2 that the Galois group of an irreducible cubic over F_0 is A_3, the alternating group, if and only if the discriminant is the square of an element of F_0. In this case we have

$$G = A_3 \supset (\epsilon),$$
$$F_0 \subset K = F_0(\theta),$$

where θ is a root of the cubic. The extension is normal but K is not necessarily an extension by radicals; that is, the cube of θ may not be in F_0. We want the correspondence to be for extensions by radicals. Now notice that $\sqrt{-D/27} = \sqrt{D}\sqrt{-3}/9$, and hence if instead of F_0 we take $F = F_0(\sqrt{-3})$, then (10.3) shows that if $\sqrt{D}$ is in F, then the right side of the following equation is in F:

$$(13.1) \qquad z^3 = \frac{-q \pm \sqrt{-D/27}}{2}.$$

Thus, corresponding to A_3 and (ϵ), we have $F = F_0(\sqrt{-3})$ and $K = F(z_1)$, respectively, where z_1 is a root of (13.1). Note that $y = z - p/3z$ shows that at least one root of the cubic is in $F(z_1)$. Furthermore, since $F_0(\sqrt{-3}) = F_0(\omega)$ for ω an imaginary cube root of 1, every root of (13.1) is in $F(z_1)$, hence every root of the cubic is in $F(z_1)$; that is, $F(z_1)$ is the root field of the cubic. Hence $F(z_1)$ is normal over F. Thus by taking F to be $F_0(\omega)$, we have accomplished two results: (1) We have a correspondence between the groups and the fields by adjunction of radicals, and (2) Equation (13.1), and hence the cubic, is normal over F.

Second, consider the case of a cubic whose Galois group has order 6. We take our cue from the previous paragraph and instead of F_0 start with $F = F_0(\omega)$. Then we have the sequences

$$G \quad \supset \quad A_3 \quad \supset \quad (\epsilon)$$
$$F_0(\omega) = F \subset F_1 = F(\sqrt{-D/27}) = F(\sqrt{D}) \subset K = F_1(z_1).$$

F_1 is obtained from F by the adjunction of a root of an equation $x^2 = b_1$, where b_1 is in F, and K is obtained from F_1 by the adjunction of a root of an equation $x^3 = b_2$, where b_2 is in F_1.

In both cases we have adapted the Galois correspondence between groups and fields to apply to a sequence of normal adjunctions by radicals.

Now consider the quartic equation using the notations of Section 12. Then for solutions by radicals we start with F_0 a field of characteristic zero that contains the coefficients of the quartic and $F = F_0(\omega, i)$; that is, F is obtained from F_0 by adjoining a cube root of 1 and a fourth root of 1. We have the following sequence of fields:

$$(13.2) \qquad F \subseteq F_1 = F(\sqrt{D}) \subseteq F_2 = F_1(x_1) \subseteq F_3 = F_2(v_1) \subseteq F_4$$
$$= F_3(v_2) \subseteq F_5 = F_4(v_3) = K,$$

where x_1 is a root of the resolvent cubic and v_1^2, v_2^2, and v_3^2 are the roots of the resolvent cubic. We shall prove the following theorem.

Theorem 13.1. In the sequence (13.2), K is the root field. Also, $F_1 \neq F_2$ unless the resolvent cubic is reducible. Moreover, K is a quadratic extension of F_3 unless $K = F_3$. Each field is a normal extension of its predecessor and the extension is by a root of an equation of the form $x^2 = b_1$ or $x^3 = b_2$.

Proof: Let K_0 be the root field of the quartic equation. Equations (12.7) show that $K \subseteq K_0$. But previous to (12.7) we showed that each v_i is a sum of roots of the quartic and hence $K_0 \subseteq K$. This shows that $K_0 = K$. Also if $F_1 = F_2$, it follows that F_2 is a quadratic extension of F. This can happen only if the resolvent cubic is reducible.

To show that K is at most a quadratic extension of F_3, look at one of the equations (12.6), namely,

$$(13.3) \qquad y^2 + \frac{u}{2} - (vy + w) = 0$$

and take $v = v_1$. The discriminant of (13.3) is

$$e = v_1^2 - 2u + 4w,$$

which is in $F_3 = F_2(v_1)$ from (12.3). Hence two roots of the quartic, say y_1 and y_2, are in $F_3(\sqrt{e})$. To show that $F_3(\sqrt{e}) = K$, we write

$$(13.4) \quad \pm\sqrt{D} = (y_1 - y_2)[y_1^2 - (y_3 + y_4)y_1 + y_3 y_4]$$
$$\cdot [y_2^2 - (y_3 + y_4)y_2 + y_3 y_4](y_3 - y_4).$$

We know that y_1 and y_2 are in $F_3(\sqrt{e})$. Also $y_3 + y_4$ and $y_3 y_4$ are in $F_3(\sqrt{e})$ since $y_3 + y_4 = -v_1$ and $y_3 y_4 = u/2 + w = (p + v_1^2)/2 - q/2v_1$ from (12.3). Hence (13.4) shows that $y_3 - y_4$ is in $F_3(\sqrt{e})$ as well as $y_3 + y_4$. Hence y_3 and y_4 are in $F_3(\sqrt{e})$ and we have completed the proof.

We have in general the following companion sequences:

$$(13.5) \quad \begin{array}{ccccccccc} & 2 & & 3 & & 2 & & 2 & \\ G \supseteq & & G_1 & \supseteq & G_2 & \supseteq & G_3 & \supseteq & (\epsilon) \\ F \subseteq F_1 = & F(\sqrt{D}) & \subseteq F_2 = & F_1(x_1) & \subseteq F_3 = & F_2(v_1) & \subseteq K = & F_3(\sqrt{e}) \end{array}$$

with x_1 a zero of the resolvent cubic and $v_1 = \pm\sqrt{x_1}$. In (13.5) two fields are equal if and only if the corresponding groups are equal. When inequality holds, the index is as indicated. Also, $G_1 = G_2$ if and only if the resolvent cubic is reducible over F. This enables us to construct an example, promised at the end of Section 2, of a quartic equation whose roots are not contructible. First, we state a theorem and then give a specific example.

Theorem 13.2. The Galois group of a quartic equation has a power of 2 as its order if and only if its resolvent cubic is reducible over F.

Proof: Since all the indices in (13.5) are 2 except that of G_2 in G_1, it follows that the order of G is a power of 2 if and only if $G_1 = G_2$; that is, the resolvent cubic is reducible. This completes the proof.

From this theorem it follows that to give an example of a quartic equation whose roots are not constructible, it is only necessary to find a quartic polynomial whose resolvent cubic is irreducible. One then starts with any irreducible cubic $f(x)$ and by means of equations (12.3) finds a quartic whose resolvent cubic is $f(x)$. For example, take $f(x) = x^3 - 3x - 1$. As noted at the end of Section 9, the discriminant of $f(x)$ is $D = 81$. Reference to (12.4) shows that the quartic $g(y) = y^4 + py^2 + qy + r$ has $f(x)$ as its resolvent cubic if and only if

$$2p = 0, \qquad p^2 - 4r = -3, \qquad \text{and} \qquad -1 = -q^2.$$

If we choose q to be -1, we see that

$$g(y) = y^4 - y + \tfrac{3}{4}$$

has $f(x)$ as its resolvent cubic. Using Section 10 and Cardan's formulas (10.6), we find that one root of $f(x) = 0$ is $x_1 = z_1 + z_2$, where $z_1^3 = (1 + \sqrt{-3})/2$, $z_2^3 = (1 - \sqrt{-3})/2$, and the choice of z_1 and z_2 is made so that $z_1 z_2 = 1$. Now, using the trigonometric form, we have

$$z_1^3 = \cos\frac{\pi}{3} + i\sin\frac{\pi}{3},$$

and a permissible choice of z_1 and z_2 is

$$z_1 = \cos\frac{\pi}{9} + i\sin\frac{\pi}{9}, \qquad z_2 = \cos\frac{\pi}{9} - i\sin\frac{\pi}{9}.$$

Thus $x_1 = 2\cos(\pi/9)$ is a root of the resolvent cubic. Since the square roots of the roots of the cubic are sums of roots of the quartic, it would follow that if the roots of **the** quartic were constructible, then the roots of the resolvent cubic would be constructible. But $\cos(2\pi/9)$ is not constructible, as was shown in Section 3. Hence $\cos(\pi/9)$ is not constructible, which leads to a contradiction. We have shown that the roots of $y^4 - y + 3/4$ are not constructible.

14.   Normal Extensions and Groups

Our mode of attack on the general problem is to show that certain subgroups of S_n do not exist and hence certain fields obtained by adjunctions of radicals to F do not exist. If we concerned ourselves only with unrestricted subgroups of S_n, we would not be able to make the exclusions needed. But we can make the desired exclusions if we require the subgroups to be normal. (See Section 10 of Chapter II.) Let us look back at the above with this in view. Both for the cubic and quartic equations G_1, being of index 2 in G, is a normal subgroup of G. (See Theorem 10.5 of Chapter II.) Coupled with this is the fact that F_1 is a normal extension of F; for if a square root is in F_1, its conjugate, being its negative, is also in F_1. The cubic equation, K, is a normal extension of F_1, since K, being the root field of the cubic, contains all the roots of the same cubic viewed as over F_1. By the same token for the quartic, F_2 is a normal extension of F_1. It would seem reasonable that G_2 would be a normal subgroup of G_1, since it plays the same role for the resolvent cubic that (e) plays for the cubic; and (e) is a normal subgroup of $G_1 = A_3$ in the case of the cubic equation.

Now we proceed to show that if an equation is solvable by radicals, its root field can be obtained by a sequence of *normal* extensions of a certain type. After that we will make the transition to normal subgroups.

Theorem 14.1.   Let F be a field of characteristic zero and K a field obtained from F by a finite sequence of adjunctions of roots of equations of the form $x^n = a$. Then there is the following sequence of fields:

$$(14.1) \qquad\qquad F \subset K_2 \subset K_3 \subset \cdots \subset K,$$

where each field is a *normal* extension of its predecessor by the adjunction of a root of an equation of the form $x^p = b$, where p is a prime and b is in the preceding field.

Proof: Certainly, if we can prove the theorem for a single equation $x^n = a$, it will follow for a sequence of such adjunctions. First, we consider the case $n = p$ and $a = 1$. Let ρ be a root, different from 1, of $x^p = 1$. Then since the order of ρ must be a factor of p, it must be p, and the roots of $x^p = 1$ are

$$1, \rho, \rho^2, \ldots, \rho^{p-1}.$$

If ρ is not in F, then $F(\rho)$ is a normal extension of F of degree p. Therefore, the theorem holds for this case. [Note that we would be in trouble if the characteristic of the field were p, for then $(x - 1)^p = x^p - 1$ and 1 is the only zero.]

Next we prove the theorem for $a \neq 1$ and $n = p$. If α is a root of $x^p = a$, where a is in F and ρ a root not 1 of $x^p = 1$, then the roots of $x^p = a$ are

$$\alpha, \alpha\rho, \alpha\rho^2, \ldots, a\rho^{p-1}.$$

Thus the root field of $x^p = a$ is obtained first by adjoining ρ to F, which by the above can be done by a sequence (14.1), and then by adjoining α to $F(\rho)$, which again is an extension of prime degree and normal, since all the conjugates of α are in $K = F(\alpha, \rho)$.

We now prove the theorem for a in F and all n. Let $n = pp_2 \cdots p_k$, where the p's are prime numbers, not necessarily distinct. Our proof is by induction on k. We have shown the theorem for $k = 1$. If $n = pm$, we assume the theorem for the exponent m. Let K_0 be the field obtained by adjoining to F all the zeros of $x^m = a$. If α is a root of $x^{pm} = a$, then $\alpha^p = \beta$ for some β in K_0. If $\beta = 1$, $K = K_0(\rho)$ for ρ a root different from 1 of $x^p = 1$, the extension is of degree p and normal; hence the theorem holds for K. If $\beta \neq 1$, then first adjoin ρ just defined, to K_0, which is a normal extension of prime degree, and then a root of $x^p = \beta$, which is another normal extension of prime degree. So, by one or two more normal adjunctions of prime degree, we can go from K_0 to K. The theorem is proved.

Now that we have the proof we can look on the sequence of extensions in slightly simpler form. To express the root field of $x^n = a$ as a sequence of normal extensions by radicals, we first do this for the roots of $x^n = 1$. Then if $a \neq 1$, we adjoin one root of $x^n = a$. It is simpler to consider F to contain the needed roots of unity in the first place. Next we make the transition to normal subgroups of the Galois group. We state and prove the following fundamental result.

Theorem 14.2. Let F be a field containing all roots of unity (or as many as are needed). Let K be an extension of F by a sequence (14.1). Then there is a sequence of subgroups G_i of G:

$$(14.2) \qquad\qquad G \supset G_1 \supset G_2 \supset \cdots \supset G_k = (e),$$

where every G_i is a normal subgroup of prime index in its predecessor.

Our proof is by a repetitive argument. In order to set up the process, it seems better first to state and prove a lemma.

Lemma 1. Let F be a field that contains all roots of unity and let $F_1 = F(\lambda)$, where λ is a zero of an irreducible polynomial $x^p - a$ over F (with a in F) for p a prime number. Let K be a normal extension of F that contains F_1. Let $G = G(K:F)$ be the Galois group of K over F, $G_1' = G(F_1:F)$, and $G_1 = G(K:F_1)$. Then G_1 is a normal subgroup of G, G/G_1 is isomorphic to G_1', and $o(G) = o(G_1)o(G_1') = o(G_1)o(G/G_1)$.

Proof: Now if σ_1 is in G_1, it is an automorphism of G that leaves fixed each element of F_1 and hence of F. Thus G_1 is a subgroup of G from the fundamental theorem of Galois theory. Let σ be any element of G. Then for each λ in F_1,

$$[\lambda\sigma][\sigma^{-1}\sigma_1\sigma] = \lambda\sigma_1\sigma = \lambda\sigma.$$

Hence $\lambda\sigma$ is unchanged by $\sigma\sigma_1\sigma^{-1}$. Since σ leaves fixed each element of F, σ takes λ into a conjugate of λ over F. But the conjugates of λ are, as we have seen above, $\rho\lambda$ for ρ some root of unity. Thus if $\lambda\sigma = \rho\lambda$,

$$\rho\lambda(\sigma^{-1}\sigma_1\sigma) = \lambda\sigma = \rho\lambda,$$

and since ρ is in F, σ and σ_1 take ρ into itself. This shows that

$$\lambda\sigma^{-1}\sigma_1\sigma = \lambda$$

and that $\sigma^{-1}\sigma_1\sigma$ is in G_1. Hence G_1 is a normal subgroup of G.

Next we show that G/G_1 is isomorphic to G_1'. Now every element of G_1 takes λ into one of its conjugates, and the image of λ determines the image of each element in F_1. (Note that F_1 is normal over F.) So we can identify ω_i in G_1' by

$$\lambda\omega_i = \rho^i\lambda$$

for $\rho \neq 1$ a pth root of unity. Similarly, the elements of G take λ into a conjugate (still over F). Thus we can define the class $[\sigma]_i$ modulo G_1 of G to be the set of all σ of G for which $\lambda\sigma = \rho^i\lambda$, and set up the correspondence

$$[\sigma]_i \leftrightarrow \omega_i.$$

This correspondence is one-to-one, since if $\lambda\sigma = \lambda\sigma'$, it follows that $\sigma'\sigma^{-1}$ is in G_1, and hence σ and σ' are in the same class modulo G_1.

To complete the proof of the isomorphism, we need to show that the correspondence is preserved under multiplication. Now $[\sigma_i]$ is the set of all σ of G such that $\lambda\sigma = \rho^i\lambda$ and $[\sigma_j]$ such that $\lambda\sigma = \rho^j\lambda$. If $\lambda\sigma_1 = \rho^i\lambda$ and $\lambda\sigma_2 = \rho^j\lambda$, then $\lambda\sigma_1\sigma_2 = \rho^i\lambda\sigma_2 = \rho^{i+j}\lambda$ shows that

$$[\sigma_i][\sigma_j] \leftrightarrow \omega_i\omega_j,$$

which completes the proof of the isomorphism. Finally, G/G_1 isomorphic to G_1' implies that $o(G/G_1) = o(G_1') = o(G)/o(G_1)$ from the corollary of Theorem 10.3 in Chapter II.

Now we return to the proof of Theorem 14.2. Let K be the sth field in the following sequence:

$$F \subset F_1 = F(\lambda_1) \subset F_2 = F_1(\lambda_2) \subset \cdots \subset F_{s-1} = F_{s-2}(\lambda_s) = K,$$

where each λ_i is a zero of an irreducible polynomial $x^{p_i} - a_{i-1}$ over F_{i-1}, each p_i is a prime number (the p's need not be distinct), F contains all the p_ith roots of unity for $1 \leq i \leq s$, and each F_i is normal over F.

To show that F_i is normal over F_{i-1}, we note that if λ_i is a zero of the irreducible polynomial $f_i(x)$ over F_{i-1} and $f(x)$ over F, $f_i(x)$ must be a factor of $f(x)$. Thus all the conjugates of λ_i over F_{i-1} are among the conjugates of λ_i over F; all of these are in F_i.

If $s = 3$, the lemma suffices to establish the theorem; that is, $G(K\!:\!F_1)$ is a normal subgroup of $G(K\!:\!F)$ and

$$\frac{G(K\!:\!F)}{G(K\!:\!F_1)} \text{ is isomorphic to } G(F_1\!:\!F),$$

and hence $G(K\!:\!F_1)$ is of index p_1 in $G(K\!:\!F)$. Now use the same result with F replaced by F_1 and F_1 by F_2 to see that $G(K\!:\!F_2)$ is a normal subgroup of index p_2 in $G(K\!:\!F_1)$ and

$$\frac{G(K\!:\!F_1)}{G(K\!:\!F_2)} \text{ is isomorphic to } G(F_2\!:\!F_1).$$

Continuing, we develop a sequence of subgroups $G(K:F)$

$$G(K:F) \supset G(K:F_1) \supset G(K:F_2) \supset \cdots \supset G(K:F_{s-1}) \supset (e),$$

where each is a normal subgroup of prime index in its predecessor. This completes the proof of the theorem.

The converse of this theorem is also true; that is, if such a sequence of groups exists, K is a normal extension by radicals; but we shall not need it to solve our problem.

Now we are in sight of our goal. Two things remain to be proved. First, we shall show that the symmetric group S_n for $n > 4$ does not have a sequence of normal subgroups having the properties of the sequence (14.2) in Theorem 14.2. Second, we shall show that there is a quintic equation that has as its Galois group the symmetric group S_5 on five symbols.

Theorem 14.3. The symmetric group S_n for $n > 4$ does not have a sequence of normal subgroups (14.2) each of prime index in its predecessor. That is, such a sequence must stop short of (ϵ).

Proof: The proof is essentially that given on p. 420 of Birkhoff and Mac Lane's *A Survey of Modern Algebra*. Suppose that

$$G = S_n \supset G_1 \supset G_2 \supset \cdots \supset G_s$$

is a sequence of subgroups of S_n ($n > 4$), where each is a normal subgroup of prime index in its predecessor. Since $n \geq 3$, S_n contains every 3-cycle, that is, contains all permutations (bcd), where b, c, and d are three positive integers between 1 and n inclusive. We want to show that G_s contains every 3-cycle and hence cannot be the identity group as in (14.2). We do this by induction on s; that is, we show that if G_i contains every 3-cycle, so does G_{i+1}.

The first step is to show that if g_1 and g_2 are in G_i, then their "commutator" $g_1^{-1}g_2^{-1}g_1g_2$ is in G_{i+1}. The commutator is certainly in G_i. Now the quotient group G_i/G_{i+1}, being of prime order, is cyclic and hence commutative. This means that $g_1g_2 = g_2g_1k$ for some k in G_{i+1}. But

$$g_1^{-1}g_2^{-1}g_1g_2 = g_1^{-1}g_2^{-1}g_2g_1k = k$$

is in G_{i+1}. So the commutator of any two elements of G_i is in G_{i+1}.

The second step is to show that for $n \geq 5$, g_1 and g_2 may be chosen so that k is any 3-cycle. Let $k = (abc)$ and choose $g_1 = (adb)$ and $g_2 = (bcf)$, where a, b, c, d, and f are five distinct numbers between 1 and n inclusive. Then

$$g_1^{-1}g_2^{-1}g_1g_2 = (abd)(bfc)(adb)(bcf) = (abc).$$

Thus the sequence of subgroups stops short of (ϵ), and the proof is complete.

Corollary. If the Galois group of a normal extension of F is the symmetric group S_n for $n > 4$, then the extension is not an extension by radicals.

An interesting and important fact in this connection is that for $n > 4$, the alternating group A_n has no proper normal subgroup except the identity. A proof of this can be found in van der Waerden, vol. I, pp. 149ff.

15. Equations Not Solvable by Radicals

In this section we assume that all fields are subfields of $\mathbf{C}$, the field of complex numbers. It follows from Section 14 that to solve the second problem of this chapter it remains to exhibit some quintic equation whose Galois group is the symmetric group. There are various ways to do this. (See the Bibliography for references to the books of van der Waerden and Birkhoff and Mac Lane.) Here we use the approach of Irving Kaplansky in his *Fields and Rings*, 2nd ed., p. 37, which gives a large class of equations not solvable by radicals. The basic theorem is the following.

Theorem 15.1. Let p be a prime number and $f(x)$ an irreducible polynomial of degree p over $\mathbf{Q}$, the field of rational numbers. If $f(x) = 0$ has exactly two imaginary roots, then the Galois group of $f(x)$ is S_p.

Proof: First, we show that $p \mid o(G)$. This is true since to form the root field of $f(x)$, we can first adjoin a root of $f(x) = 0$. Thus p divides the degree of the root field over $\mathbf{Q}$. Since the degree of the root field is the order of G, it follows that p divides $o(G)$.

Now we showed in Theorem 12.5 of Chapter II that if $p \mid o(G)$, then G has an element of order p, that is, a p-cycle. Furthermore, G contains a transposition, for let x_1 and $x_2 = \bar{x}_1$ be the conjugate imaginary roots of $f(x) = 0$. Then let σ be a linear transformation of the root field K onto itself, which leaves all elements of $\mathbf{Q}$ fixed and is defined by

$$x_i \sigma = \bar{x}_i, \qquad i = 1, 2, \ldots, p.$$

Now from Theorem 4.3, σ is an isomorphism of $F(x_1, x_2, \ldots, x_p)$ onto itself, that is, an automorphism of K. Thus we have shown that G contains a transposition, since σ leaves fixed every element of $\mathbf{Q}$ and interchanges x_1 and x_2, leaving fixed all other roots.

The proof will be complete when we have established the following lemma.

Lemma 2. Let p be a prime number. If a subgroup G of S_p contains a transposition and a p-cycle (that is, an element of order p), then $G = S_p$.

Proof: We may without loss assume that the p-cycle is $\alpha_1 = (1\ 2\ 3 \cdots p)$ and the transposition $\beta_2 = (1\ 2)$. It helps to write α_1 in the form

$$\alpha_1 = (1\ 2)(1\ 3) \cdots (1\ p).$$

We begin by performing two computations. First,

$$\beta_2\alpha_1\beta_2 = (1\ 3)(1\ 4)(1\ 5)\cdots(1\ p)(1\ 2).$$

We define $\alpha_2 = \beta_2\alpha_1\beta_2$ and note that α_2 is obtained from α_1 by making $(1\ 3)$ the first transposition and permuting cyclically the transpositions. The second computation leads to

$$\alpha_2^{-1}\beta_2\alpha_2 = (1\ 3) = \beta_3.$$

At this point one can say, "Repeat this process to see that the transpositions $(1\ 4), (1\ 5), \ldots, (1\ p)$ are all in G"; and hence by Exercise 26 after Section 6 in Chapter II, $G = S_p$.

If this procedure leaves you with some misgivings, we can find a more systematic approach by writing, in place of α_1,

$$\pi_1 = \tau_2\tau_3\cdots\tau_p,$$

where

$$\tau_2 = (1\ a_2),\ \tau_3 = (1\ a_3), \ldots, \tau_p = (1\ a_p)$$

and $(a_2\ a_3\cdots a_p) = (2\ 3\cdots p)$; that is, the a_i form in order some cyclic permutation of $2, 3, \ldots, p$. Then define π_2 by

$$\pi_2 = \tau_2\pi_1\tau_2 = \tau_3\tau_4\cdots\tau_p\tau_2.$$

We want to show by analogy with the case for α_1 and β_2 that

(15.1) $\pi_2^{-1}\tau_2\pi_2 = \tau_3;$ that is, $\tau_2\pi_2 = \pi_2\tau_3.$

This is done by the following calculations:

$$\tau_2\pi_2 = (1\ a_2)(1\ a_3)\cdots(1\ a_p)(1\ a_2) = (a_2a_3\cdots a_{p-1}a_p),$$
$$\pi_2\tau_3 = (1\ a_3)(1\ a_4)\cdots(1\ a_p)(1\ a_2)(1\ a_3) = (a_3a_4\cdots a_pa_2).$$

The equality (15.1) is established by noting that the two cycles on the right are equal.

Thus we have shown that if π_1 and τ_2 are in G, so are π_2 and τ_3, for all π_2 and τ_2. Thus, if $(1\ 2\ 3\cdots p)$ and $(1\ 2)$ are in G, so are $(1\ 3), (1\ 4), \ldots, (1\ p)$ in turn. One completes the argument by citing the exercise indicated above. This completes the proof of the lemma and establishes Theorem 15.1.

From Theorem 15.1 it is easy to exhibit specific equations having the symmetric group for their Galois group. For instance,

$$f(x) = x^5 - 10x + 5$$

is irreducible over $\mathbf{Q}$ by Eisenstein's criterion. As you will be asked to show in Exercise 3, $f(x) = 0$ has exactly two imaginary roots. Thus the Galois group of $f(x)$ is in S_5, and hence the equation is not solvable by radicals over $\mathbf{Q}$.

As an addendum it is worth noting that H. B. Mann has considered solvability by radicals over finite fields in vol. 29(1974), pp. 551–54, of the *Journal of Algebra.* Now, in a sense, every solution of an equation in a finite field F is expressible as a radical, since every element of F is a zero of $x^q = x$, where q is the number of elements of F. But suppose that in the polynomial

$$f(x) = x^n + c_1 x^{n-1} + c_2 x^{n-2} + \cdots + c_{n-1} x + c_n,$$

the coefficients are considered to be variables, and one asks if the zeros of $f(x)$ can be expressed in terms of radicals over a polynomial function of the coefficients. Then the problem becomes more difficult. Mann has derived explicit expressions that give the roots of every irreducible polynomial of degree n over F, a finite field F, in terms of one radical of a polynomial function of the c_i, provided that n is not a multiple of the characteristic of F. The restriction that the characteristic not divide n is at least partially to avoid inseparability of the field. (See Section 16 of Chapter III.)

Notice that of fundamental importance in this chapter are sequences of fields and groups each contained in or containing its predecessor. Since a field is a special case of an ideal, we see again how fundamental is the idea of a Noetherian ring. (See Section 5 of Chapter V.)

The history of solutions of cubic equations is a long one, going back at least as far as Archimedes in the third century B.C. Omar Khayyam in the twelfth century seems to have been the first to give a method of solving a cubic equation with positive roots. Construction of the regular pentagon was given by Euclid. For further details on these and other subjects, see the books by Coolidge and Struik listed in the Bibliography.

Exercises

1. Let $F = \mathbf{Q}(\sqrt{-3})$. Show that every polynomial $x^3 - a$, where a is a real number, is normal over F.

2. If in Exercise 1 the number a is imaginary, is the polynomial still normal over F?

3. Prove that exactly two of the roots of $x^5 - 10x + 5$ are imaginary.

4. Let q be a prime number. For what integers c does Theorem 15.1 imply that the Galois group of the following equation is the symmetric group S_5?

$$x^5 - cqx + q = 0.$$

5. In Example 3 of Section 10 in Chapter II we showed that

$$H = \{\epsilon, (1\ 2)(3\ 4), (1\ 3)(2\ 4), (1\ 4)(2\ 3)\}$$

is a normal subgroup of A_4. Let y_1, y_2, y_3, and y_4 denote the zeros of an irreducible quartic polynomial $f(x)$ over $\mathbf{Q}$. Show that $y_1 y_2 + y_3 y_4$ is left fixed by each element of H, and find the other two similar expressions in the y_i left fixed by each element of H.

6. Let K be the root field of the polynomial $f(x)$ in Exercise 5 and F the subfield of K that consists of all those elements of K left fixed by all elements of H. What are the possibilities for the degree of F over $\mathbf{Q}$? For each of these, find the degree of K over $\mathbf{Q}$.

7. To what extent is a quartic equation determined by its resolvent cubic?

8. In the proof of the Lemma after Theorem 14.2, how many different automorphisms of K over F take λ into itself? Why?

9. Prove that if the resolvent cubic of a quartic equation is normal, then the Galois group of the quartic is of order 12 or less.

10. Prove that if the Galois group of a quartic equation is S_4, then its discriminant is not a square of a rational number.

11. Suppose that the Galois group of a quartic equation is of order 4 or less. What does this imply about the resolvent cubic?

12. Find the order of the Galois group of $y^4 - y + \frac{3}{4}$. (See the end of Section 13.)

13. Show that the order of the Galois group of every irreducible polynomial over $\mathbf{Q}$ of degree n is divisible by n.

14. Show that S_3 has a subgroup H of prime index with H not a normal subgroup.

15. Let $G = S_4$. Find a sequence of groups

$$G \supset G_1 \supset G_2 \supset G_3 \supset (\epsilon)$$

such that each G_i is a normal subgroup of its predecessor and of prime index. Let $f(x)$ be a quartic irreducible polynomial over $\mathbf{Q}$, whose Galois group is S_4. Write a sequence of subfields of the root field of $f(x)$.

16. We showed that if for the sequence $F \subset F_1 \subset K$ if K is normal over F, it is normal over F_1. Is it also true that if K is normal over F_1, it must be normal over F? Why or why not?

17. Suppose that a cubic polynomial over $\mathbf{Q}$ has two imaginary roots. Show that the root field K of the polynomial is not a quadratic extension of any cubic normal extension of $\mathbf{Q}$.

18. A group G is called *solvable* if there is a sequence of subgroups

$$G \supset G_{k-1} \supset G_{k-2} \supset \cdots \supset G_1 = (\epsilon),$$

where each G_i is a normal subgroup of its predecessor and each G_i/G_{i-1} is cyclic. Show that the group G in Theorem 14.2 is solvable.

19. If a group G is solvable, show that there is a sequence (14.1) in which each subgroup is of prime index in its predecessor.

20. Let K be the root field of a polynomial over a field F. Two intermediate fields F_1 and F_2 are called *conjugate* if there is an element σ of $G(K:F)$ such that $(F_1)\sigma = F_2$. Prove that F_1 and F_2 are conjugate if and only if $\sigma^{-1}G_1\sigma = G_2$, where $G_i = G(K:F_i)$, $i = 1, 2$.

16. Galois

Evariste Galois (1811–1832) was born in Bourg-la-Reine, near Paris. His life was shorter and more tragic than Abel's; he died in a duel. His lack of recognition, coupled with the loss of two of his manuscripts at the hands of mathematicians of high repute, made him bitter and full of pride, which manifested itself in scorn and snobbery. Apparently he failed an important examination at the École Polytechnique because he refused to answer a question that he judged ridiculous. After his death it was recognized that although Lagrange, Gauss, and Abel made much progress in the theory of equations, none of them found the fundamental connection between equations and groups, that is, that to each equation corresponds a group that reflects the essential character of the equation. Also he made important contributions to the theory of elliptic integrals; there he seemed to have had in his possession results that Riemann obtained 55 years later.

In 1897, in the introduction to Galois's brief collected works, Emile Picard, then president of the Mathematical Society of France, wrote (freely translated): "It appears, alas, that this unhappy young man sadly paid the ransom of his genius ... It is not without emotion that one reads the scientific will of a young man of twenty years written on the eve of his death. Galois was without doubt an equal among the great mathematicians of the century. None surpassed him in originality and depth of conception."

This is no mean tribute, since in many respects the nineteenth century was the Golden Age of mathematics.

Bibliography

BIRKHOFF, G., and S. MAC LANE. *A Survey of Modern Algebra*, 3rd ed. New York: Macmillan, 1965.

COOLIDGE, J. L. *The Mathematics of Great Amateurs*. New York: Dover, 1963.

EVES, H. *An Introduction to the History of Mathematics*, rev. ed. New York: Holt, Rinehart and Winston, 1964.

HALL, M. *The Theory of Groups*. New York: Macmillan, 1959.

HARDY, G. H., and E. M. WRIGHT. *Introduction to the Theory of Numbers*, 4th ed. New York: Oxford University Press, 1960.

HERSTEIN, I. N. *Topics in Algebra*. Lexington, Mass.: Xerox, 1964.

JONES, B. W. *Linear Algebra*. San Francisco: Holden-Day, 1973.

KAPLANSKY, I. *Fields and Rings*, 2nd ed. Chicago: University of Chicago Press, 1972.

LEVEQUE, W. J. *Topics in Number Theory*, 2 vols. Reading, Mass.: Addison-Wesley, 1956.

MCCOY, N. H. *Introduction to Modern Algebra*, rev. ed. Boston: Allyn and Bacon, 1968.

MCCOY, N. H. *Rings and Ideals*. Buffalo, N.Y.: Mathematical Association of America, 1948.

NIVEN, I., and H. S. ZUCKERMAN. *Introduction to the Theory of Numbers*. New York: Wiley, 1965.

POLLARD, H. *Theory of Algebraic Numbers*. New York: Mathematical Association of America, 1948.

STRUIK, D. J. *Concise History of Mathematics*, 3rd rev. ed. New York: Dover, 1967.

VAN DER WAERDEN, B. L. *Modern Algebra*, vols. I and II (English translation). New York: Ungar, 1949 and 1950.

Answers and Partial Answers
to Selected Exercises

Chapter *I*

Section 2

1. (a) Binary and commutative.   (b) Binary but not commutative.
   (d) $\sqrt[a]{b}$ is not usually defined unless a is a positive integer. If it were to be defined as $b^{1/a}$, the operation would still not be a binary one.
   (e) Not a binary operation. (The answers to the other parts are left to you.)

2. Only one of the S's in Exercise 1 is a group.

7. Since $c^3 = 1$, S is a group.

9. The smallest group that contains R is the set $\{x\}$ in radians, where x is an integer. This is an infinite additive group.

10. The numbers 0 and 1 are the additive and multiplicative identities of $\mathbf{Z}[x]$, respectively. The elements of $\mathbf{Z}[x]$, whose reciprocals are in $\mathbf{Z}[x]$, are 1 and -1.

12. To prove property 5, let $a \circ b = e = a \circ c$. Let $\bar{a}$ be *an* inverse of a. Then, using the associative property we have

$$\bar{a} \circ e = \bar{a} \circ (a \circ b) = (\bar{a} \circ a) \circ b = e \circ b = b.$$

Similarly, $\bar{a} \circ e = c$, showing that $b = c$.

18. If $h \circ g = g \circ h$ and $h' \circ g = g \circ h'$ for all g, then $(h' \circ h) \circ g = h' \circ g \circ h = h' \circ h \circ g$, and hence the product of two elements of H is in H. Since $e \circ g = g \circ e$ for all g, e is in H. To complete the proof, one must show that if h is in H, then $\bar{h}$ is in H.

Section 4

1. $A^{-1} = \begin{bmatrix} -2 & 1 \\ \frac{3}{2} & -\frac{1}{2} \end{bmatrix}$, $\quad (BA)^{-1} = \begin{bmatrix} -7 & 11 \\ \frac{9}{2} & -7 \end{bmatrix}$.

The solution X of $AX = B$ is $X = A^{-1}B$.

3. The inverse does not exist.

4. (b) S is not an additive group but is a multiplicative group. (The answers to the other parts are left to the reader.)

7. $X = \begin{bmatrix} 2a & 2b \\ -a & -b \end{bmatrix}$ for all a and b.

11. Use the form of the inverse of A given in Theorem 4.1.

14. Define A as in (4.7) and A' and A'' similarly. Then

$$A(A' + A'') = \begin{bmatrix} a_1 & a_2 \\ b_1 & b_2 \end{bmatrix}\begin{bmatrix} a_1' + a_1'' & a_2' + a_2'' \\ b_1' + b_1'' & b_2' + b_2'' \end{bmatrix} = \begin{bmatrix} r_1 & r_2 \\ s_1 & s_2 \end{bmatrix},$$

where $r_1 = a_1(a_1' + a_1'') + a_2(b_1' + b_1'')$, $r_2 = a_1(a_2' + a_2'') + a_2(b_2' + b_2'')$, $s_1 = b_1(a_1' + a_1'') + b_2(b_1' + b_1'')$, $s_2 = b_1(a_2' + a_2'') + b_2(b_2' + b_2'')$.

Then compute $AA' + AA''$.

15. The identity vector for addition is $(0, 0)$ and the inverse of (a_1, a_2) is $(-a_1, -a_2)$.

18. Yes.

20. (a) holds but (b) has no meaning.

23. Use (4.2) and see that $(a_1a_1' + a_2b_1')(b_1a_2' + b_2b_2') - (b_1a_1' + b_2b_1')(a_1a_2' + a_2b_2') = a_1b_2(a_1'b_2' - a_2'b_1') - b_1a_2(-a_2'b_1' + a_1'b_2') + a_1b_1a_1'a_2' + a_2b_2b_1'b_2' - a_1b_1a_1'a_2' - a_2b_2b_1'b_2'$.

25. If B is nonsingular, $(x_1, x_2)B = (0, 0)$ implies that $(x_1, x_2) = (0, 0)B^{-1} = (0, 0)$. If B is singular, call it $B = \begin{bmatrix} a_1 & a_2 \\ b_1 & b_2 \end{bmatrix}$ with $a_1b_2 - b_1a_2 = 0$. Then two solutions of $(x_1, x_2)B = (0, 0)$ are $(b_2, -a_2)$ and $(b_1, -a_1)$. These are not both $(0, 0)$ unless B is the zero matrix. If B is the zero matrix, every ordered pair is a solution.

Section 8

1. Suppose that we write $f(x) = n_f(x)/d_f(x)$ with a similar notation for $g(x)$. Then

$$\begin{bmatrix} n_f(x) \\ d_f(x) \end{bmatrix} = \begin{bmatrix} 3 & 4 \\ 2 & 3 \end{bmatrix}\begin{bmatrix} x \\ 1 \end{bmatrix}, \qquad \begin{bmatrix} n_g(x) \\ d_g(x) \end{bmatrix} = \begin{bmatrix} 2 & 1 \\ 3 & 2 \end{bmatrix}\begin{bmatrix} x \\ 1 \end{bmatrix},$$

$$\begin{bmatrix} n_{fg}(x) \\ d_{fg}(x) \end{bmatrix} = \begin{bmatrix} 3 & 4 \\ 2 & 3 \end{bmatrix}\begin{bmatrix} n_g(x) \\ d_g(x) \end{bmatrix} = \begin{bmatrix} 3 & 4 \\ 2 & 3 \end{bmatrix}\begin{bmatrix} 2 & 1 \\ 3 & 2 \end{bmatrix}\begin{bmatrix} x \\ 1 \end{bmatrix} = \begin{bmatrix} 18 & 11 \\ 13 & 8 \end{bmatrix}\begin{bmatrix} x \\ 1 \end{bmatrix}.$$

This implies that $fg(x) = (18x + 11)/(13x + 8)$.

5. $(x_1, x_2)\tau^2 = (x_2, -x_1 - x_2)\tau = (-x_1 - x_2, x_1)$ and $(-x_1 - x_2, x_1)\tau = (x_1, x_2)$, which shows that $\tau^3 = e$, the identity.

7. $\varphi = n\pi$ for n an integer.

10. $\begin{bmatrix} \cos\varphi & \sin\varphi \\ -\sin\varphi & \cos\varphi \end{bmatrix}\begin{bmatrix} \cos\varphi' & \sin\varphi' \\ -\sin\varphi' & \cos\varphi' \end{bmatrix} = \begin{bmatrix} \cos(\varphi + \varphi') & \sin(\varphi + \varphi') \\ -\sin(\varphi + \varphi') & \cos(\varphi + \varphi') \end{bmatrix},$

since from trigonometry $\cos\varphi\cos\varphi' - \sin\varphi\sin\varphi' = \cos(\varphi + \varphi')$ and $\cos\varphi\sin\varphi' + \sin\varphi\cos\varphi' = \sin(\varphi + \varphi')$.

12. If $a_1 = \cos\varphi$, $a_2 = \sin\varphi$, $b_1 = -\sin\varphi$, $b_2 = \cos\varphi$; then, from (8.1), we should verify: $x_1^2 + x_2^2 = (x_1\cos\varphi - x_2\sin\varphi)^2 + (x_1\sin\varphi + x_2\cos\varphi)^2$. The coefficient of x_1^2 on the right is $\cos^2\varphi + \sin^2\varphi = 1$, and similarly for the coefficient of x_2^2. Also the coefficient of x_1x_2 is 0.

15. The answer to the question is "no," for $S \cap T$ may not even be a group. Why?

18. The set of reflections does not form a multiplicative group.

19. Consider $L = \begin{bmatrix} \frac{1}{2} & \frac{1}{4} \\ 3 & -\frac{1}{2} \end{bmatrix}$. Its determinant is -1 and its square the identity matrix. Yet it is not of the form of L_2 in (8.4).

$$\begin{bmatrix} \cos\varphi & \sin\varphi \\ \sin\varphi & -\cos\varphi \end{bmatrix}\begin{bmatrix} \cos\varphi' & \sin\varphi' \\ \sin\varphi' & -\cos\varphi' \end{bmatrix} = \begin{bmatrix} \cos(\varphi' - \varphi) & \sin(\varphi' - \varphi) \\ -\sin(\varphi' - \varphi) & \cos(\varphi' - \varphi) \end{bmatrix}.$$

22. From Section 8, the first matrix corresponds to a reflection in a line that makes an angle of $\varphi/2$ with the x-axis and the second matrix to a reflection in a line making an angle of $\varphi'/2$ with the x-axis. Hence the angle between the lines of reflection is $\frac{1}{2}(\varphi' - \varphi)$, which is half the angle $(\varphi' - \varphi)$.

Section 9

2. S_3 has three subgroups of order 2 and one of order 3.

5. The transformations of the octic group that leave unchanged the orientation of a square are ϵ, ρ, ρ^2, and ρ^3.

8. Only the elements ϵ and ρ^2 of the octic group are commutative with all its elements.

9. For the octic group there are five elements of order 2 and two of order 4. One group of order 4 is $\{\epsilon, \rho, \rho^3, \rho^2\}$. Every other group of order 4 must be composed of ϵ and elements of order 2. Now if a subgroup contains α and ρ, then it is the entire octic group. Hence if a subgroup contains α and $\alpha\rho$, it must contain $\alpha\alpha\rho = \rho$ and hence be the octic group. A second subgroup of order 4 is $\{\epsilon, \alpha\rho, \alpha\rho^3, \rho^2\}$. There are in all three subgroups of order 4.

13. The set of even integers is a cyclic additive group. The set of odd integers is not.

17. There are six rotations of a regular hexagon through multiples of $60°$, call them ϵ, ρ, ρ^2, ρ^3, ρ^4, and ρ^5. If we number the vertices of the hexagon from 1 to 6 in order, a reflection α is the interchange of 1 and 2, 3 and 6, and 4 and 5. So, if

$$\alpha = \begin{pmatrix} 1 & 2 & 3 & 4 & 5 & 6 \\ 2 & 1 & 6 & 5 & 4 & 3 \end{pmatrix} \quad \text{and} \quad \rho = \begin{pmatrix} 1 & 2 & 3 & 4 & 5 & 6 \\ 2 & 3 & 4 & 5 & 6 & 1 \end{pmatrix},$$

computation shows that $\alpha\rho = \rho^5\alpha$. From this it may be shown that the group of symmetries of the regular hexagon has order 12 and consists of

$$\alpha^i \rho^j \qquad \text{where } i = 0 \text{ or } 1 \text{ and } j = 0, 1, 2, 3, 4, 5,$$

with $\alpha^2 = \epsilon = \rho^6$ and $\alpha\rho = \rho^5\alpha$.

Section 11

1. Consider the set $S = \{ax + by + cz\}$ where a, b, and c are fixed integers and x, y, and z range over all integers. Now, $ax + by + cz + (ax' + by' + cz') = a(x + x') + b(y + y') + c(z + z')$ shows that S is closed under addition. It is associative, since addition of integers is, and because of the distributive property. The additive identity is $0 \cdot a + 0 \cdot b + 0 \cdot a$, and the additive inverse of $ax + by + cz$ is $a(-x) + b(-y) + c(-z)$.

3. The group of Section 11 is not generated by a single element, since the square of each element is I. The only proper subgroups beside (I) are those three of order 2 consisting of I and one of C, V, and O.

4. No.

5. Let $g = a_1 c_1 + a_2 c_2 + \cdots + a_n c_n$, and let $c_1 = qg + r$, where q is an integer and $r = 0$ or $0 < r < g$. Then $r = c_1 - q(a_1 c_1 + a_2 c_2 + \cdots + a_n c_n)$ shows that r is in S. Since g is the least positive integer in S, r must be zero. Thus g is a factor of c_1. Similarly, g can be shown a factor of every c_i.

8. If 1 is the g.c.d. of a and b, Corollary 1 shows that $ax + by = 1$ is solvable for integers x and y. If $ax + by = 1$ for integers x and y, then every common divisor of a and b must be a factor of 1. Hence 1 is the g.c.d.

Chapter II

Section 3

1. The Klein four-group.

3. No.

6. Let σ be an isomorphism between G and G' and σ' one between G' and G''. Thus if g is in G, $g\sigma\sigma'$ is a unique element of G''. If g'' is in G'', there is a unique g' in G' such that $g'\sigma' = g''$ and a unique g in G such that $g\sigma = g'$. Hence g'' determines g by means of $g\sigma\sigma' = g''$. To show property (1.1), compute $(g_1 g_2)\sigma\sigma' = [(g_1\sigma)(g_2\sigma)]\sigma' = (g_1\sigma\sigma')(g_2\sigma\sigma')$. So $\sigma\sigma'$ is an isomorphism between G and G''.

9. Let g be an element of order 3 and h an element of G not a power of g. Then (why?) the elements of G are e, g, g^2, h, hg, and hg^2. Now, h^2 must be one of these six elements. If $h^2 = e$, then calculation shows that the six powers of hg are the elements of G. If $h^2 = g$ or $h^2 = g^2$ and $h \neq g$, the six powers of h are the elements of G.

12. G and H are each of order 2. Any two groups of order 2 are isomorphic. Why?

16. Transformations and matrices are isomorphic if the basis is fixed.

21. One isomorphism is $s \leftrightarrow r^5$. There are other isomorphisms.

27. Let σ be the correspondence and e the identity of G. Then $(eg)\sigma = g\sigma = g'$ implies that $(e\sigma)(g\sigma) = (e\sigma)g'$ for all elements of G'. Hence $(e\sigma) = e'$ is the identity element of G'. Let g' be any element of G'; then $g\sigma = g'$ determines g in G. Let $g'_0 = (g^{-1})\sigma$ and see that $e' = g'_0 g'$ and hence that g'_0 is the inverse of g' in G'.

Section 4

1. One solution is $(0, 2)$ and the solutions of the homogeneous set of equations are $\{x(2, -3)\}$. Hence the solutions are the coset

$$(0, 2) + H, \qquad \text{where } H = \{x(2, -3)\}.$$

3. The two left cosets are H and gH, where g is not in H. The right cosets are H and Hg. Furthermore, $Hg = gH$. Why?

5. Besides H, the left cosets are $(1\ 3)(2\ 4)H = \{(1\ 3)(2\ 4), (1\ 4)(2\ 3)\}$, $(1\ 2\ 3\ 4)H = \{(1\ 2\ 3\ 4), (2\ 4)\}$, $(1\ 3)H = \{(1\ 3), (1\ 4\ 3\ 2)\}$. What are the right cosets?

8. The coset $A + H$ is the set of coordinates of the points on the line through (a_1, a_2) parallel to the line from the origin to the point (b_1, b_2). Also, $(a_1, a_2) + x(b_1, b_2) = r(a_1, a_2) + s(a_1 + b_1, a_2 + b_2)$, where $s = x$ and $r = 1 - x$.

10. From Lagrange's theorem the only subgroups of the octic group have orders 1, 2, 4, or 8.

13. (a) $c_1 e^{2x} + c_2 e^{-2x} - \frac{1}{5} \cos x$.

16. Suppose that a and b are in N_g. Then $abg = agb = gab$ shows that ab is in N_g. Also $ag = ga$ implies $ga^{-1} = a^{-1}g$. The left cosets of N_g are of the form $N_g, g_2 N_g, g_3 N_g, \ldots$, where each g_i is chosen not to be in any of the previous cosets.

18. It was shown that the transformations are closed under composition. The identity transformation is $\sigma_{1,0}$. To find the inverse of $\sigma_{a,b}$, call it $\sigma_{c,d}$ and determine c and d. To show that H is a subgroup, note that $x\sigma_{1,b} = x + b$ and hence $\sigma_{1,b}\sigma_{1,c} = \sigma_{1,b+c}$. The coset aH is $\{\sigma_{a,b}\}$, where a is fixed and b varies over all real numbers.

Section 6

1. (a) The order is 6.   (e) The order is 3.

2. To start the proof, note that $(a\ b)(c\ d)(a\ b\ c) = (a\ c\ d)$ implies that $(a\ b)(c\ d) = (a\ c\ d)(a\ c\ b)$.

6. No, it does not make any difference.

8. The even permutations of the octic group are the symmetries of a rectangle that is not a square.

11. Let $g_1, g_2, \ldots, g_s$ be the even elements of G, with $e = g_1$. Let h be an odd permutation of G. Then $g_i h$ are all odd permutations of G and every odd permutation of G is $g_j h$ for some j. Why?

12. No.

21. See Exercise 9 after Section 3.

23. If the order of s is t, then $t \mid p$ and hence $t \neq 1$ implies that $t = p$. So p is the order of s and $p \mid r$.

26. If 1, a, and b are distinct positive integers between 1 and n inclusive, then $(1 \ a)(1 \ b)(1 \ a) = (a \ b)$. From this, one can complete the proof.

Section 7

2. Write the elements of the Klein four-group as follows: $e = a_1, a_2, a_3, a_4$, where $a_4 = a_2 a_3$. Then form the following table:

	$e = a_1$	a_2	a_3	a_4	Representation
σ_{a_2}	a_2	a_1	a_4	a_3	$(1\ 2)(3\ 4)$
σ_{a_3}	a_3	a_4	a_1	a_2	$(1\ 3)(2\ 4)$
σ_{a_4}	a_4	a_3	a_2	a_1	$(1\ 4)(2\ 3)$

5. Here the operation is addition and hence $x\sigma_n = x + n$.

7. In this case $g_i \sigma_g \sigma_{g'} = (gg_i)\sigma_{g'} = g'gg_i = \sigma_{g'g}$. So $\sigma_{gg'} = \sigma_g \sigma_{g'}$. If G is not Abelian, the order of g and g' is changed. Instead of an isomorphism we have a kind of reverse isomorphism.

9. The regular representation of p consists of m/r cycles of length r. Thus for the regular representation to be odd, $n!/r$ must be odd and r even. Now since $r \leq n$, $n!/r$ is even if $n \geq 4$. If $n = 3$ and $r = 2$, then $3!/2 = 3$, and the regular representation is an odd permutation. See σ_{a_4} for S_3 in the above section.

10. No. Why?

Section 8

2. Use the notation of Example 1 and note that $a\sigma_1^2 = (ar)\sigma_1 = ar^2$, $a\sigma_1^3 = (ar^2)\sigma_1 = arr^2 = a$. So $\sigma_1^3 = \epsilon$, the identity automorphism. Also

$\sigma_2^2 = \epsilon$. So we set up the correspondence $\sigma_1 \leftrightarrow r$, $\sigma_2 \leftrightarrow a$. We need to show that $\sigma_1\sigma_2 = \sigma_2\sigma_1^2$. Now

$$a\sigma_1\sigma_2 = (ar)\sigma_2 = ar^2 \qquad \text{and} \qquad a\sigma_2\sigma_1^2 = a\sigma_1^2 = (ar)\sigma_1 = ar^2.$$

Similarly, one can show that $r\sigma_1\sigma_2 = r\sigma_2\sigma_1^2$.

3. To see that (e) is the center of S_4, call N the center. First, $(a\ b)(a\ c) = (a\ b\ c)$ shows that no transposition is in N. Why? Second, $(a\ b\ c)(a\ d) = (a\ b\ c\ d)$ and $(a\ d)(a\ b\ c) = (a\ d\ b\ c)$ shows that N contains no 3-cycle. Third, $(a\ b)(c\ d)(b\ c) = (a\ c\ d\ b)$ and $(b\ c)(a\ b)(c\ d) = (a\ b\ d\ c)$ shows that N contains no permutation $(a\ b)(c\ d)$. Finally, $(a\ b\ c\ d)(b\ c) = (a\ c\ d)$ and $(b\ c)(a\ b\ c\ d) = (a\ b\ d)$ shows that N contains no permutation that is a four-cycle. This shows that (e) is the center of S_4. This does *not* show that (e) is the center of A_4.

5. Denote the generators of the octic group G by r and a with $r^4 = e = a^2$ and $ra = ar^3$. From the table in Section 9 of Chapter I, the only elements of G of order 4 are r and r^3. All other elements except e are of order 2. So the possibilities are

$$r\sigma = r \text{ or } r^3 \qquad \text{and} \qquad a\sigma = r^2, a, ar, ar^2, \text{ or } ar^3.$$

So $r\sigma = r^t$ for $t = 1$ or 3 and $a\sigma = r^2$ or ar^v, where v is one of 0, 1, 2, 3.

First, $a\sigma = r^2$ is not possible, since $r^2\sigma = r^2$ and the correspondence is one-to-one. All other possible values of $a\sigma$ work, for $ra\sigma = r^tar^v$ and $ar^3\sigma = ar^{v+3t}$, and also $r^ta = ar^{3t}$. Hence there are eight automorphisms. Four of these are inner automorphisms. Which?

9. $\varphi(n)$.

Section 11

1. (a) S is the set of positive rational numbers. The correspondence is a homomorphism, since $|r_1|\,|r_2| = |r_1r_2|$ and r determines $|r|$ uniquely. The kernel of K is the set $\{1, -1\}$. G/K is the set of classes where $[a] = [b]$ if and only if $a = b$ or $a = -b$.
   (c) The correspondence is a homomorphism and the kernel is the number 1.

2. 1(c) is an automorphism. 1(e) is not. The others may or may not be.

5. Let the generators of the octic group G be r and a with $r^4 = e = a^2$. Take $G_1 = \{e, a\}$, $G_2 = \{e, a, r^2, r^2a\}$. Then G_1 is not a normal subgroup of G, for $r\{e, a\}r^3 = \{e, r^2a\}$.

7. Yes.

12. The key is to note that $(a\ c)(a\ b)(c\ d)(a\ c) = (a\ d)(c\ b)$. The rest of the proof is left to you.

17. Now $(ac)^{-1}bc = c^{-1}a^{-1}bc$. If $a^{-1}b$ is in S, we want $c^{-1}a^{-1}bc$ to be in S. Thus $c^{-1}sc$ must be in S for all s in S and c in G. Hence $a \equiv b \pmod{S}$ implies that $ac \equiv bc \pmod{S}$ if and only if S is a normal subgroup of G. For $ca \equiv cb \pmod{S}$ we do not need the condition.

19. If K is normal, $hkh^{-1} = k_1$ for some k_1 in K. Thus $hkh^{-1}k^{-1} = k_1k^{-1}$ is in K. The rest is up to you.

20. Every g in G determines a unique inner automorphism. The correspondence $g \leftrightarrow \sigma_g$ is a homomorphism because

$$x\sigma_g\sigma_{g'} = (g^{-1}xg)\sigma_{g'} = g'^{-1}g^{-1}xgg' = (gg')^{-1}x(gg') = x\sigma_{gg'}.$$

The kernel of the homomorphism is the set of g such that $x\sigma_g = \epsilon$, the identity automorphism. This means that $g^{-1}xg = x$ for all x, that is, that g is commutative with all x in G. Thus the kernel of the homomorphism is the center of G.

23. Using the notation of the answer to Exercise 5, see that ar and ar^2 are both of order 2, but $(ar)(ar^2) = r$, which is not of order 2. The proof asked for is left to you.

24. For every k_1, k_2, g_1, and g_2, $(g_1k_1)(g_2k_2)$ is in g_3K for some g_3 in G, where g_3K is independent of k_1 and k_2. Taking $k_1 = k_2 = e$, we see that g_1g_2 is in g_3K, and hence $g_3K = g_1g_2K$. Then $g_1k_1g_2k_2 = g_1g_2k_3$ for some k_3 in K implies that $k_1g_2k_2 = g_2k_3$, which implies that $g_2^{-1}k_1g_2 = k_3k_2^{-1}$ is in K. Hence K must be a normal subgroup of G.

25. No.

28. $(aN)(bN) = (bN)(aN)$ implies that $ab = (bn_1)(an_2)$ for some n_1 and n_2 in N. Hence $b^{-1}ab = n_1an_2$. Since N is normal, $n_1a = an_3$, for some n_3 in N. Hence $b^{-1}ab = an_3n_2$, which shows that $a^{-1}b^{-1}ab$ is in N.

32. Given $g^{-1}hg = h^i$, we want to prove $g^{-k}hg^k = h^t$ for $t = i^k$, by induction. It holds for $k = 1$. Assume it for k and then compute

$$g^{-k-1}hg^{k+1} = g^{-1}h^tg = (g^{-1}hg)^t = h^{it} = h^{t'}, \qquad \text{where } t' = i^{k+1}.$$

This establishes the second equation of the exercise. Taking $k = m$, we see that $h^{i^m} = h$ and hence that $i^m - 1 \equiv 0 \pmod q$, where q is the order of h. If $i \neq 1$, this implies that $m \mid q - 1 = \varphi(q)$, since q and m are primes.

Section 12

2. Let the octic group be defined by $r^4 = e = a^2$, $ra = ar^3$. The center is $\{e, r^2\}$. The conjugates of r are r and r^3; of a are a and ar^2; of ar are ar and ar^3. So $2 + 2 + 2 + 2 = 8$.

5. No. Give an example.

7. To show the symmetric property of the equivalence relation, note that $g = x^{-1}hx$ for x in G is equivalent to $h = (x^{-1})^{-1}gx^{-1}$.

9. To prove that the relation described in Theorem 12.5 is an equivalence relation, note that $\alpha\sigma^0 = \alpha$, $\alpha\sigma^i = \beta$ implies that $\alpha = \beta\sigma^{p-i}$, and $\alpha\sigma^i = \beta$ with $\beta\sigma^j = \gamma$ imply that $\alpha\sigma^{i+j} = \gamma$.

Chapter III

Section 2

3. Closure under division is shown by $1/(a + b\sqrt{3}) = (a - b\sqrt{3})/(a^2 - 3b^2)$, where the denominator is not zero, since 3 is not the square of a rational number.

4. No, because $(a + b\sqrt[3]{2})^2$ is not of the form $c + d\sqrt[3]{2}$.

5. Closure for multiplication is shown by $(a + bx)(c + dx) = ac + 3bd + (bc + ad)x$. To show closure for division, one needs to prove that $a^2 - 3b^2 \equiv 0 \pmod 5$ implies that a and b are divisible by 5. This follows from noting that if a and b are not both divisible by 5, then $a^2 - 3b^2 \equiv 0 \pmod 5$ only if 5 divides neither a nor b. In this case $a^2 \equiv 1$ or 4 $\pmod 5$ and $-3b^2 \equiv 2$ or 3 $\pmod 5$ and for no pairing of these is $a^2 - 3b^2 \equiv 0 \pmod 5$.

9. Here $x^2 + 1 \equiv (x + 2)(x + 3) \equiv 0 \pmod 5$, but neither $x + 2$ nor $x + 3$ is congruent to zero modulo 5.

12. No, since $a^2 - 3b^2 = 0$ for $a = \sqrt{3}$ and $b = 1$.

16. Let w' be the other root of $x^2 + x + 1 = 0$. To show closure under division, compute $1/(a + bw) = (a + bw')/(a^2 - ab + b^2)$ and show that $a^2 - ab + b^2 = 0$ for a and b rational only if $a = b = 0$.

17. Note that $a + bi + \sqrt{2}(c + di)$ may be written $r + is$, where $r = a + c\sqrt{2}$ and $s = b + d\sqrt{2}$. Then to show closure under division, note that $(r + is)(r - is) = r^2 + s^2$, which is not zero unless $r = s = a = c = b = d = 0$.

19. No, because $(a + b\sqrt{2})\sigma = a + b\sqrt{3}$ implies that $(a - b\sqrt{2})\sigma = a - b\sqrt{3}$ and, taking the product, $(a^2 - 2b^2)\sigma = a^2 - 3b^2$. The first equation shows that $(1)\sigma = 1$ and the last, taking $a = 3$ and $b = 2$, that $(1)\sigma = -3$; thus the image of 1 under σ is not unique.

20. Since (1)$\sigma = 1$, then (2)$\sigma = 1 + 1 = 2$ and (3)$\sigma = 3$. Hence the only automorphism of $\mathbf{Z}_3$ is the identity automorphism.

24. No. Why?

25. Let w be a root of $x^2 + x + 1 = 0$ and let $P = (a + bw\sqrt[3]{2} + cw^2\sqrt[3]{4}) \cdot (a + bw^2\sqrt[3]{2} + cw\sqrt[3]{4})$. Show that $P = a^2 - 2bc + \sqrt[3]{2}(-ab + 2c^2) + \sqrt[3]{4}(-ac + b^2)$ and $(a + b\sqrt[3]{2} + c\sqrt[3]{4})P = a^3 + 2b^3 + 4c^3 - 6abc = Q$. If Q were zero for rational numbers a, b, and c not all zero, we could multiply each term by the least common multiple of the denominators and have $Q = 0$ for a, b, and c integers not all zero. Also, we could without loss assume that 1 is the g.c.d. of a, b, and c. The proof that $Q = 0$ is impossible for integers a, b, and c not all zero is completed by showing that $Q = 0$ implies first that a is an even integer, then b even, and finally c even. This is a contradiction.

Section 5

1. In $\mathbf{Z}_{11}$ the product is $3x^5 + 6x^3 + 6x^2 - x - 3$.

3. There is closure under addition, multiplication, and subtraction. The set has no divisors of zero, since it is a subset of the field of complex numbers. [One can also show this directly by proving that if $(a + bi)(c + di) = 0$, either $a = b = 0$ or $c = d = 0$.]

5. Exercise 16 of the previous section shows that the set $\{a + bw\}$ for a and b rational forms a field. This means that we do not need to check many of the properties of an integral domain, but we do need to establish closure properties.

7. S is isomorphic to the set $\{x + yi\}$ for x and y integers.

9. $fg(x) = 0$ if x is nonnegative or $-4x$ if x is negative.

11. Since D is ordered, then for a and b in D with $b \neq 0$, exactly one of the following holds: $ab = 0$, $ab > 0$, $ab < 0$. In the first case $(a, b) = 0'$, where $0'$ is the zero element of the quotient field. In the second case $(a, b) > 0'$. In the third case $(-a)b > 0$ shows that $(-a, b) > 0'$ and hence that $-(a, b) > 0'$. Hence property 1 holds for the quotient field. To prove property 2 for addition, we start with $(a, b) > 0'$ and $(c, d) > 0'$ and seek to show that $(ad + bc, bd) > 0'$. The last inequality is equivalent to $abd^2 + b^2cd > 0$; this is true, since $(a, b) > 0'$ is equivalent to $ab > 0$ and $(c, d) > 0'$ to $cd > 0$, while d^2 and b^2 are both positive elements of D.

13. There are 6 units.

14. The units of $\mathbf{Q}$ are the nonzero elements of $\mathbf{Q}$.

15. The associative and commutative properties for multiplication follow directly. Every pair (c, c) with $c \neq 0$ is an identity for multiplication; all such are equivalent. If $ab \neq 0$, the inverse of (a, b) is (b, a), since $(a, b)(b, a) \cong (ab, ab) \cong (1, 1)$.

16. Yes.

18. (iii) Now $a > 0$ implies that $a^{-1} > 0$, for $aa^{-1} = 1$ shows that a^{-1} is neither zero nor negative. Similarly, $b^{-1} > 0$. Hence $a^{-1}b^{-1} > 0$. Thus $0 < a < b$ implies that $aa^{-1}b^{-1} < ba^{-1}b^{-1}$, which implies that $b^{-1} < a^{-1}$.

20. To prove property iv, see that if c is in D^+, then $(a - b)c$ is in D^+ and hence $ac - bc$ is in D^+, which implies that $ac > bc$. For property v, begin by noting that if $c < 0$, then $-c > 0$.

27. Yes, for suppose that $b < 0$. Then property v shows that $(-b) > 0$, which establishes property 1. Also, properties 2 and ii are equivalent.

28. $\{c + di\}$ for c and d rational numbers.

30. Yes.

31. The field S is isomorphic to the field of rational numbers.

34. No.

Section 7

2. Let S be some nonempty set of integers $n \geq s$. Let $T = \{n - s + 1\}$ for all n in S. Every element of T is a positive integer. Since the set of positive integers is well-ordered, T has a least element. Call it $n_t - s + 1$, where n_t is in S. Now n_t is the least element in S, for if $n' < n_t$ were in S, $n' - s + 1$ would be in T.

6. First compute $u_{4s+4} = u_{4s+3} + u_{4s+2} = 2u_{4s+2} + u_{4s+1} = 3u_{4s+1} + 2u_{4s}$. Now $u_4 = 3$ implies, taking $s = 1$, that u_8 is divisible by 3. Then take $s = 2$ to see that u_{12} is divisible by 3 and so complete the proof by an induction argument.

9. By the binomial theorem, $(1 + 1/n)^n = 1 + n(1/n) + K$, where K is a positive number. Hence $(1 + 1/n)^n > 2$ for all integers $n > 1$.

12. Let $k = a/d$, where a and d are relatively prime integers and $d > 0$. Consider the set T of all integers $dx - a$ that are positive for integers x. There is a one-to-one correspondence between the elements of T and S, since $dx - a > 0$ if and only if $x > a/d$. Let T' be any nonempty subset of T and S' the corresponding subset of S. Since the set of positive integers is well-ordered, T' has a least element. Call it t_0 and let x_0 be

determined by $t_0 = dx_0 - a$. Then x_0 is the least element of S'. This shows that S is well-ordered.

13. Yes, by the following argument: Let S be a nonempty set of integers less than some integer k. Define the set $T = \{k - s\}$ for all s in S. This is a set of positive integers and is nonempty, since S is. Hence T has a least element, call it $k - s_0$, for s_0 in S. Thus $k - s_0 \le k - s$ for all s in S. Hence $s \le s_0$ for all s in S, and s_0 is the maximum element of S.

17. Consider the set $S = \{60t\}$ for t in T. This is a set of positive integers and hence has a least element s. Then $s \le 60t$ for all t in T and $s = 60t_0$ for some t_0 in T. Hence $t_0 = s/60$ is the least element in T.

18. If r is an integer, Exercise 13 gives the answer. Assume that r is not an integer and let G be the set of all integers greater than r and S the set of all integers less than r. Since the set of reals is ordered, every integer is in one of S and G. Now, the second assumption implies that G is not empty and also that S is not empty, since it shows the existence of an integer $-s$ such that $-s > -r$. Let $W = \{g - s\}$ for all g in G and s in S. Now W is not empty and all its elements are positive integers. Hence it has a least element. Call it $g_0 - s_0$. Since $g_0 - s_0$ is an integer and positive, $g_0 - s_0 \ge 1$. If $g_0 - s_0 = 1$, then $s_0 + 1 = g_0$ is in G. Hence s_0 is the maximum integer less than r. If $g_0 - s_0 > 1$, then $g_0 - (s_0 + 1) > 0$. Now $s_0 + 1$ is not in S, since $g_0 - s_0$ is the smallest number in W. Hence $s_0 + 1$ is in G and we are back in the previous case.

Section 9

1. Since $x^3 + 2x^2 + 2x + 4 = (x + 1)(x^2 + x + 1) + 3$, we have divisibility only in $\mathbf{Z}_3$.

5. No.

6. To prove part 2, note that $ar = b$ and $as = c$ for integers r and s implies that $b + c = ar + bs = a(r + s)$.

7. From Theorem 9.1 there are integers q and r such that $r = 0$ or $0 < r < b$. In the former case $a - bq = 0$ shows that the result holds. In the latter case there are two possibilities. First, $0 < r < b/2$, in which case $-b < 2r \le b$, as required. Second, if $b/2 < r < b$, then $a = b(q + 1) + (r - b)$. In this case $-b/2 < r - b < 0$, and the desired result holds for $q + 1$ in place of q and $r - b$ in place of r. Uniqueness follows.

10. The units among the Gaussian integers are 1, -1, i, and $-i$.

11. Suppose that $f(x) = x^2 + 1$ and $g(x) = 3x + 2$. Then since $q(x)$ and $r(x)$ in $f(x) = q(x)g(x) + r(x)$ of Theorem 9.1p are uniquely determined,

the leading coefficient of $q(x)$ must be $\frac{1}{3}$, which is not an integer. One requirement that would make the theorem hold for polynomials with integral coefficients would be that $g(x)$ have leading coefficient 1. There are other requirements that would do.

13. Using the method of proof of Theorem 9.5, let $n(s + t\sqrt{-5} - [u + w\sqrt{-5}]) = (s - u)^2 + 5(t - w)^2$. If $s = t = \frac{1}{2}$, it follows that $|s - u| \geq \frac{1}{2}$ and $|t - w| \geq \frac{1}{2}$ for all integers u and w. Hence the n-value is not less than $\frac{1}{4} + \frac{5}{4} = \frac{6}{4} > 1$.

Section 11

1. (a) 7.

3. (a) From Theorem 9.5, the Gaussian integers form a Euclidean domain with $n(\alpha) = \|\alpha\|$. The first step in the Euclidean algorithm process is to find Gaussian integers α and ρ such that

$$2 + 11i = \alpha(1 + 3i) + \rho, \qquad \text{with } \|\rho\| < \|1 + 3i\|;$$

that is, $(2 + 11i)/(1 + 3i) = \alpha + \rho/(1 + 3i)$. We want to find a Gaussian integer α so that $\|(2 + 11i)/(1 + 3i) - \alpha\| < 1$; that is, $\|\frac{1}{2}(7 + i) - \alpha\| < 1$. Such an α is $\alpha = 3$. So we compute

$$2 + 11i = 3(1 + 3i) + (-1 + 2i), \qquad \text{where } \|-1 + 2i\| = 5 < \\ \|1 + 3i\|.$$

Now, for the second step in the process, note that $(1 + 3i)/(-1 + 2i) = 1 - i$, and hence

$$1 + 3i = (-1 + 2i)(1 - i) + 0.$$

Hence $-1 + 2i$ is a g.c.d. of $1 + 3i$ and $2 + 11i$.

5. $x^4 + x^2 + 1 = (x^2 + 1)^2 - x^2 = (x^2 + 1 - x)(x^2 + 1 + x)$ in $\mathbf{Q}$ and $\mathbf{R}$. In $\mathbf{C}$, $x^2 - x + 1 = [x - \frac{1}{2}(1 + \sqrt{-3})][x - \frac{1}{2}(1 - \sqrt{-3})]$. In $\mathbf{Z}_3$, $x^2 - x + 1 = x^2 + 2x + 1 = (x + 1)^2$. Also, $2x^3 + x^2 - 2x - 6 = (2x - 3)(x^2 + 2x + 2)$. In $\mathbf{Z}_3$, $x^2 + 2x + 2$ is a prime polynomial.

8. Suppose that $ax_1 + by_1 = c = ax_0 + by_0$. Then $a(x_1 - x_0) = b(y_0 - y_1)$. Since a and b are relatively prime, $x_1 - x_0 = kb$ for some integer k. Thus $y_0 - y_1 = ak$. So $x_1 = x_0 + kb$ and $y_1 = y_0 - ka$.

10. Suppose that $f(x)$ and $g(x)$ were relatively prime. Then $f(x)h(x) + g(x)k(x) = 1$ for some polynomials $h(x)$ and $k(x)$. Replacing x by c gives a contradiction.

23. Suppose that $2(x + y\sqrt{-5}) = 1 - \sqrt{-5}$. Then $2x = 1$ and $2y = -1$, which is impossible for integers x and y. To show that 3 is a prime, one

uses the same method as for 2 except that one has to take into account that $9 = x^2 + 5y^2$ has a solution $x^2 = 4$, $y^2 = 1$. The same method is used to prove that $1 + \sqrt{-5}$ and $1 - \sqrt{-5}$ are primes.

Section 13

1. (b) $\frac{2}{3}$. (e) Since there are no positive roots, the only possibilities are -1, -2, -3, -6, $-\frac{1}{2}$, $-\frac{3}{2}$. But if $x = a$ is a root for a an integer, the form of the equation shows that a is divisible by 3, eliminating -1 and -2. Similarly, -3 and -6 can be eliminated, leaving only the last two to check. A quicker way to show that there are no rational roots is to use Eisenstein's criterion for irreducibility.

2. Now $-1 \pm \frac{13}{2} = (-2 \pm 13)/2$ cannot be a zero of $f(x)$, since the numerator is not a factor of 12. So there are no rational zeros.

4. All rational roots of $x^2 - c = 0$, for c an integer, are integers.

6. $f(y + 1) = [(y + 1)^p - 1]/y$. In $(y + 1)^p$ all but the first and last terms are divisible by p. The coefficient of y in $(y + 1)^p$ is p. Thus Eisenstein's criterion can be applied.

10. (b) 2.

11. (b) 4.

12. Substitution shows that $cx_1 + d = cx_2 + d = 0$, which implies that $c = d = 0$ if $x_1 \neq x_2$.

Section 15

1. (b)

	1	x	x^2	$x+1$	x^2+1	x^2+x	x^2+x+1
1	1	x	x^2	$x+1$	x^2+1	x^2+x	x^2+x+1
x	x	x^2	$x+1$	x^2+x	1	x^2+x+1	x^2+1
x^2	x^2	$x+1$	x^2+x	x^2+x+1	x	x^2+1	1
$x+1$	$x+1$	x^2+x	x^2+x+1	x^2+1	x^2	1	x
x^2+1	x^2+1	1	x	x^2	x^2+x+1	$1+x$	x^2+x
x^2+x	x^2+x	x^2+x+1	x^2+1	1	$1+x$	x	x^2
x^2+x+1	x^2+x+1	x^2+1	1	x	x^2+x	x^2	$x+1$

2. (b) $x - 1$.   (c) $-x^2 + x$.

5. The quadratic over F' is $(x - \sqrt{2} - \sqrt{3})(x - \sqrt{2} + \sqrt{3}) = (x - \sqrt{2})^2 - 3$.

9. Suppose that $a^3 \equiv 2 \pmod 7$. Then $a^6 \equiv 4 \pmod 7$, which contradicts Fermat's theorem. Hence $x^3 = 2$ has no roots in $\mathbf{Z}_7$. Thus $x^3 - 2$ is irreducible in $\mathbf{Z}_7$. If $c^3 = 2$, then $2c$ and $4c$ are also roots of $x^3 - 2 = 0$ in $\mathbf{Z}_7(c)$. These are the three roots.

11. $g(y) = 3y^2 + 2y - 5$, which factors in $\mathbf{Q}$. The splitting field is $\mathbf{Q}(\sqrt{-3}, \sqrt{-11})$.

Section 17

1. $(I + N)^{1/3} = I + (\frac{1}{3})N + (\frac{1}{3})(-\frac{2}{3})(\frac{1}{2})N^2$ by formal expansion. Thus a cube root of $I + N$ should be $I + (\frac{1}{3})N - (\frac{1}{9})N^2$. If we cube this expression, we have $[I + (\frac{1}{3})N]^3 + 3[I + (\frac{1}{3})N]^2[-(\frac{1}{9})N^2] = I + N$.

3. First, there is not a square root of the form $aI + bN$ for a field of characteristic 2 since $(aI + bN)^2 = a^2I \neq I + N$ unless $N = (0)$. The following is a matrix N over $\mathbf{Z}_2$ such that $I + N$ has no square root. Suppose that

$$N = \begin{bmatrix} 0 & 1 \\ 0 & 0 \end{bmatrix}; \qquad \text{then } N^2 = (0) \quad \text{and} \quad I + N = \begin{bmatrix} 1 & 1 \\ 0 & 1 \end{bmatrix}.$$

Now if $\begin{bmatrix} a & b \\ c & d \end{bmatrix}^2 = I + N$, we would have $a^2 + bc = 1 = bc + d^2$ and $ab + bd = 1$. The first two equations imply that $a^2 = d^2$ and hence $a + d = 0$, which denies the third equation.

5. $(1)\sigma = 1$ implies that $(n)\sigma = n$ for all integers n (mod m). Hence the only automorphism is the identity.

7. $(x^p - 1) = (x - 1)^p$ in $\mathbf{Z}_p$.

9. Try products.

12. Since we are dealing with polynomials, there is no trouble with convergence. Since $f(x) = f(x - c + c)$, we can write $f(x)$ as a sum of powers of $(x - c)$ with coefficients in F', an extension of F that contains c. So we have

$$f(x) = a_0 + a_1(x - c) + a_2(x - c)^2 + \cdots + a_k(x - c)^k + \cdots$$
$$+ a_n(x - c)^n.$$

The expression terminates, since $f(x)$ is of degree n, and the coefficients are in F'. Then, using the formulas for the derivatives, we have

$$f^{(k)}(x) = (k!)a_k + q_k(x)(x - c),$$

where $q_k(x)$ is a polynomial in x with coefficients in F'. This implies that

$$f^{(k)}(c) = (k!)a_k; \qquad \text{that is, } a_k = f^{(k)}(c)/(k!).$$

Chapter **IV**

Section 3

1. (b) The sum does not exist.   (c) (4, 6, 23).

2. (c) Yes, since $a^3 + b^3 = 0$ *and a* and *b* real implies that $a = -b$ and similarly for *c* and *d*. This shows closure under addition and multiplication by a scalar. However, over the field of complex numbers one does not have a vector space.

3. (b) $\alpha = (\frac{1}{3}, \frac{2}{3}, \frac{4}{3})$.

4. (b) There is no solution.   (c) There is a solution.

6. No.

7. The set of solutions of each equation is a vector space. Why? The set of solutions of both equations is the intersection of the set of solutions of the first equation with the set of solutions of the second. (Recall that the intersection of two sets is the set of elements common to the two sets.)

12. To show that S_2 spans V^3, note the following: $(1, 1, 1) - (1, -1, 1) = (0, 2, 0)$; $(1, 1, 1) - (-1, 1, 1) = (2, 0, 0)$; $(1, -1, 1) + (-1, 1, 1) = (0, 0, 2)$. Thus the vectors $(0, 1, 0)$, $(1, 0, 0)$, and $(0, 0, 1)$ are in S_2, which shows that S_2 spans V^3.

13. Yes.

16. $\alpha + (-1)\alpha = 1 \cdot \alpha + (-1)\alpha = [1 + (-1)]\alpha = 0 \cdot \alpha = \theta$, using in succession properties 5, 4, and 6 of a vector space.

17. See Theorem 3.1 in Chapter I.

Section 5

1. (d) They are linearly independent.   (e) They are linearly independent.

3. The vector $(-1, 5)$ is the resultant of $r\alpha$ and $s\beta$. It remains to show this on the graph.

5. The conditions imply that $(c, d) = x(a, b)$ and $(e, f) = y(a, b)$ are both solvable for *x* and *y*.

7. We need to solve: $0.40x + 0.70y = c$, for *x* and *y* nonnegative real numbers whose sum is 1, to get 1 gallon of the mixture. We have $0.4x + 0.7(1 - x) = c$, which is equivalent to $7 - 10c = 3x$. This is solvable for *x* with $0 \le x \le 1$ provided that $7 - 10c \ge 0$, and $7 - 10c \le 3$, that

is, $0.7 \geq c \geq 0.4$. (The limits on c can also be seen by common sense.) We are involved with linear combinations.

10. Let the vectors α and β correspond to the points A and B. If r and s are both positive, the point corresponding to $r\alpha + s\beta$ is on the line AB between A and B. What happens otherwise?

12. (a) Since the given vectors are triples, they are all in V^3. Hence at most three vectors are needed to span the space. The first two vectors are linearly independent. To see if the first three are, look at

$$x(1, 2, 3) + y(1, 0, 5) + z(1, 1, 4) = (0, 0, 0).$$

This yields the equations $x + y + z = 0$, $2x + z = 0$, and $3x + 5y + 4z = 0$. Replacing z by $-2x$ in the first and third equations gives us $-x + y = 0$ and $-5x + 5y = 0$, which hold if and only if $x = y$. The third triple is a linear combination of the first, and we may eliminate it from consideration. Then, to try the last triple, we look at

$$x(1, 2, 3) + y(1, 0, 5) + z(3, 2, 1) = (0, 0, 0).$$

Here the only solution is $x = y = z = 0$ and the three vectors are linearly independent. Hence the vector space is of dimension 3 and is the same as V^3.

14. (b) S and T are isomorphic, since each has dimension 2. A basis for S is $(1, 1, 0)$, $(0, 0, 1)$ and for T: $(1, 0)$, $(0, 1)$. So we set up a one-to-one correspondence between the basis vectors, for example,

$$(1, 1, 0) \leftrightarrow (0, 1) \quad \text{and} \quad (0, 0, 1) \leftrightarrow (1, 0).$$

The S and T in part (c) are not isomorphic.

16. Since 3 is the dimension of the space spanned by the four vectors, three must be linearly independent and the fourth a linear combination of the three. Assume that α, β, and γ are linearly independent and $\delta = r\alpha + s\beta + t\gamma$, for r, s, t in the field. Then $\theta = a\alpha + b\beta + c\gamma + d\delta = (a + rd)\alpha + (b + sd)\beta + (c + td)\gamma$. This holds if and only if the coefficients of α, β, γ are all zero. Hence

$$W = \{(-rd, -sd, -td, d)\} = \{d(-r, -s, -t, 1)\},$$

where r, s, t are fixed and d is any element of the field. So W is of dimension 1.

19. $(a, b, c) = \frac{1}{2}(-a + b + c)(0, 1, 1) + \frac{1}{2}(a - b + c)(1, 0, 1) + \frac{1}{2}(a + b - c)(1, 1, 0)$.

22. $r\alpha_1 + s\alpha_2 = (rx_1 + sx_2, ry_1 + sy_2, rz_1 + sz_2)$. Then $a(rx_1 + sx_2) + b(ry_1 + sy_2) + c(rz_1 + sz_2) = r(ax_1 + by_1 + cz_1) + s(ax_2 + by_2 + cz_2) = 0$.

Section 7

1. $B + C = \begin{bmatrix} 8 & 7 & 4 \\ 4 & 0 & 5 \end{bmatrix}.$

2. $DA = [17 \ 29].$

3. $A + A^2 + A^3 = N$, where N is the matrix of order 3, each of whose elements is 1.

6. The inverse of $\begin{bmatrix} 1 & 4 \\ 0 & 1 \end{bmatrix}$ is $\begin{bmatrix} 1 & -4 \\ 0 & 1 \end{bmatrix}.$

8. (b) To find the inverse of the matrix, one can look at the following equation:

$$\begin{bmatrix} -1 & 1 & 1 \\ 1 & -1 & 1 \\ 1 & 1 & -1 \end{bmatrix} \begin{bmatrix} x \\ y \\ z \end{bmatrix} = \begin{bmatrix} r \\ s \\ t \end{bmatrix}.$$

The inverse will be the matrix of coefficients of r, s, t in the solution for x, y, z. We want to solve the following set of equations: $-x + y + z = r$, $x - y + z = s$, $x + y - z = t$. Adding the first two equations gives $z = \frac{1}{2}(r + s)$. Similarly, $x = \frac{1}{2}(s + t)$ and $y = \frac{1}{2}(r + t)$. Hence the inverse matrix is

$$\begin{bmatrix} 0 & \frac{1}{2} & \frac{1}{2} \\ \frac{1}{2} & 0 & \frac{1}{2} \\ \frac{1}{2} & \frac{1}{2} & 0 \end{bmatrix}.$$

11. Let S be the set of solutions of $\xi A = 0$. Use Theorem 2.2 and note that if $\xi_1 A = \xi_2 A = 0$, then $(\xi_1 + \xi_2)A = 0$; this shows closure under addition. Finally, $\xi A = 0$ implies that $(c\xi)A = 0$ for every c in the field. Hence there is also closure under multiplication by a scalar.

12. No.

14. If $A = (a_{ij})$ and $B = (b_{ij})$, then $\text{tr}(AB) = \sum a_{ij}b_{ji} = \sum b_{ji}a_{ij} = \text{tr}(BA)$, where the sums are over i and j between 1 and n inclusive.

15. Let $A = (a_{ij})$, $B = (b_{ij})$, $C = (c_{ij})$, and see that the element in ith row and jth column of $A(B + C)$ is

$$\sum a_{ik}(b_{kj} + c_{kj}) + \sum a_{ik}b_{kj} + \sum a_{ik}c_{kj},$$

where the sums are over k between 1 and n inclusive.

19. We show that the product of two matrices is nonsingular if and only if each is nonsingular. If A and B are nonsingular, then $(AB)(B^{-1}A^{-1}) = I$ shows that AB is nonsingular. Conversely, if AB is nonsingular, then

$ABC = I$ for some matrix C. Then A and C are both nonsingular. Further-more, $BC = A^{-1}$ implies that $B = A^{-1}C^{-1}$ and $BCA = I$, which shows that B is nonsingular.

Section 9

1. The determinants of the first and last matrices are 23 and 0, respectively.

3. $n!$

5. One basis is $(1, 0, 1)$ and $(0, 1, 1)$.

7. $\det(CAC^{-1}) = (\det C)(\det A)(\det C^{-1})$. Use of the result of Exercise 6 will complete the proof.

10. We can use polar coordinates: $x_1 = r_1 \cos \varphi_1$, $y_1 = r_1 \sin \varphi_1$, $x_2 = r_2 \cos \varphi_2$, $y_2 = r_2 \sin \varphi_2$. From a figure you can see that the area of the triangle with vertices $(0, 0)$, (x_1, y_1), (x_2, y_2) is the absolute value of $\frac{1}{2}[r_1 r_2 \sin(\varphi_1 - \varphi_2)] = \frac{1}{2}[r_1 r_2 \sin \varphi_1 \cos \varphi_2 - r_1 r_2 \sin \varphi_2 \cos \varphi_1] = \frac{1}{2}[y_1 x_2 - x_1 y_2]$. Hence the area of the parallelogram is the absolute value of $x_1 y_2 - x_2 y_1$.

13. If the three points are collinear, the matrix of order 3 in the lower right corner is singular (see Exercise 9). Thus in the determinant the coefficient of $x^2 + y^2$ is 0, which implies that the determinant is a linear function of x and y.

16. One can show this result by writing out the matrices involved. Alge-braically it may be shown as follows. First, notice that $I_{ij}(k) = I + M_{ij}(k)$, where $M_{ij}(k)$ is a matrix all of whose elements are zero except for a k in the jth row and ith column. Then $B - A = AM_{ij}(k)$. Now the element in the rth row and sth column of $AM_{ij}(k)$ is $\sum a_{rt}m_{ts}$, where the sum is over t and $M_{ij}(k) = (m_{rs})$. But $m_{ts} = 0$ unless $t = j$ and $s = i$, and $m_{ji} = k$. Hence the sum is zero except for $t = j$ and $s = i$, when it re-duces to $a_{rj}k$ for $1 \leq r \leq n$. Therefore, B and A are the same except that in order to get the ith column of B, we add k times the jth column of A to its ith column.

17. We show that the second operation of Theorem 9.2 does not change the vector space spanned by the rows of the matrix. Let $\rho_1, \rho_2, \ldots, \rho_n$ denote the rows of the matrix and S the vector space which they span. Let T be the space spanned by the set $\rho_1, \rho_2 + k\rho_1, \rho_3, \ldots, \rho_n$. To show that T and S are the same space, consider

$$x_1\rho_1 + x_2\rho_2 + x_3\rho_3 + \cdots + x_n\rho_n$$
$$= (x_1 - kx_2)\rho_1 + x_2(\rho_2 + k\rho_1) + x_3\rho_3 + \cdots + x_n\rho_n.$$

This shows that $S \subseteq T$. Similarly, one can show that $T \subseteq S$.

18. Consider x_1 as a variable and see that the determinant is a polynomial $f(x_1)$ of degree $n - 1$ or less in x_1 with coefficients that are polynomials in the x_i for $2 \leq i \leq n$. Since the determinant is zero if x_1 is replaced by any x_i with $i \neq 1$, we see that $x_1 - x_i$ is a factor of $f(x_1)$ for each $i \neq 1$. We use the same process for each x_j instead of x_1 and see that the determinant is kP, where $P = \prod (x_i - x_j)$ for $1 \leq j < i \leq n$ and k is a function of the x_i. Since P is of degree $n - 1$ in each x_i, k must be independent of the x_i. Now one term of the determinant is the product of the elements along the principal diagonal, that is, $x_2 x_3^2 x_4^3 \cdots x_n^{n-1}$. If we look at the product P, we see that by taking the first x_i in each difference $x_i - x_j$ we get the same product of powers of the x_i. This shows that k above must be 1.

Section 10

1. $C^{-1}MC = C'^{-1}MC'$ for all M if and only if $C'C^{-1}M = MC'C^{-1}$ for all nonsingular matrices M. This is true if and only if $C'C^{-1}$ is in the center of the multiplicative group of nonsingular matrices of order n.

4. The first row of a nonsingular matrix over $\mathbf{Z}_p$ can be any ordered pair of elements of $\mathbf{Z}_p$ except the null vector. Hence the number of possibilities for the first row is $p^2 - 1$. The second row can be any ordered pair that is not dependent on the first row, that is, which is not a multiple of the first row by an element of $\mathbf{Z}_p$. Hence the number of matrices in $M_2^*(F)$ for $F = \mathbf{Z}_p$ is $(p^2 - 1)(p^2 - p)$. The rest is left to you.

Chapter **V**

Section 3

3. To prove Corollary 1, note that if $r \not\equiv 1 \pmod 4$, the corollary is a restatement of part 1 of the theorem. Suppose that $r \equiv 1 \pmod 4$. Then $\sqrt{r} = 2\alpha - 1$ and $a + b\sqrt{r} = a + b(2\alpha - 1) = x + y\alpha$, where $x = a - b$ and $y = 2b$. Now if a and b are both integers or both halves of odd integers, then x and y are integers. Conversely, if x and y are both integers, then a and b are both integers or halves of odd integers.

4. To factor 3, write $3 = \frac{1}{2}(a + b\sqrt{-11})\frac{1}{2}(c + d\sqrt{-11})$ and try to determine integers a, b, c, d with $a \equiv b \pmod 2$ and $c \equiv d \pmod 2$. Then also $3 = \frac{1}{2}(a - b\sqrt{-11})\frac{1}{2}(c - d\sqrt{-11})$ and

$$9 = \tfrac{1}{4}(a^2 + 11b^2)\tfrac{1}{4}(c^2 + 11d^2).$$

Now if neither factor of 3 is to be a unit, it follows that $a^2 + 11b^2 = 12 = c^2 + 11d^2$. This is possible only if $a^2 = b^2 = c^2 = d^2 = 1$. So if we try $a = b = 1$, we see that $3/[\frac{1}{2}(1 + \sqrt{-11})] = \frac{1}{2}(1 - \sqrt{-11})$, which is an algebraic integer.

6. Use the same kind of argument as in Section 1.

7. If $r \not\equiv 1 \pmod 4$, then the units of $\mathbf{Q}(\sqrt{r})$ are $a + b\sqrt{r}$, where $a^2 - rb^2 = 1$. If $r \equiv 1 \pmod 4$, the units are $\frac{1}{2}(a + b\sqrt{r})$, where $a^2 - rb^2 = 4$. In both cases a and b are integers. It remains to find for what negative r, b must be zero.

9. If a and b are integers and $\sqrt{r}$ is not rational, then $(a + b\sqrt{r})^k$ can be written $a_k + b_k\sqrt{r}$, where a_k and b_k are integers. If $(a + b\sqrt{r})(a - b\sqrt{r}) = \pm 1$, then $(a_k + b_k\sqrt{r})(a_k - b_k\sqrt{r}) = (\pm 1)^k$. From this the answer can quickly be deduced.

11. Define $f(x)$ and $g(x)$ as in the proof of Theorem 3.2 and

$$F(x) = \prod (x - e_i d_j).$$

We want to define $z_j = x/e_j$ and hence need to avoid $e_j = 0$. First, deal with the case when some zeros of $f(x)$ and $g(x)$ may be zero and then exclude them from further consideration. The rest of the proof is like that of Theorem 3.2.

13. (b) When we use the method of proof of the fundamental theorem on symmetric functions, we have $h_0 = \sum x_1^4$, $b = 4$, $k = 1$, and the summation, and all those below are over all permutations of the x_i. Then

$$c_1^4 = (x_1 + x_2 + x_3)^4 = (x_1 + x_2)^4 + 4(x_1 + x_2)^3 x_3 + \cdots$$
$$= h_0 + 4\sum x_1^3 x_2 + 6\sum x_1^2 x_2^2 + 12\sum x_1^2 x_2 x_3.$$

Now

$$\sum x_1^3 x_2 = \left(\sum x_1^2\right)\left(\sum x_1 x_2\right) - \sum x_1^2 x_2 x_3 = (c_1^2 - 2c_2)c_2 - c_1 c_3,$$

$$\sum x_1^2 x_2^2 = \left(\sum x_1 x_2\right)^2 - 2\sum x_1^2 x_2 x_3 = c_2^2 - 2c_1 c_3.$$

Hence

$$c_1^4 = h_0 + 4(c_1^2 - 2c_2)c_2 - 4c_1 c_3 + 6c_2^2,$$
$$h_0 = c_1^4 - 4c_1^2 c_2 + 2c_2^2 + 4c_1 c_3.$$

14. (b) The answer is the same as that for 13(b) minus $4c_4$.

19. To take the step from $k = 2$ to $k = 3$, write

$$\sum x_1^3 = \left(\sum x_1^2\right)\left(\sum x_1\right) - \sum x_1^2 x_2 = s_2 c_1 - \left[\left(\sum x_1\right)\left(\sum x_1 x_2\right) - 3c_3\right],$$

where the sums are over all permutations of the subscripts. Thus $s_3 = s_2c_1 - c_1c_2 + 3c_3$ and $s_3 - c_1s_2 + c_2s_1 - 3c_3 = 0$. (This and the answer for $k = 4$ can be used to check the answers of Exercise 13.)

20. Suppose that $g(x)$ is a reducible polynomial and write $g(x) = f(x)h(x)$, where f and h are of positive degrees r and s, respectively. Then

$$p(x) = x^n g\left(\frac{1}{x}\right) = \left[x^r f\left(\frac{1}{x}\right)\right]\left[x^s h\left(\frac{1}{x}\right)\right]$$

shows that $p(x)$ is reducible.

21. Let $b = s/t$, where 1 is the g.c.d. of s and t. Then if rs^2/t^2 is to be an integer, t^2 must be a factor of rs^2. Since 1 is the g.c.d. of s and t, t^2 must divide r. This almost completes the proof.

Section 5

1. $\mathcal{D}\mathcal{C} = (3)$.

4. $\mathcal{D} + \mathcal{C}$ contains $12 = (6 + \sqrt{-3}) + (6 - \sqrt{-3})$ and also 7. Hence it contains 5 and 2 and therefore 1. Hence $\mathcal{D} + \mathcal{C} = \mathbf{R}$, where $\mathbf{R}$ is the ring of algebraic integers in $\mathbf{Z}[\sqrt{-3}]$.

5. Take $\mathcal{E} = (3, c + \sqrt{-5})$ and determine c properly.

9. $\mathcal{T}$ cannot divide (x), since (x) is a prime ideal. Why?

13. Let $\mathcal{A} + \mathcal{B} = \mathbf{R}$ and see that $\mathcal{A}\mathcal{C} + \mathcal{B}\mathcal{C} = \mathcal{C}$. This shows that $\mathcal{A} \supseteq \mathcal{C}$.

14. To show that $\mathcal{B} = (4, x)$ is a prime ideal, suppose that $\mathcal{D}\mathcal{E} = \mathcal{B}$. Since $\mathcal{D}$ and $\mathcal{E}$ contain $\mathcal{B}$, they contain 4 and x. If $\mathcal{D} \neq \mathbf{R}$, then $\mathcal{D}$ contains no odd integer. If all the integers in $\mathcal{D}$ are multiples of 4, then $\mathcal{D} = \mathcal{B}$ and $\mathcal{E} = \mathbf{R}$. Hence the only possibility is $\mathcal{D} = \mathcal{E} = (2, x)$. But then $\mathcal{D}\mathcal{E} = (4, 2x, x^2)$, which does not contain $\mathcal{B}$, since $\mathcal{D}\mathcal{E}$ does not contain x. Hence $\mathcal{B}$ is a prime ideal. But it is not maximal.

19. Here we may follow the pattern of the proof of Theorem 4.1. The norms (that is, the squares of the absolute values) of the Gaussian integers are positive integers and hence are well-ordered. Thus any set of Gaussian integers has an element (or elements) of least norm. Call such an element μ. Now by Theorem 9.5 of Chapter III the Gaussian integers form a Euclidean domain. Furthermore, μ is not only an element of S but (see the proof of Theorem 4.1) is a factor of each element of S. Thus for every σ in S, there is a τ in S such that $\mu\tau = \sigma$. So $S = (\mu)$.

21. Let $\mathcal{A}, \mathcal{B}$, and $\mathcal{C}$ be three ideals over $\mathbf{R}$. Then $\mathcal{A}(\mathcal{B} + \mathcal{C})$ consists of all sums of elements of the form $a(b + c)$, where a, b, and c are in $\mathcal{A}, \mathcal{B}$, and $\mathcal{C}$, respectively. Now $a(b + c) = ab + ac$, since $\mathbf{R}$ is distributive. Hence all sums of terms $a(b + c)$ consists of all sums $ab + ac$.

Section 7

5. We want to find integers a and b such that $\mathscr{C}\mathscr{D} = (3)$, where $\mathscr{C} = [3, a + \sqrt{-5}]$ and $\mathscr{D} = [3, b + \sqrt{-5}]$. Now $\mathscr{C}\mathscr{D} = [9, 3(a + \sqrt{-5}), 3(b + \sqrt{-5}), ab - 5 + (a + b)\sqrt{-5}]$. Try $a = 1 = -b$ so that 3 will divide every element of $\mathscr{C}\mathscr{D}$. Then $1 + \sqrt{-5} + 1 - \sqrt{-5} = 2$, which shows that 3 is in $\mathscr{C}\mathscr{D}$.

8. First notice that if m is not a prime number, then (m) is certainly not a prime ideal. Thus assume that m is a prime. Then choose $\mathscr{C} = [m, c + \sqrt{-5}]$ and $\mathscr{D} = [m, c - \sqrt{-5}]$ and see that $\mathscr{C}\mathscr{D} \subseteq \mathscr{A}$, since m is a factor of $c^2 + 5$. But $\mathscr{C}\mathscr{D}$ also contains $2mc$. If m and $2c$ are relatively prime, we can choose k so that $2ck \equiv 1 \pmod m$, which implies that m is in $\mathscr{C}\mathscr{D}$ and hence that $\mathscr{C}\mathscr{D} = \mathscr{A}$. Now $m \mid c^2 + 5$ shows that m and c are relatively prime unless $m = 5$. One must consider separately the cases $m = 2$ and $m = 5$.

14. No.

15. If $\alpha = 1 + \sqrt{-5}$, then $\Delta(\alpha) = -2\sqrt{-5}$ and $\Delta^2(\alpha) = -20$.

17. Let $\gamma_1, \gamma_2, \ldots, \gamma_n$ be a linearly independent set of elements of $\mathscr{A}$ (over **Q**). Since the α_i form an integral basis, then

$$\gamma_i = \sum_j c_{ij}\alpha_j, \quad 1 \leq i \leq n,$$

for integers c_{ij}. From this it follows that $\gamma_{ik} = \sum_j c_{ij}\alpha_{jk}$ and hence $\Delta(\gamma) = C\,\Delta(\alpha)$, where $C = (c_{ij})$. Since C has integral elements and is not singular, its determinant must be not less than 1 in absolute value. This implies that $|\det \Delta^2(\gamma)| \geq |\det \Delta^2(\alpha)|$.

First, if the β_i form an integral basis, we see that not only is $|\det \Delta^2(\beta)| \geq |\det \Delta^2(\alpha)|$ from the above, but, interchanging the roles of α and β, we see that the inequality holds in the other direction, too. Hence the determinants of $\Delta^2(\beta)$ and $\Delta^2(\alpha)$ are equal in absolute value. Second, if the β_i do not form an integral basis, then there is a linearly independent set γ_i such that $|\det \Delta^2(\gamma)| < |\det \Delta^2(\beta)|$. If the squares of the determinants of $\Delta(\beta)$ and $\Delta(\alpha)$ were equal in absolute value, then $|\det \Delta^2(\gamma)|$ would be less than $|\det \Delta^2(\alpha)|$, which denies what we showed above.

18. If $\mathscr{A}$ is not the zero ideal and contains $\mathscr{B}$, then $\mathscr{A}$ is a divisor of $\mathscr{B}$ and $\mathscr{D}\mathscr{A} = \mathscr{B}$. If $\mathscr{D}\mathscr{A} = \mathscr{D}'\mathscr{A}$, then Theorem 7.2 implies that $\mathscr{D} = \mathscr{D}'$.

Section 9

2. Note that S is closed under addition. To see that it is a right ideal, let M be a matrix of S and T a matrix of R. Then MT has a zero first row and hence is in S.

5. Suppose that $(N + c)^k \subseteq N$ for some positive integer k. Then every element of $N + c$ is nilpotent and hence c is nilpotent. Why? Thus c is in N and $N + c = N$.

7. One can find an example in which $A \cap B$ is neither a right ideal nor a left ideal.

8. Now $r_1\sigma = r_2\sigma$ if and only if $(r_1 - r_2)\sigma = 0'$, the zero element of R'. That is, two elements of R have the same image in R' if and only if their difference is in the kernel K This means that r_1 and r_2 have the same image in R' if and only if $r_1 + K = r_2 + K$.

10. Denote the homomorphism by σ and see that for each r in R, $r\sigma = r'$ for some unique r' in R'. Define τ by $(r + K)\tau = r'$. Note that since τ is a transformation of the coset, if r_1 is in $r + K$, then $(r_1 + K)\tau = (r + K)\tau = r'$. We want to show that τ is an isomorphism. To show τ is a homomorphism, first notice that it takes each element of R/K into a unique element of R'. Second, $[(r_1 + K) + (r_2 + K)]\tau = r_1\sigma + r_2\sigma = (r_1 + K)\tau + (r_2 + K)\tau$. Third, one can show similarly that the correspondence preserves products. Hence τ is a homomorphism. To show it an isomorphism, see that if $(r + K)\tau = 0'$, then $r + K$ is in K, and hence $r + K$ is the zero coset of K.

11. Theorem 8.5 follows from Theorems 8.3 and 8.4.

14. Let 1 denote the identity element of R' and σ the homomorphism of Z onto R'. Then $1\,\sigma = 1$, $2\,\sigma = 2, \ldots, t\sigma = t$, for $t < m$. What is $m\sigma$? It must be $0'$, since it cannot be equal to any of the previous images and R' has m elements. Thus the kernel of σ is (m) and R' is isomorphic to $Z/(m)$.

16. To find an example of an ideal that has property P without being maximal, note from Theorem 8.6 that $R/\mathscr{A}$ must have infinitely many elements. So, choose $R = Z[x]$, $\mathscr{A} = (x)$. Then $R/\mathscr{A}$ is isomorphic to Z, which is an integral domain but not a field.

21. It is not circular reasoning.

24. (a) Since $\mathscr{A}$ is a right ideal, $ar = a'$ is in $\mathscr{A}$ for every r in R. Hence $(ar)\sigma = (a\sigma)(r\sigma) = a'\sigma$, which implies that $(\mathscr{A}\sigma)(r\sigma)$ is in $\mathscr{A}\sigma$.
(b) No, for R' might be a field.

Chapter **VI**

Section 3

3. If we let $2x = y$, the equation becomes $y^3 - 3y + 1 = 0$.

4. (b) If the lengths of the sides are denoted by a, b, c, then from trigonometry the area of the triangle is $A = \sqrt{s(s - a)(s - b)(s - c)}$, where $2s = a + b + c$, and the altitude is $2A/a$.

(e) The vertices of the regular 7-sided polygon in the unit circle are the roots of $x^7 - 1 = (x - 1)(x^6 + x^5 + x^4 + x^3 + x^2 + x + 1) = 0$. If we let $x + 1/x = y$, we see that

$$\frac{x^7 - 1}{x - 1} = x^3(y^3 + y^2 - 2y - 1)$$

and the cubic in y has no rational zeros.

5. One method of solution is to use the law of cosines twice, where m is the length of the median. Then $m^2 = (a^2/4) + b^2 - ab \cos C$ and $c^2 = a^2 + b^2 - 2ab \cos C$.

8. Yes.

9. A construction of $\sqrt{c}$, for c a positive real number, is as follows: Construct a semicircle with diameter $1 + c$ and let A be a point 1 unit from one end of the diameter. Then $\sqrt{c}$ is the length of the line segment from A to the semicircle perpendicular to the diameter.

12. If the quadratic polynomial is $ax^2 + bx + c$, by the formula, its zeros are $(-b \pm \sqrt{d})/2a$, where $d = b^2 - 4ac$. Then construct $\sqrt{-d}$, $-b/2a$, and $\sqrt{-d}/2a$ and see that the roots are $(-b/2a) \pm i\sqrt{-d}/2a$.

14. One can solve the equation of the line for x or y and substitute in the equation for the circle to get a quadratic equation in x or y with complex coefficients. Call it $ax^2 + bx + c = 0$. Then the construction is as described in the answer to Exercise 12.

16. The degree of the intermediate field over F would have to be a factor of the prime degree of K over F.

20. Yes.

21. No.

Section 4

1. $\sigma_2^2 = \sigma_1$, the identity automorphism, $\sigma_6^2 = \sigma_4$, $\sigma_6^3 = \sigma_7$, $\sigma_6^4 = \sigma_1$, $\sigma_2\sigma_6 = \sigma_5$, $\sigma_2\sigma_6^2 = \sigma_3$, $\sigma_2\sigma_6^3 = \sigma_8 = \sigma_6\sigma_2$.

3. See the answer to Exercise 9 of Section 9 in Chapter I. There is one more subgroup of order 4 and two more of order 2. One of the latter two is $H = \{\sigma_1, \sigma_4\}$. Now H takes each of the following into itself: $1, x_2x_3, x_2^2, x_2^3x_3$. These four elements are linearly independent over $\mathbf{Q}$ and span the field F corresponding to H.

4. See the multiplication table of the octic group in Section 9 of Chapter I and the answer to Exercise 3. The group $H = \{\sigma_1, \sigma_4\}$, being the center

of the octic group, is also normal. No other subgroup of order 2 is a normal subgroup, as may be seen by trial. All three subgroups of order 4 are normal.

9. Take $c = 1 + \sqrt{2}$ and $c' = 1 - \sqrt{2}$. Then $x_2^2 = 1 + \sqrt{2}$ and $x_3^2 = 1 - \sqrt{2}$ imply that x_2 is real and x_3 imaginary. Hence x_3 cannot be in $\mathbf{Q}(x_2)$.

10. In doing the exercise, it is helpful to notice that x_3 is not in $\mathbf{Q}(x_2)$.

12. To show $(0)\sigma = 0$, note that $(a)\sigma = (a + 0)\sigma = (a)\sigma + (0)\sigma$, which implies that $(0)\sigma = 0$.

13. The theorem gives information only about automorphisms of $\mathbf{Q}$. But from the proof, $(1)\sigma = 1$ implies that $(n)\sigma = n$ first for all integers n and then for all rational numbers, just as in the proof of Theorem 4.2.

*15. Let $h(x)$ be the irreducible polynomial over $\mathbf{Q}(\alpha)$ such that $h(\beta) = 0$. Call its degree u. Then the degree of $\mathbf{Q}(\alpha, \beta)$ over $\mathbf{Q}$ is ru. Similarly, call $k(x)$ the irreducible polynomial over $\mathbf{Q}(\beta)$ such that $k(\alpha) = 0$ and call its degree v. Then $ru = vs$. Since s and r are relatively prime, $r \mid v$ and $s \mid u$. But $u \le s$ and $v \le r$, which shows that $u = s$ and $v = r$. This completes the proof.

Section 7

1. Those less than 25 are 2, 3, 4, 5, 6, 8, 10, 12, 15, 16, 17, 20, and 24.

3. By Section 3, the roots of any cubic equation whose roots are all real can be constructed using the instrument I if every angle 3θ can be constructed using this instrument. Under the conditions imposed, each cubic equation for an intermediate field has three real roots. Hence I gives a construction.

4. No.

5. Trial shows that 6 is a generator of $\mathbf{Z}_{17}^*$ and hence $\lambda = \sigma_6$ is a generator of G. Then $r(\lambda^2)^k = r(\sigma_6^{2k}) = r^{6^{2k}} = r^{2k}$, since $6^2 \equiv 2 \pmod{17}$. Construction of a table for $2^k \pmod{17}$ shows that $t_1 = s_1$.

7. Since x_2 and x_3 are roots of $(x^2 - c)(x^2 - c') = x^4 - 2x^2 - 1$, where $c = 1 + \sqrt{2}$, the conjugates of x_2 are $x_2, -x_2, x_3, -x_3$. Since x_3 satisfies an equation of degree 2 over $\mathbf{Q}(x_2)$, namely, $x_3^2 = 2 - x_2^2$, the conjugates of x_3 are x_3 and $-x_3$ over $\mathbf{Q}(x_2)$. We wish to choose a rational number b so that the following are distinct:

$$x_2 + bx_3, \quad -x_2 + bx_3 \qquad x_3 + bx_3, \quad -x_3 + bx_3$$
$$x_2 - bx_3, \quad -x_2 - bx_3 \qquad x_3 - bx_3, \quad -x - bx_3.$$

Since x_2 and x_3 are linearly independent over $\mathbf{Q}$ and b is in $\mathbf{Q}$, it follows that none of the four terms on the left is equal to any term on the right. We can avoid equality between two on the right if we make $1 + b$, $-1 + b$, $1 - b$, $-1 - b$ distinct. This can be accomplished, for instance, by $b = 2$. This choice also avoids equality of any two on the left of the line. So we may take $x_4 = x_2 + 2x_3$.

8. We calculate only the product $s_1 s_1'$, since it is the most difficult. First, by writing out the expressions for s_1 and s_1' or by other means, we see that no product of powers of r is $r^{17} = 1$. Hence $s_1 s_1' = \sum a_i r^i$ summed over $1 \le i \le 16$. But $s_1 s_1'$ is a symmetric function of the powers of r and hence $s_1 s_1' = c \sum r^i$ for some integer c. Since $\sum r^i = 0$ for $0 \le i \le 16$, it follows that $s_1 s_1' = -c$. On the other hand, the product $s_1 s_1'$ has $8 \cdot 8 = 64$ terms. This shows that $16c = 64$ and hence that $c = 4$. Hence s_1 and s_1' are the roots of the equation $x^2 + x - 4 = 0$.

10. Each G_i is a normal subgroup of its predecessor, since its index is 2. The same result would not necessarily hold if the index were some prime other than 2.

Section 12

2. If $z_1 z_2 = -p/3$, then $z_1 - p/3z_1 = -p/3z_2 + z_2$. Hence $y = z - p/3z$ is the same for $z = z_1$ as for $z = z_2$.

3. If z is real in (10.3), then y is real. $D < 0$ implies that two roots of the cubic are imaginary. Hence when z is imaginary, the corresponding y's must be imaginary also. Why?

4. For the equation $x^3 - 3x + 1 = 0$, the discriminant is $81 = D$. Hence $2z^3 = -1 \pm \sqrt{-3}$ and $y = z + 1/z$. Now $\frac{1}{2}(-1 - \sqrt{-3})\frac{1}{2}(-1 + \sqrt{-3}) = 1$ shows that we need consider only one of the two signs $\pm$. Hence we may take only $z_1^3 = \frac{1}{2}(-1 + \sqrt{-3})$, for instance, and have as the values of y:

$$z_1 + \frac{1}{z_1}, \quad z_1\omega + \frac{1}{z_1\omega}, \quad z_1\omega^2 + \frac{1}{z_1\omega^2}, \quad \omega^2 + \omega + 1 = 0.$$

Then, using (9.3) with y_i in place of θ_i, we have, since D is the square of a rational number:

$$y_2 = \frac{y_1^2 + y_1}{1 - 2y_1} \quad \text{and} \quad y_3 = \frac{y_1^2 - 2y_1}{1 - 2y_1}.$$

6. Yes.

8. (a) We see, using Section 11, that $\cos 3\theta = \pm\frac{1}{2}$. So we may choose $\theta = 20° = \pi/9$ radians. Hence a root of the cubic is $(\sqrt{-4p/3})\cos 20° = 2\cos 20°$. The other two roots are $2\cos 140°$ and $2\cos 260°$.

9. Since the roots of the equation $4x^3 - 3x - \cos 3\theta = 0$ are $\cos \theta$, $\cos(\theta + 2\pi/3)$, and $\cos(\theta + 4\pi/3)$, the second is a linear combination of the powers of the first if and only if the equation is normal. Computation shows that the discriminant is $27(1 - c^2)/16$, where $c = \cos 3\theta$. Hence a necessary and sufficient condition for the equation to be normal is that $3(1 - c^2)$ be the square of a rational number. One such value of c is $\frac{1}{2}$. (See Exercise 8a.)

12. Using (12.5), we see that the resolvent cubic is $x^3 + 4x^2 - x - 4$, which has three rational zeros. It is not hard to show that the quartic has no rational zeros. (However, the quartic is the product of $x^2 + x + \frac{1}{2}$ and $x^2 - x + 5/2$ and thus is reducible.)

16. If $z_2 = -p/3z_1$, then $y = z_1 - p/3z_1 = z_1 + z_2$. If in place of z_1 we have ωz_1, then, to preserve the relation $z_1 z_2 = -p/3$, we must choose $\omega^2 z_2$ in place of z_2. Therefore, another root of the cubic is $z_1\omega + z_2\omega^2$. Similarly, the third root is $z_1\omega^2 + z_2\omega$.

19. Suppose that the roots are as given. If we denote by c_1 the sum of the roots and by c_2 the sum of the products of pairs of the roots, we compute

$$\frac{-\theta^3 + 3\theta - 1}{(1 - \theta)\theta} = c_1, \qquad \frac{-\theta^3 + 3\theta^2 - 1}{(1 - \theta)\theta} = c_2.$$

Then it turns out to be true that $c_2 - c_1 = -3$. So if we let $c = -c_1$, we have $c_2 = -(c + 3)$. The product of the roots is -1, showing that the constant term is 1. Hence if the roots are as given, the equation is $x^3 + cx^2 - (c + 3)x + 1 = 0$.

Suppose that the equation is as given and θ is one root. Then substitution shows that $1/(1 - \theta)$ is also a root. It immediately follows that the third root must be $(\theta - 1)/\theta$. Why?

21. If we call $g(x)$ the resolvent cubic: $x^3 + 2px^2 + (p^2 - 4r)x - q^2$, then $g'(x)$, the derivative, is $3x^2 + 4px + (p^2 - 4r)$. Now, a look at the graph of a cubic equation whose roots are all real and whose constant term is negative shows that two of its roots will be negative if and only if $g'(x) = 0$ has a negative root. But the roots of $g'(x) = 0$ are $(-2p \pm \sqrt{p^2 + 12r})/3$. So the derivative has a negative zero if and only if either $p > 0$ or $(2p)^2 < p^2 + 12r$. The latter inequality is equivalent to $p^2 < 4r$. This shows that a resolvent cubic with three real roots has two negative ones unless the coefficient of x^2 is negative and that of x is positive.

Two comments should be made: (1) If $p < 0$ and $p^2 - 4r > 0$, one can see more directly than the above that no root is negative, but the converse of this statement does not follow so directly; (2) all of the above is predicated on the assumption that the resolvent cubic has three real roots. Whether or not *this* condition is met can be determined by computing the discriminant of the cubic or using the derivative. Neither of these is very difficult in a numerical case.

Section 15

3. If $f(x) = x^5 - 10x + 5$, then $f'(x) = 5x^4 - 10$ and $f'(x) = 0$ has exactly two real roots: $\pm\sqrt[4]{2}$. This shows that $f(x)$ has exactly one or three real roots. To see that the number of real roots is actually three, note that $f(0) = 5$ and $f(\sqrt[4]{2}) = -8\sqrt[4]{2} + 5 < 0$.

5. The following three expressions are left fixed by H:

$$w_1 = y_1y_2 + y_3y_4, \qquad w_2 = y_1y_3 + y_2y_4, \qquad w_3 = y_1y_4 + y_2y_3.$$

6. From Section 12 we can compute the following:

$$-v_1^2 = w_2 + w_3, \qquad -v_2^2 = w_1 + w_3, \qquad -v_3^2 = w_1 + w_2.$$

This shows that $\mathbf{Q}(w_1, w_2, w_3) \supseteq \mathbf{Q}(v_1^2, v_2^2, v_3^2)$. Similarly, one can show that $\mathbf{Q}(w_1, w_2, w_3) \subseteq \mathbf{Q}(v_1^2, v_2^2, v_3^2)$. If F is of degree 3 over $\mathbf{Q}$, the fundamental theorem of Galois theory shows that $[K:F] = 4$, and hence K is of degree 12 over $\mathbf{Q}$. (Why?) (Note that F is the root field of the resolvent cubic.) The other possibilities are left to you.

8. Only the identity automorphism takes λ into itself.

11. This implies that the resolvent cubic is reducible. Notice that the resolvent cubic can be reducible while the quartic is irreducible. For example, let $f(x) = x^4 + 10x^2 + 5x + 55$. This quartic is irreducible by Eisenstein's criterion. But the resolvent cubic is $y^3 + 20y^2 - 120y - 25$, which has 5 as one of its zeros.

12. Since the discriminant of the resolvent cubic and thus that of the quartic is 81, in the sequence (13.2), $F_1 = F$ and $[K:F] \mid 12$. Now $[F_2:F_1] = 3$ and $4 \mid [K:\mathbf{Q}]$. Hence $[K:\mathbf{Q}] = 12$.

14. For instance, $H = \{\epsilon, (1\ 2)\}$.

15. Now $S_4 \supset A_4 \supset H_1 \supset H_2 \supset (\epsilon)$, where $H_1 = \{\epsilon, (1\ 2)(3\ 4), (1\ 3)(2\ 4), (1\ 4)(2\ 3)\}$ and $H_2 = \{\epsilon, (1\ 2)(3\ 4)\}$. Then the corresponding sequence of fields is $\mathbf{Q} \subset F_1 \subset F_2 \subset F_3 \subset F_4 = K$, where $F_1 = \mathbf{Q}(\sqrt{D}) \neq \mathbf{Q}$, since the discriminant is not a rational square and F_2 is the field of Exercise 6. The rest is up to you.

List of Special Symbols

Z   Domain of integers, **Z*** same with zero omitted, 2

$\mathbf{Z}_p$ Field of integers modulo p, 115

Z[t]   Domain of polynomials in t with integer coefficients, 137

$\cap, \cup$ Intersection and union of sets, 29

$\subseteq$ Is a subset of, 24

$\subset$ Is a proper subset of, 24

$\phi(n)$ Euler phi function, 67

$\equiv \pmod{m}$ Congruence modulo m, 53

Index